AF581871

Historical Landscape Evolution

Historical Landscape Evolution

Editors

Bellotti Piero
Alessia Pica

Basel • Beijing • Wuhan • Barcelona • Belgrade • Novi Sad • Cluj • Manchester

Editors

Bellotti Piero
Department of Earth Sciences,
University of Rome
Rome, Italy

Alessia Pica
Department of Earth Sciences,
Sapienza University of Rome
Rome, Italy

Editorial Office
MDPI
St. Alban-Anlage 66
4052 Basel, Switzerland

This is a reprint of articles from the Special Issue published online in the open access journal *Land* (ISSN 2073-445X) (available at: https://www.mdpi.com/journal/land/special_issues/historical_landscape_evolution).

For citation purposes, cite each article independently as indicated on the article page online and as indicated below:

Lastname, A.A.; Lastname, B.B. Article Title. *Journal Name* **Year**, *Volume Number*, Page Range.

ISBN 978-3-0365-9909-0 (Hbk)
ISBN 978-3-0365-9910-6 (PDF)
doi.org/10.3390/books978-3-0365-9910-6

Contents

 land

Article

A Cultural Landscape Emerges: Analyzing the Evolution of Two Historic North Pole Expedition Bases in Virgohamna, Svalbard, from Trash to a Protected Cultural Heritage Site

Anne Cathrine Flyen [1,2]

1 Building Department, Norwegian Institute for Cultural Heritage Research, 0155 Oslo, Norway; anne.flyen@niku.no; Tel.: +47-90040152

2 Department of Architecture and Technology, Norwegian University of Science and Technology, 7491 Trondheim, Norway; annecfly@stud.ntnu.no

Abstract: The identification and preservation of cultural landscapes worthy of protection is a challenging task, as their significance is often not immediately apparent. Analyzing the process through which a site or landscape became a heritage site and understanding the historical context and the factors that contributed to its designation allows making informed decisions on the management and preservation of the site. To provide research-based knowledge, this paper aims to analyze the transformation of the degraded remains of the North Pole expedition bases of Andrée and Wellman in Virgohamna, Svalbard, into a protected historic landscape and the subsequent emergence of the site as a current popular tourist destination. Virgohamna serves as an illustrative case for examining the heritagization process of cultural heritage sites in Svalbard. This article adopts a multidisciplinary approach, drawing upon case studies, mapping and categorizing the historic and current landscape, the condition and vulnerability assessment of historic remains, behavior studies on visiting tourists and guides on-site, expert interviews, and document studies. The findings highlight the complex processes that have influenced the making of the cultural heritage landscape in Virgohamna, the enduring narrative associated with the site, and the need for continued efforts to ensure the preservation and dissemination of its historical significance. Analyzing the process through which Virgohamna has become a heritage site and understanding the historical context and the factors that have contributed to its designation as a heritage site has the potential to enhance comprehension regarding historical importance and heritage values. Furthermore, it might facilitate engaging stakeholders and formulating management approaches and provide insights for policy suggestions. The comprehensive examination serves as a foundation for responsible and sustainable heritage management, ensuring the preservation and promotion of Virgohamna's cultural heritage for present and future generations.

Keywords: cultural heritage; cultural landscape; expedition site; heritage values; heritagization; Svalbard; tourism

Citation: Flyen, A.C. A Cultural Landscape Emerges: Analyzing the Evolution of Two Historic North Pole Expedition Bases in Virgohamna, Svalbard, from Trash to a Protected Cultural Heritage Site. *Land* **2023**, *12*, 1481. https://doi.org/10.3390/land12081481

Academic Editors: Bellotti Piero and Alessia Pica

Received: 26 June 2023
Revised: 20 July 2023
Accepted: 21 July 2023
Published: 25 July 2023

1. Introduction

The identification and preservation of cultural landscapes worthy of protection is a challenging task, as their significance is often not immediately apparent [1]. Research on heritagization processes and the criteria for determining the worthiness of heritage sites okfor protection has been a subject of scholarly investigation (e.g., [1,2]). Understanding the factors that contribute to the recognition and preservation of cultural heritage is essential for effective conservation practices and responsible management of these sites [3]. This study focuses on the case of Virgohamna, an Arctic landscape in the Svalbard archipelago, which was initially described by Fridtjof Nansen as a desolate and eerie place [4]. Despite its initial characterization, Virgohamna later gained recognition as a cultural heritage site due to the presence of two historical expedition bases: the staging area for Swede Salomon August Andrée's attempts to reach the North Pole by balloon in 1896 and 1897 and the base camp of American Walter Wellman's airship expeditions in 1906, 1907, and 1909. According to Polar historian Arlov [5], the renowned polar ventures undertaken by Andrée and Wellman constitute captivating chapters in the history of Svalbard, and the expeditions have left enduring cultural landmarks that rank among the most frequented on the archipelago.

The term "heritagization" refers to the procedural aspect of cultural heritage, that cultural heritage is something that occurs and is happening [6]. Løkka et al. describe it as a phenomenon being defined as cultural heritage within a specific context, thereby acquiring a new status [7]. It is this process that underlies the concept of "heritagization". Swensen upholds that mechanisms in our contemporary society determine what is worth preserving and why [8]. The process of heritagization is also defined as the formal process of heritage making. However, the development of cultural environments in Virgohamna goes beyond the formal selection and preservation of elements as a cultural heritage site. In this article, the term heritagization encompasses all the additional aspects, including physical development, intangible narratives, and the way visitors interact with the cultural environment.

The study of the heritagization of the cultural heritage landscape in Virgohamna is significant. This unique site encompasses a rich historical tapestry that stretches back to the 1600s and includes early whaling activities, as well as the expedition bases of Andrée and Wellman [9]. The earliest traces of whaling activities from the 1600s provide a window into the Arctic's maritime history and its economic importance at that time [10]. The expedition bases of Andrée and Wellman shed light on the intense race among explorers to conquer the North Pole and the challenges they faced in their quests. These elements also annotate the pioneering developments in aeronautics during that era [11]. By examining the process of heritagization in Virgohamna, valuable insights into the historical significance of this cultural landscape will be gained. Furthermore, the development of early aeronautics represented by these bases showcases the human desire to push boundaries and expand scientific knowledge.

Exploring the heritagization of Virgohamna is important for managerial reasons. Delving into the process through which the site became a heritage site will provide valuable insights that inform effective management and conservation practices. Understanding the historical context and the factors that contributed to its designation allows making informed decisions about the site's preservation.

Knowledge of the site's heritage values is another important aspect to consider. Studying its history and significance can make it possible to identify the key elements that make Virgohamna significant, whether it be its cultural, historical, or natural values. This knowledge helps prioritize conservation efforts and ensures that the site's most important heritage values are safeguarded as long as possible for future generations.

The research field of the heritagization of cultural landscapes and cultural heritage sites is a dynamic and evolving area of study, actively exploring various aspects of heritagization, contributing to a deeper understanding of its complexities.

Conceptual frameworks to analyze heritagization processes often incorporate multidisciplinary approaches, drawing on fields such as discourses of heritage [12,13], critical heritage studies [14,15], landscape research [16–18], and anthropology [19,20]. They provide a theoretical foundation for understanding the social, cultural, ecological, and political dimensions of heritagization.

According to O'Hare [21], the character of a site originates from historical interactions between the natural and cultural components within the landscape. A significant focus of research has been on identifying and understanding the heritage values and criteria that inform the selection and recognition of cultural landscapes and heritage sites. This involves exploring the different dimensions of value (e.g., [22–26]), the meaning of places [27], and how places shape identity [2]. Researchers have examined how these values are assessed, negotiated, and prioritized within various cultural contexts.

Discussing cultural heritage sites and identity refers to the conceptualizations, perceptions, and understandings related to the cultural, social, and historical aspects of a specific place or site [2,28,29]. It encompasses the beliefs, values, narratives, and symbolic meanings associated with the site's significance and its role in shaping individual and collective identities [30]. The concept of identity reflects the connections people establish with the cultural heritage site, how they interpret its past, and how it contributes to their sense of belonging, heritage, and cultural identity.

The importance of community engagement and participation in heritagization processes has gained considerable attention. Researchers have explored the role of local communities [31,32] in shaping and managing cultural landscapes and heritage sites. This research emphasizes the significance of participatory approaches, inclusive decision making, and empowering local communities in heritage conservation.

There is growing recognition of the need for sustainable heritage management practices. Research in this area explores strategies for balancing conservation, tourism, and sustainable development [33]. It addresses the challenges of managing visitor pressure [34], promoting community benefits, and ensuring the long-term viability and resilience of cultural landscapes and heritage sites [35].

In Svalbard, the pursuit of knowledge-based management regarding cultural and natural heritage sites is an esteemed objective, particularly in response to the escalating tourism levels and anticipated impacts of climate change [9,36].

While there is a wealth of national and international literature of historical and archaeological information concerning Svalbard's cultural history available (e.g., [5,37–44]), the accessibility of knowledge-based information specifically related to heritagization is limited. This holds true for the cultural heritage site in Virgohamna as well, despite the fact that archaeological research and thorough investigations have been conducted on the expeditions of Andrée and Wellman, with particular emphasis on the comprehensive study of Wellman's base [11,39,45–47].

In order to provide research-based knowledge, Virgohamna serves as an illustrative case for examining the heritagization process of cultural heritage sites in Svalbard. This paper aims to analyze the transformation of the degraded remains of the expedition bases of Andrée and Wellman into a protected historic landscape and the subsequent emergence of Virgohamna as a current popular tourist destination.

The analyses encompassed four key areas:

1. Policy documents and legal framework:
    a. To what extent has the overarching policy of the authorities influenced the management of cultural heritage, and what implications does it have on the cultural environment of Virgohamna?
    b. How has the evolution of cultural heritage legislation specific to Svalbard impacted the cultural environment?
2. Physical transformations of the region:
    a. Which natural forces have caused degradation within the site?
    b. In what ways have human activities influenced the condition of the site?
3. Narratives and cultural heritage value:
    a. What narratives have existed and currently exist about Virgohamna?
    b. Which heritage values can be associated with, or can be ascribed to, the site?
4. Impact of tourism:
    a. How has tourism influenced the perception and interpretation of the site?
    b. In what manner has tourism affected the utilization of the site?

Today, the historic environment is being worn down by trampling visitors, and the paper offers suggestions for measures that can preserve heritage values and at the same time allow visitors to experience this magnificent environment. An analysis of the heritagization process of Virgohamna can contribute to a deeper understanding of its historical significance, identify heritage values, engage stakeholders, develop management strategies, and inform policy recommendations. This comprehensive examination serves as a foundation for responsible and sustainable heritage management, ensuring the preservation and promotion of Virgohamna's cultural heritage for present and future generations.

This article is structured as follows: To answer how the transformation and heritagization of Virgohamna were studied, the Materials and Methods section is organized with subsections displaying the case study area and outlining the mixed methods used for data collection and analysis. The results obtained are presented under the heading Results organized with subsections in accordance with the four key areas introduced above: (1) Policy documents and legal framework; (2) Physical transformations of the region; (3) Narratives and cultural heritage value; and (4) Impact of tourism, including a short description of the historic landscape of the site. Next is the Discussion section reasoning with the findings and debating further development of the historic site. Finally, a short section with concluding remarks completes the article.

2. Materials and Methods

This article adopts a multidisciplinary approach, drawing upon archival research, historical analysis, and field observations. Archival research involved examining expedition records, trappers' personal diaries, and published and unpublished official reports to gain insights into the historical context and significance of the expedition bases and the process leading to the listing of the site. Field observations were conducted in Virgohamna, documenting the current state of the historic landscape, visitor activities, and potential threats to its preservation. Interviews with local authorities and heritage professionals provided valuable perspectives on the management of the site and visitor experiences. The different methods were selected according to which research questions they were to elucidate:

Mapping and categorizing landscape and historic landscape were used to analyze the physical transformation of the region and help answer the following research questions: (1a) Which natural forces have caused degradation within the site? (1b) In what ways have human activities influenced the condition of the site? (3b) Which heritage values can be associated with, or can be ascribed to, the site?

Condition and vulnerability assessment of historic remains was used to analyze the physical transformation of the region and help answer the following research questions: (1a) Which natural forces have caused degradation within the site? (1b) In what ways have human activities influenced the condition of the site? (3b) Which heritage values can be associated with, or can be ascribed to, the site? (4a) How has tourism influenced the perception and interpretation of the site? (4b) In what manner has tourism affected the utilization of the site?

Behavior studies on visiting tourists and guides were used to analyze the physical transformation, the narratives and cultural heritage values, and the impact of tourism and help answer the following research questions: (2a) Which natural forces have caused degradation within the site? (2b) In what ways have human activities influenced the condition of the site? (3a) What narratives have existed and currently exist about Virgohamna? (3b) Which heritage values can be associated with, or can be ascribed to, the site? (4a) How has tourism influenced the perception and interpretation of the site? (4b) In what manner has tourism affected the utilization of the site?

Expert interviews and document studies were used to help answer all eight research questions.

2.1. Case Study Area

The case study area Virgohamna is situated on the northern side of Danskøya (Danes Island) on the north-western coast of Spitsbergen, the largest island in the Svalbard archipelago. Danskøya is surrounded by icy waters and characterized by a rugged and remote Arctic landscape, with snow-capped mountains, glaciers, and vast expanses of tundra. The region experiences harsh Arctic weather conditions, including long, cold winters and relatively cool summers.

In this part of Svalbard, the coastline features a captivating archipelago with small fjords, straits, and islands, surrounded by Alpine rock formations and glacier fronts plunging into the fjords. The bedrock mainly consists of metamorphic rocks such as granite, gneiss, and migmatite. The landscape is shaped by the glaciers, continuing eroding the bedrock and leaving the mountains with pointed ridges [48]. The inland region, however, is challenging to access due to its rugged mountains and glaciers. When Wilhem Barentz's expedition discovered Svalbard in 1596, the north-western corner of present-day Spitsbergen was the second landing, next to Bjørnøya [42]. The expedition's resulting map only depicted this specific north-western section of Spitsbergen [49].

The sea ice conditions are among the most favorable in Svalbard, making it a popular hunting ground throughout the centuries [42]. Furthermore, it stands out as one of the most northerly regions worldwide that remains free of ice during summertime [42]. This factor played a significant role in choosing Virgohamna as a base for expeditions to the North Pole.

The area's remote location in the far north is an inherent attraction for many visitors. Moreover, it offers breathtaking scenery, abundant wildlife, and a rich concentration of cultural sites spanning various periods in Svalbard's history [9,42,49,50]. Large cruise ships, limited to a few landing spots, smaller expedition vessels, and small private boats make Northwest Spitsbergen a primary destination. Consequently, this region has been and continues to be most frequented during the summer, placing significant strain on its cultural environments [36,49].

Virgohamna derives its name from "Virgo's harbor", Andrée's expedition ship, and is a small inlet nestled within the strait of Danskegatet [51]. This picturesque location is shielded by sheer cliffs and scattered islets, creating a protective embrace for the harbor. The site itself is situated on a rocky beach at the foot of the cliffs, facing north to the green and icy sea (see Figures 1–4).

Figure 1. To the left: Maps showing Svalbard with Longyearbyen, the capital of Svalbard, and Virgohamna, both located on Spitsbergen, the largest island of the archipelago. To the right: Virgohamna on Danskøya (Danes Island) in detail. Map by A.C. Flyen, reproduced from Norwegian Polar Institute with permission.

Figure 2. Virgohamna in 1906 seen toward west. In the foreground, remnants of Andrée's expedition base can be seen, to the right is Pike's House, and in the background, Wellman's base with the airship hangar and several buildings is shown. Photo: Wilse, A.B./Norsk Folkemuseum [52].

Figure 3. Tourists visiting Virgohamna in 2014 with the historic remnants from Dutch whaling station Hardingers Kokerij, Pike's House, and Andrée's and Wellman's expedition bases. Photo: Anne-Cathrine Flyen/NIKU.

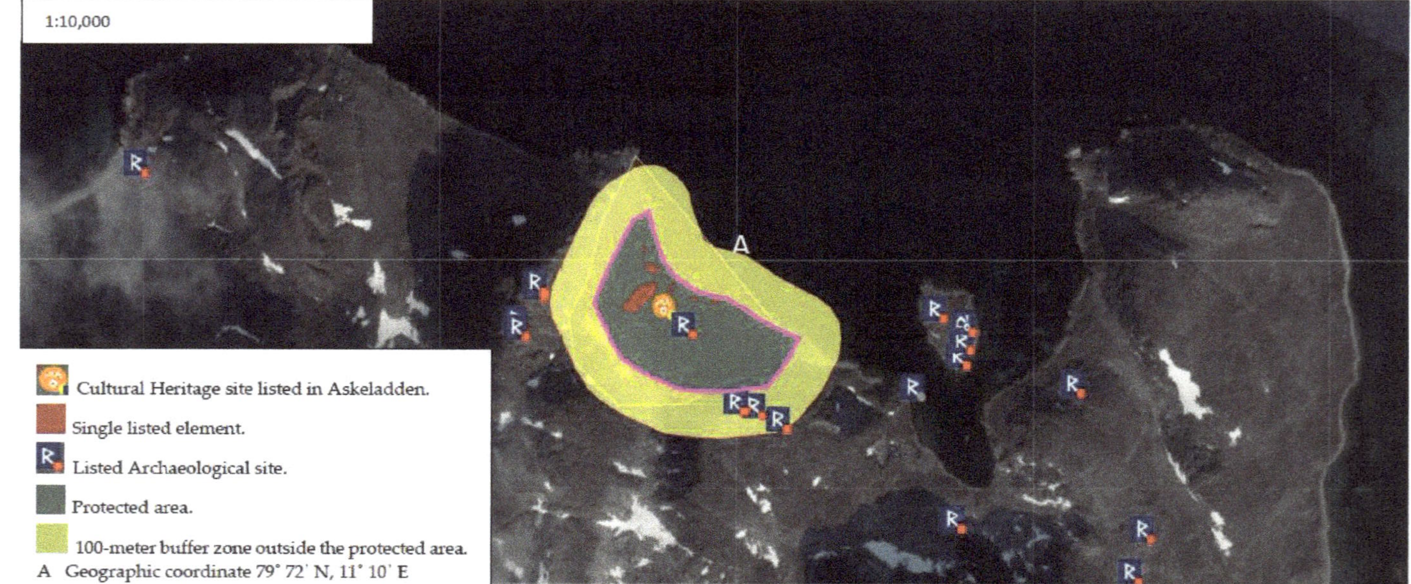

Figure 4. The protected area in Virgohamna, as it is registered in the Norwegian cultural heritage database, named "Askeladden". The information in Askeladden is only available to the public cultural heritage administration. However, most of the information (except for the aerial photo in the background) can be retrieved from the publicly accessible service "Kulturminnesøk" [53]. The area is marked on the aerial photograph, which includes the protected cultural environment and a buffer zone of 100 m surrounding it in all directions. All cultural heritage sites and cultural environments in Svalbard have such a 100-meter buffer zone, which is regulated as strictly as the single element or site itself [54]. Map by A.C. Flyen, reproduced from Askeladden with permission.

2.1.1. Harlinger Kokerij

The stony beach of Virgohamna preserves the remnants of a Dutch whaling station known as Harlinger Kokerij. The station was established in 1636 in what was then referred to as Houker Bay [48] including houses and lard ovens. However, within a few years, the fjords and coastal areas of Svalbard saw a sharp decline in whale populations, necessitating a shift toward hunting them in the open sea [55]. According to Johansen et al. [48], German physician and naturalist Friedrich Martens observed the decay of Harlinger Kokerij in 1671, describing how the remnants of the former whaling station were cowered by layers of formidable ice.

Today, the remains of the houses are still visible in Virgohamna, as are foundations for three double lard ovens including ramps on the side for facilitating the transport of blubber to and from the cookware (see Figure 5). At least 12 graves are situated east of the ovens, and historical relics from the daily life of the station, like remnants from the whalers' chalk pipes, are still lying on the beach.

Figure 5. Remains of the blubber ovens from the Dutch whaling station. The stone painted with number 1 is part of the information structure made by the Governor of Svalbard. Photo: Anne-Cathrine Flyen/NIKU.

2.1.2. Pike's House

Englishman Arnold Pike wintered in the bay with his skipper Kræmer and six Norwegians from September 1888 until May 1889, in a wooden dwelling transported from Norway [42,56,57]. According to Chapman [57] who retells from Pike's daily journals, the wintering, along with the voyage preceding and following their stay in Virgohamna aboard the ship Seggur, was an adventurous expedition centered around hunting pursuits. Pike is regarded as the first wintering tourist in Svalbard [58]. The hut was left on the shores of Danes Island and was later used by Andrée and Wellman [59] as part of their expedition bases.

The house was constructed on the remnants of the ancient Harlringer Kokerij (see Figure 4). Both the name and the cookery itself had been forgotten, and for a brief period, the place was known as Pike's Bay [51,59].

According to Bjerck et al. [59], Pike's House was relocated to Barentsburg in 1925. Trapper Waldemar Kræmer wrote in his diary in 1925 that his family had previously acquired the house from Pike [60]. Since Pike's winter stay in 1988–1989, the house had been used by wintering trappers. Many trappers had also helped themselves to materials from the house. As the house could not remain undisturbed, Kræmer wrote in his diary that he demolished the building and took the materials to Grønnfjorden. Around 1925, there were several coal mines in operation there, and he exchanged the materials for "provisions and oil" [60] (p. 38). What happened to the materials afterward is unknown. Today, the stone foundations, some sills and sleepers, and a pipe leading out of the ruins are telling the story of Pike as one of the first tourists wintering in these waters (see Figures 6 and 7).

Figure 6. Pike's House to the left. Remnants of Andrée's balloon house to the right, and tourists visiting the commemorative monument of Andrée's expedition. Photo: Wilse, A.B./Museene for kystkultur og gjenreisning i Finnmark IKS [61].

2.1.3. Andrée's Expedition Base

With the aim of reaching the North Pole aboard the hot air balloon "Örnen" (The Eagle), Swedish engineer Salomon August Andrée established his expedition base in Virgohamna during the summer of 1896 [42]. According to Arlov [42], a dedicated structure for housing the balloon and a separate facility to produce hydrogen gas, necessary for inflating the balloon, were constructed in the bay (see Figure 8). Unfortunately, the weather conditions were unfavorable that season, leading to the cancellation of the journey to the Pole. Andrée returned the following year for another attempt to leave the base which this time succeeded. A fleet of cruise ships anchored in the fjord off Danskøya to witness the departure. The balloon headed north, disappearing behind glaciers and mountains [42]. For 33 years, the fate of the expedition and its participants remained a mystery until the remains of the expedition and its members were discovered on Kvitøya [42].

Figure 7. The remnants of Pike's House in Virgohamna. Photo: Anne-Cathrine Flyen/NIKU.

Presently, the remnants of Andrée's expedition base in Virgohamna are clearly visible but scarce. While parts of the hydrogen gas production facility persist, only a few planks from the balloon house remain. The location where the balloon house once stood is discernible, and at the center of the circular foundation, a stone monument stands as a tribute to Andrée and his comrades (see Figure 9). Scattered around the area, lies remnants of the balloon house and the gas plant, including nails, bolts, wires, fragments of planks, fasteners, and piles of rusted iron filings. These scattered artifacts serve as poignant reminders of the remarkable journey undertaken by Andrée and his team.

2.1.4. Wellman's Expedition Base

Through his meticulous investigations and analyses, Polar researcher Capelotti has carefully documented and described various aspects of the Wellman base [11,39,45–47]. Capelotti [11] describes the expedition base providing a detailed account of the constructions. Wellman's headquarters, the largest building after the hangar, had ample space to accommodate a diverse range of individuals, including 40 scientific staff, engineers, aeronauts, mechanics, sailors, and workmen. Among the various structures present, there was a machine shop, a boiler house, a steam engine, a steam pump, and a shed specifically designed to house the gas apparatus used for inflating the dirigible. Notably, the airship hangar itself was an immense structure, featuring nine sturdy arches using timber for the floor construction salvaged from Andrée's abandoned balloon house [59] (see Figure 10). According to Capelotti's account [11] (pp. 12–16), Wellman brought an impressive quantity of building materials, including three or four hundred tons of timber and iron, which were utilized in the construction of the hangar and other supplementary buildings. Additionally, the expedition carried 125 tons of sulfuric acid and 75 tons of scrap iron filings for producing hydrogen for powering the airship.

Figure 8. The balloon "Örnen" within the balloon house in Virgohamna. The hydrogen gas production facility in front; Pike's House to the left (partly visible). Photo: Grenna Museum– Andréexpeditionen Polarcenter [62].

Figure 9. Historic remains from the expedition base of Andrée. Part of the hydrogen gas production facility to the right. At the very back of the picture, the Andrée monument is visible looking like a small cairn. Remnants from Pike's House to the left. Photo: Anne-Cathrine Flyen, NIKU.

Figure 10. The airship hangar in Virgohamna showcases the pioneering work of Wellman and his crew in the field of airship technology. The photograph taken in September 1906 captures the presence of engines, the hangar, and a basket gondola attached beneath the airship, providing space for the crew and machinery. Photo: Wilse, A.B./Norsk Folkemuseum [63].

Today, the remnants of Wellman's expedition base are strikingly evident in the flat yet rocky landscape along the shores of Virgo Bay (see Figure 11). The colossal wooden beams, once part of the airship hangar, now rest upon the ground, serving as a testament to the grandeur that once occupied this site. Amidst the foundations of stone and scattered woodwork, the ruins of other structures emerge, revealing glimpses of their former glory. The area is marked by the presence of numerous iron barrels and vast quantities of scrap iron, occupying a substantial space. A considerable quantity of fragmented ceramic pipes adds to the mosaic of artifacts strewn across the site. In this remarkable location, even remnants of the airship wreck itself can be found, including fuel tanks, the frame within the airship's nacelle, and fragments of cloth, offering haunting echoes of the airship's ill-fated voyage.

Figure 11. Historic remains from the expedition base of Wellman. Rusty barrels dominate the scenery; to the right, remnants of one of the houses; to the left, parts of the enormous airship hangar. Picture from July 2014. Photo: Anne-Cathrine Flyen, NIKU.

2.2. Data Collection

2.2.1. Mapping and Categorizing Landscape and Historic Landscape

The mapping and categorization of natural and cultural landscapes form the basis of understanding the spatial context and characteristics of cultural heritage sites. This contributes to answering the following research questions: (1a) Which natural forces have caused degradation within the site? (1b) In what ways have human activities influenced the condition of the site? (3b) Which heritage values can be associated with, or can be ascribed to, the site? The method involves a systematic survey and analysis of the physical features, landforms, vegetation, and historical elements present in the landscape. To carry out this method, field surveys are conducted to document the physical features and topography of the landscape. This includes a brief identification of landforms guided by NiN [17] cultural elements after Hagen et al. 2014 [64]. While conducting fieldwork, certain challenges arose as parts of Harlinger Kokerij and Wellman's base were covered by snow, hindering a comprehensive examination of the entire area. To overcome this obstacle, measurements were collected and subsequently compared and contrasted with surveys conducted by the Governor of Svalbard [58] and researchers Capelotti [11] and Hagen et al. [65] and the database of the Directorate for Cultural Heritage called Askeladden. Unpublished historical documents from the archive of the Governor of Svalbard, maps from the Norwegian Polar Institute [66], photographs from the Norwegian Polar Institute [67], the digital photo archive maintained by the Directorate for Cultural Heritage [68], and DigitaltMuseum [69] are also examined to gather additional information about the historic site and its evolution over time. DigitaltMuseum is a joint database of collections in Norwegian and Swedish art and cultural history museums. Within DigitaltMuseum, the digital photos maintained by the Svalbard Museum, Norges Arktiske Universitetsmuseum, Museene for kystkultur og gjenreisning i Finnmark IKS, Tromsø Museum, Sør-Troms Museum, Grenna Museum, Norsk Folkemuseum, and Norsk Teknisk Museum have been studied.

2.2.2. Condition and Vulnerability Assessment of Historic Remains

This method focuses on assessing the condition and integrity of both the cultural environment and the individual historic structures within it as well as conducting vulnerability assessments for the site and single heritage structures. This contributes to answering the following research questions: (1a) Which natural forces have caused degradation within the site? (1b) In what ways have human activities influenced the condition of the site? (3b) Which heritage values can be associated with, or can be ascribed to, the site? (4a) How has tourism influenced the perception and interpretation of the site? (4b) In what manner has tourism affected the utilization of the site? The aim is to identify deterioration, damage, or potential risks that may affect their resilience and longevity and to chart their susceptibility to human activities. Since the historic remnants in Virgohamna consist of ruins, remnants of construction materials, and artifacts, the condition assessment is conducted as a general evaluation. The vulnerability assessments follow the guidelines presented in Hagen et al.'s [64] handbook, which outlines a methodology for assessing vulnerability to human activities.

2.2.3. Behavior Studies on Visiting Tourists and Guides

These methods contribute to answering the following research questions: (2a) Which natural forces have caused degradation within the site? (2b) In what ways have human activities influenced the condition of the site? (3a) What narratives have existed and currently exist about Virgohamna? (3b) Which heritage values can be associated with, or can be ascribed to, the site? (4a) How has tourism influenced the perception and interpretation of the site? (4b) In what manner has tourism affected the utilization of the site? This approach entails the observation and mapping of visitor behavior, preferences, and patterns within the site. It offers valuable insights into the impact of tourism on the site's physical and social environment, as well as visitor interactions with cultural heritage structures and information. These observations also provide insights into how guides function within the group and how they inform and monitor tourists. Additionally, it sheds light on the effectiveness of the information system implemented by the Governor of Svalbard to manage site operations. During the observation period, a total of 6 medium-sized expedition cruise ships, typically carrying between 12 and 500 passengers, visited Virgohamna. These ships disembarked groups of 10 to 15 tourists per guide. The total number of visiting groups during the observation period was 39. In addition, a private boat visited the site with 6 individuals.

2.2.4. Expert Interviews

This method contributes to answering all the research questions. Interviews are a valuable method for gathering in-depth information and insights from key individuals or stakeholders. Semi-structured in-depth expert interviews were conducted with the regional cultural heritage management in Svalbard (The Governor) and at the directorate level in Oslo. At the regional level, five in-depth interviews were conducted, while at the directorate level, two interviews took place. The advisors at the regional level are periodically replaced due to fixed-term positions. Therefore, all five interviews were conducted with advisors/cultural heritage managers who had previously worked in the regional management of cultural heritage in Svalbard (at the Governor's office). Both advisors at the directorate level had or had been responsible for the management of cultural heritage in Svalbard. The interviews were not conducted on-site.

2.2.5. Document Studies

This method contributes to answering all the research questions. Document studies involve the examination and analysis of various historical documents, archival records, literature, and photographs related to the cultural heritage site. Studies were conducted on documents from the archives of the Governor of Svalbard and the Norwegian Directorate for Cultural Heritage. They also involved studying the photographic databases as referred to in Section 2.2.1. This method provides valuable insights into the historical context, significance, and evolution of the site, as well as the traditional construction techniques, materials, and cultural practices associated with it.

3. Results

The Results section presents the findings and outcomes of the study, focusing on the four key areas highlighted in the Introduction section including a short description of the historic landscape.

3.1. The Historic Landscape of Virgohamna

Virgohamna is a bay on Amsterdam Island in the far northwest of Spitsbergen. It is surrounded by up to 100-meter-high mountain walls to the south, and to the north, it is separated from Amsterdam Island by Danskegattet. The landscape in Virgohamna can be categorized as either a fjord landscape or a coastal plain, but detailed information is currently lacking precision. If the sea depth is significant, it will be defined as a fjord landscape extending up the valley slopes. There are small islands and some flat land in the area. Virgohamna's rocky and moss-covered terrain is scattered with cultural heritage sites that bear witness to significant periods of Svalbard's history from the 17th to the early 20th century. These include remnants of European whaling from the 17th century, a foundation from the first tourist to overwinter in Svalbard, remnants of Andrée's scientific and adventurous expedition with a hot air balloon to the North Pole in the late 19th century, and remnants of Wellman's large base for his three attempts to reach the North Pole by airship in the early 20th century. The cultural landscape also encompasses the development of tourism in Svalbard, as Virgohamna has been and still is a beloved tourist destination, evident from signs and well-trodden paths indicating this activity. The cultural heritage sites in Virgohamna are located partly on a rocky north-facing beach with sparse vegetation, including patches of moss. These sites have undergone significant deterioration, particularly the wooden structures, metal parts, and iron filings in Wellman's base. The historical remnants of whaling activities are difficult to perceive and easy to overlook. Additionally, it is challenging to avoid stepping on loose cultural heritage sites as paths pass through and partially overlap them. This makes the cultural heritage sites in Virgohamna highly vulnerable to visitors.

3.2. Policy Documents and Legal Framework

Until 1920, no nation had any form of sovereignty over the islands, then known as Spitsbergen. Consequently, there was no overarching responsibility for historical remains [39]. In 1920, the Svalbard Treaty was signed, and after addressing land ownership issues, implementing mining regulations, and establishing a treaty text and comprehensive legal framework [70] for the ratification and enforcement of the treaty, Norway obtained sovereignty over the archipelago in 1925 [42]. Even in this initial legal framework, called the

Svalbard Act, a provision regarding the preservation of cultural heritage was included [70], stating that "general regulations" could be issued for "ancient remains".

3.2.1. Overarching Goals and Guidelines

Despite the possibilities outlined in the first Svalbard Act, no regulations specifically related to cultural heritage and preservation were introduced until 1974 [71]. In this legislation, all cultural heritage sites and environments dating back to before 1900 were automatically protected [72]. Accordingly, parts of the cultural environment in Virgohamna, including the historical remains of Harlingers Kokerij, Pike's House, and Andrée's expedition base, were automatically preserved, while Wellman's expedition base fell outside the scope of protection. In 1992, new regulations extended the scope of automatic preservation to include all cultural heritage, both movable and immovable, predating 1 January 1946 [73]. In consequence, Wellman's expedition base was also automatically protected. These regulations were later replaced by the Svalbard Environmental Protection Act, which came into effect in 2002 and was revised in 2012 [74]. The cultural heritage chapter is incorporated into this act as part of a comprehensive environmental legal framework, integrating cultural heritage considerations throughout the environmental protection legislation.

A united Norwegian Parliament has agreed upon the overall goals of Norwegian Svalbard policy for several decades. These goals, formulated in the 1980s and remaining unchanged since then, enjoy broad political consensus [75]. The overarching goals are as follows (author's translation from Norwegian) [76–80] (p. 12):

- "Consistent and firm enforcement of sovereignty.
- Proper compliance with the Svalbard Treaty and ensuring its adherence.
- Preservation of peace and stability in the region.
- Conservation of the unique wilderness and nature of the area.
- Maintenance of Norwegian communities in the archipelago".

It is also emphasized that Svalbard possesses a distinct natural and cultural heritage, which Norwegian authorities have a particular responsibility to preserve. Environmental protection, therefore, constitutes a fundamental aspect of Norwegian Svalbard policy, and all economic activities, resource utilization, and research must operate within the framework established for the conservation of Svalbard's natural environment and cultural heritage [79].

At the same time, conscious efforts have been made to facilitate three specific industries in order to sustain activity in the archipelago and create job opportunities that support a community-oriented environment. These three industries are coal mining, research and education, and tourism [79]. Since 2016, coal mining has been phased out, and other economic activities, primarily tourism, have been developed [81].

Comprehensive reports in the shape of White Papers to the Norwegian Parliament (Storting) on Svalbard have historically been presented approximately every five to ten years [78–80,82]. The latest White Paper on Svalbard was presented in 2016 [80]. These reports have provided guidance for the development of the archipelago for several years, and the comprehensive review has contributed to coordinated development within the framework set by the objectives of Svalbard policy. All the White Papers specifically mention cultural heritage sites and environments, setting high goals, including [79] (pp. 52–53) (author's translation from Norwegian):

- "Based on its internationally significant natural and cultural heritage, Svalbard shall be one of the best-managed wilderness areas in the world.
- Environmental considerations shall prevail over other interests in cases of conflict between environmental protection and other interests, within the framework of treaty and sovereignty considerations.
- The extent of wilderness areas shall be maintained.
- Flora, fauna, and valuable cultural heritage sites shall be preserved as close to their natural state as possible, allowing natural ecological processes and biological diversity to develop with minimal human impact".

Furthermore, there have been White Papers within the fields of environmental protection and cultural heritage preservation for Norway as a whole, which also address the cultural environments in Svalbard [83,84]. In particular, St. Meld. nr. 16 Living with Cultural Heritage [83] (pp. 59–60) focuses on cultural heritage in greater detail and emphasizes Norway's international obligations in Svalbard as a steward of international cultural heritage (author's translation from Norwegian):

> *"Norway bears the responsibility for managing an important national and international cultural heritage on Svalbard, representing the activities of many nations. The government will maintain a restrictive practice with regard to activities and interventions that could harm or diminish the value of the cultural heritage on the archipelago.*
>
> *Outside settlements, cultural heritage sites should remain undisturbed as scattered traces of human use of the Arctic landscape for over 400 years.*
>
> *It is neither practically feasible nor desirable to implement measures against natural deterioration for all cultural heritage sites. However, for certain cultural heritage sites of great historical or experiential value, preventive measures may be considered".*

3.2.2. Local Legislation

The Norwegian Ministry of Climate and Environment holds the primary responsibility for overseeing environmental protection management in Svalbard, including cultural heritage protection, as well as coordinating environmental management efforts [85]. This ministry plays an important role in developing and implementing policies related to environmental conservation and sustainability on the archipelago.

In the realm of cultural heritage management in Svalbard, the Norwegian Directorate for Cultural Heritage assumes the authority and overall responsibility [85]. The Directorate is tasked with safeguarding and preserving the cultural heritage sites and artifacts found on the islands. They play an important role in formulating guidelines, conducting research, and coordinating activities related to cultural heritage protection.

The Governor of Svalbard serves as the regional cultural heritage authority and carries out the day-to-day administrative tasks associated with cultural heritage management. This includes preparing cases for protection and exemption, conducting inspections, and enforcing regulations [85]. The Governor works closely with the Norwegian Directorate for Cultural Heritage to ensure effective implementation of cultural heritage policies in Svalbard.

The management of cultural heritage sites on Svalbard follows the overarching policy established by legislation, all the aforementioned White Papers, international treaties which Norway has endorsed concerning cultural heritage, and official guidelines specifically developed for Svalbard, as stipulated in the current cultural heritage plan. Since 1994, a cultural heritage plan has been in place for the management of cultural heritage on Svalbard, and the plan has been expanded and renewed in two subsequent versions [49,50,71]. The cultural heritage plans are developed by the Governor of Svalbard to serve as a tool ensuring predictability and long-term management of Svalbard's cultural heritage [71]. In the 2000 cultural heritage plan [50], a prioritization of the 50 most important cultural environments was conducted, which was expanded to include the 100 most important cultural heritage sites and environments on Svalbard in the 2013 plan [71]. These sites are intended to be protected through predictable and long-term management. Currently, the Governor of Svalbard is working on a revision of the existing plan [86].

In 2000, the Ministry of Environment launched the Regulation on Area Protection and Traffic Control in Virgohamna, Svalbard. The cultural environment in Virgohamna, including a safety zone of 100 m on all sides of the cultural environment, was already automatically protected under the Svalbard Environmental Act [74]. However, the increasing tourism exerted a heavy toll on the cultural environment and individual monuments. Therefore, the Governor found it necessary to impose restrictions to reduce the impact. The regulation involved introducing restrictions on access and requiring all visitors to obtain permission from the Governor for landing [87]. This way, the Governor could maintain control over who and how many people landed and provide specific guidelines on where visitors were allowed to move within the cultural environment. A separate system was established with maps and brief descriptions [88], as shown in Figures 12 and 13. The map illustrates an absolute ban on access in parts of the area where particularly vulnerable historic remains are located. The accompanying description explicitly states that it is not permitted to step on or move building components, building materials, objects, or other historical remains.

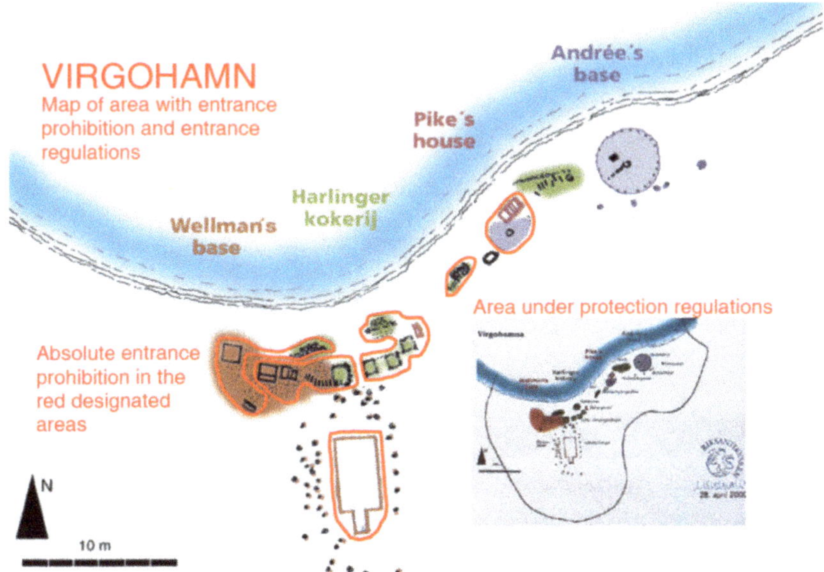

Figure 12. Virgohamna: Map of the area indicating prohibited entry and entrance regulations made by the Governor of Svalbard [88].

3.2.3. Interview with Local and Overarching Administration

The interviews with the administration (former advisors to the Governor's Office and the Directorate for Cultural Heritage) were divided into two main sections: a general overview of the management of cultural environments in Svalbard and a more specific focus on the management of the cultural environment in Virgohamna. The advisors who were interviewed had worked with cultural environments in Svalbard over the past four decades, representing a significant time span. In general, the advisors emphasized that the cultural heritage plans for Svalbard were valuable and important tools for management. Particularly, it was highlighted that these plans are crucial in compensating for the lack of continuity in the advisory role at the Governor's Office, where there is relatively high turnover due to fixed-term positions. Several interviewees were uncertain about the significant impact of overarching political guidelines on cultural heritage management. However, they specifically emphasized the importance of incorporating cultural heritage regulations into comprehensive environmental legislation when the Svalbard Environmental Protection Act was enacted in 2001 [74]. This led to an increased focus on cultural heritage considerations and their integration with the management of natural environments. All the advisors also noted that there are numerous tasks and limited staff to handle them. Many also believed that there was a lack of overarching systems to monitor the deterioration, wear, and development of cultural heritage sites. Regarding the management of the cultural environment in Virgohamna, all the advisors highlighted the informational booklet primarily showcasing the historical background of the present cultural environment. The advisors believed that this booklet had been and still is crucial in shaping visitors' perception of the site, and many purchased it to prepare for their visit and familiarize themselves with the historical context. Thus, according to the interviewed advisors, many visitors were well-prepared regarding the historical narrative but lacked an understanding of how to physically interact with the existing elements. As tourism has increased in Svalbard, the cultural environment in Virgohamna has suffered more wear and tear, necessitating the implementation of restrictions. However, since the cultural environment is highly significant and popular as a visitor site, the Governor's Office and the directorate did not want to completely close it off to visitors. Therefore, a dedicated information system with maps and descriptions was developed to guide visitors and restrict access to the most vulnerable historical remains [88]. Restrictions were also imposed on group sizes, and all groups were required to be accompanied by a guide. This approach aimed to provide comprehensive information and ensure oversight of tourists. The advisors believed that this information system had served its purpose effectively. All the advisors agreed that both trappers and tourists had contributed to the development of the cultural environment and the heritagization of Virgohamna over time. This includes conscious actions by trappers to extract materials, building components, and objects for use elsewhere on Svalbard, as well as souvenir collecting by tourists. Unintentionally, visitors walking around the site have also contributed to trampling and degradation shaping today's heritage environments.

Figure 13. Virgohamna: Map and text expanding the map of the area indicating prohibited entry and entrance regulations made by the Governor of Svalbard [88].

3.2.4. Observations of Visitors

On-site observations during casework indicated that the guidelines appeared to be somewhat confusing and unclear. The system initially established a trail to be followed in parts of the area around Wellman's base. This area holds a significant number of historical remains, such as wooden structures from the buildings, iron shavings, rusted barrels, and various metal parts that cannot withstand being stepped on (see Figures 12 and 13). However, it turned out that old paths used by visitors in the past, which were no longer allowed, were much more distinct in the terrain than the new approved paths. This led most tourists to believe that the most visible paths were the ones permitted. As a result, the tourists followed the old forbidden paths. Both the old and new paths passed directly through the historical remains, making it impossible to walk on the path (either the old unauthorized one or the new authorized one) without stepping on protected artifacts. As a result, it was not possible to comply with the Governor's requirements, despite following the Governor's other guidelines. The observations also revealed that several guides climbed onto the blubber ovens, which are slightly elevated in the landscape, to better address the tourists. This was done despite all the blubber ovens being clearly marked as off-limits on the map provided by the Governor.

3.2.5. Key Findings

The analysis of the policy documents and the legal framework surrounding the heritagization process of Virgohamna, as well as the interviews with the administration of the historic environment, revealed several key findings. The overarching policy of the authorities clearly states that cultural heritage protection is an important responsibility of the Norwegian administration of Svalbard. The evolution of cultural heritage legislation specific to Svalbard has also impacted the cultural environment of Virgohamna. The legislation has undergone changes over time, with increasing recognition of the historical and cultural significance of the region. Specific management plans have been developed that describe guidelines for the preservation of cultural environments, and 100 particularly prioritized cultural environments have been defined to receive extra attention from the authorities. The cultural environment in Virgohamna is among these. This recognition has resulted in strengthened protection measures and regulations aimed at preserving the heritage values of Virgohamna. However, challenges such as limited enforcement mechanisms and ambiguity in certain provisions seem to have affected the effectiveness of the legislation. Travel restrictions have been implemented for Virgohamna, along with a specific arrangement for visitors aimed at preserving particularly vulnerable parts of the cultural environment. Even so, these guidelines are somewhat unclear and difficult to follow. Additionally, it appears that the guidelines are not being followed by the guides.

3.3. Physical Transformations of the Cultural Heritage Site

Cultural environments and cultural heritage in polar regions are facing increasing pressure (e.g., [36,89–92]). The effects of climate change combined with growing tourism contribute significantly to this situation (e.g., [93–97]). The impacts of climate change and its subsequent effects are stronger the further north one goes, making the Arctic particularly vulnerable [98]. This has also influenced tourism, and the concept of "last chance tourism" reflects the desire to experience natural phenomena, wildlife, and cultures before they disappear due to climate change (e.g., [99–102]). In Svalbard, these climate effects are highly evident [103], while tourism has seen a significant increase in the past 10–15 years, except for the two pandemic summers in 2020 and 2021 [104]. In Virgohamna, natural decay and human-induced decay have reinforced each other.

There are few descriptions of how natural decay has affected the cultural environment in Virgohamna. However, the high-Arctic nature and climate are harsh on vulnerable buildings, typically constructed with simple foundations and designed to last only a few seasons. These structures often feature a significant amount of wood and are located along the coast, often at the base of mountains or slopes. The extreme weather conditions, including strong winds, heavy snowfall, freezing temperatures, and fluctuating sea ice, pose significant challenges to the durability and longevity of these buildings [49,50,71]. Weather and wind erosion pose strong destructive forces on standing buildings in Svalbard [49,105]. If the buildings are partially decayed, for example, due to polar bear damage or trappers' plundering, the forces of wind can take a firm grip. Once gaps or vulnerabilities occur, the buildings rapidly deteriorate. Coastal erosion is a significant problem in Svalbard [92,93,106], and the cultural environments in the archipelago are often located very close to the coastline, just like in Virgohamna. The relentless pounding of waves and the erosive action of ice can lead to the gradual retreat of the shoreline, putting coastal structures and heritage sites at risk. The combination of storm surges, sea ice dynamics, and thawing permafrost exacerbates the process of coastal erosion. This ongoing erosion poses a constant threat to the preservation of cultural sites in Svalbard, including those in the Virgohamna area. Even slower processes such as rot and rust take hold more effectively when buildings and building components are no longer protected by roofs or sheltering walls. Although it was long assumed that rot barely existed in Svalbard, research in recent years has shown that the timber is also degraded by wood-decaying fungi [90,91,107].

It seems that wind forces are particularly strong in the bay. Trappers who have spent winters in the buildings of Virgohamna recall several periods of winter winds reaching hurricane strength [60,108]. Norwegian trapper Paul Bjørvik recounts a hurricane during Christmas in 1908 that toppled Wellman's hangar [108]. When Nansen visited Virgohamna in 1912, Wellman's airship hangar had blown down, according to Nansen, because trappers had helped themselves to the steel supports that held the structure standing. The gas facility was still there, but all metal parts had been taken. The residential house was still intact, and Nansen described it as large and spacious but unwelcoming, with a large room in the middle surrounded by various rooms along the outer walls. He mentioned the presence of a bathroom with a bathtub, a kitchen, and a darkroom for developing photographs [4].

Tourist A. Schibsted recounts a trip to Svalbard, then known as Spitsbergen, in 1908. His letters have been published, and he describes Andrée's balloon house as a "pile of beams" [109] (p. 39) and explains that the balloon house collapsed already a year after the expedition left Virgohamna in the balloon and disappeared in the fog. This tourist trip was 11 years later, and Schibsted also mentions "a giant mark carved into the mountain" [109] (p. 39). The mark depicted the Masonic symbol, and Schibsted mentions that Andrée, who was a high-ranking Freemason back in Sweden, had carved the mark himself. Furthermore, Schibsted talks about all his fellow tourists eagerly and meticulously collecting souvenirs from the ruins. He describes how they took with them nails, pieces of pipes, scraps of fabric, timber, and ropes [109].

Wellman hired trappers to overwinter in the expedition base to look after equipment and buildings. It was well-known that trappers helped themselves to materials, building components, and machine parts, gradually dismantling and taking everything usable. This was considered a tradition and a necessity in these remote areas, where every man and woman had to fend for themselves, and it has become part of the historical development of trappers' cabins [42,60,71,110]. Trapper Bjørvik writes in his diary in 1908 that he and another trapper overwintered in Wellman's base to "look after Wellman's things" [110] (p. 3). Bjørvik also writes that the two of them were taking over from two other trappers who had overwintered since 1907 and that he himself had overwintered there from 1906 to 1907 [108,110]. Several Svalbard trappers and also polar explorer Fridtjof Nansen mention the plundering of materials, metals, and parts of machinery carried out by trappers, as well as tourists who visited the area right from the start of the expeditions and took souvenirs with them [4,39,60,108,109,111]. Nansen visited Virgohamna to find a part for his damaged boat engine and points out that he himself contributes to the decay. He also describes how hunters have removed fittings and latches from doors and windows, causing them to no longer seal properly, and how this contributes to the natural decay of the buildings in Virgohamna [4].

A few years after Wellman had left Virgohamna, the buildings were no longer in use by trappers and hunters, and they rapidly decayed. However, cruise ships continued visiting the site, and over the years, thousands of tourists have continued to trample upon the historical remains and take souvenirs with them. Today, there are no standing buildings remaining, but there are still numerous historical traces. A significant amount of timber from buildings, especially from the large lattice arches of Wellman's airship hangar, lies scattered around. Like in other parts of the archipelago, the timber has been heavily degraded by biological activity, primarily by wood-decay fungi [90,91]. This makes the wood highly vulnerable and susceptible to damage, leading to its collapse. Similarly, there are substantial amounts of scrap metal in the form of metal shavings used for gas extraction, as well as metal barrels and remnants of machinery and metal pipes from the airship itself. All the metal is affected by rust. Climate change is expected to further enhance several of these degradation processes, including the growth of wood-decay fungi, rust attacks, and strong gusts of wind [91,93].

Pike's House was dismantled and removed, but the bottom sills remain. These have been decayed by rot and are not very resilient to pressure. The remains of Andrée's expedition are not as extensive as Wellman's, but there is part of a wooden structure remaining from gas production, also heavily decayed.

The remnants of Harlinger's Kokerij mostly consist of low mounds and elevations from the blubber ovens where whale blubber was melted into oil. These are primarily susceptible to human-induced decay. During field work, tourists were observed trampling in the ruins and among the historic remains, and guides were observed standing on the remains of the blubber ovens talking to the tourists. This demonstrates that both natural and human-induced decay persists.

3.4. Narratives and Cultural Heritage Value

3.4.1. Narratives

The narrative of Virgohamna can be understood as the story or account of the historical, cultural, and natural significance of the landscape. In the historic context of polar exploration, Andrée and Wellman are considered pioneers for their attempts to conquer the North Pole using the airway [11]. While many explorers focused on ground expeditions or sea voyages, Andrée and Wellman ventured into the uncharted territory of using aircraft to reach the pole. Their endeavors represented a groundbreaking approach to polar exploration, pushing the boundaries of what was previously considered possible [11]. Though their attempts were ultimately unsuccessful, their pioneering spirit and willingness to explore new avenues in the quest for the North Pole left a lasting impact on the history of polar exploration [112].

The history of polar exploration is marked by heroic expeditions. Toward the end of the 19th century, the Arctic and Antarctic regions, especially the poles, remained uncharted territories on the world map [42]. Many men courageously ventured into the great unknown, hoping to be the first to reach the pole [113]. Some were adventurers, while others were scientists, and many expeditions included researchers [42]. Figures such as William Edvard Parry, John Franklin, Adolf Erik Nordenskiöld, Fridtjof Nansen, Roald Amundsen, and Umberto Nobile loom large in the history of Arctic exploration [42]. In Antarctica and the sub-Antarctic regions, notable figures include James Weddell, James Clark Ross, Carl Anton Larsen, Henrik Bull, Carstens Borchgrevink, Adrien de Gerlache, Robert Falcon Scott, Otto Nordenskjöld, William Speirs Bruce, Ernest Shackleton, and Roald Amundsen, among scientists, explorers, and adventurers [114–117]. However, by the late 19th and early 20th centuries, few had explored the possibilities of reaching the pole by air [11,42]. In this regard, Andrée and Wellman were the pioneers. Later successful and unsuccessful attempts to reach the North Pole by air built upon the achievements of these two pioneers [11]. This included, for example, Amundsen and Ellsworth's 1925 expedition with Dornier Wal flying boats as a preparation for the airship expedition in 1926, Byrd and co-pilot Bennet's claim of being the first to reach the North Pole in their Fokker F VII plane in 1926, which was later rejected and deemed false [112,118], and Amundsen, Ellsworth, and Nobile's successful overflight of the North Pole in 1926 [4,112,119].

Among his contemporary journalist colleagues and scientists, Wellman was accused of engaging in pseudo-science and being more concerned with creating headlines in newspapers than achieving scientific accomplishments [11]. After all, he failed in all his attempts and was portrayed by Nansen as something of a fool [4]. Interestingly, even though Andrée also failed in his goal of flying over the North Pole, Nansen depicted him as a hero [4]. Nansen describes the site and the historical remains of the two expeditions as "the cabin of the Swedish hero-tragedy and the noisy mess of the American magazine- and aerial humbug." [4] (p. 193) (author's translation). According to Diesen and Fulton [120], the airship attempt in 1906 garnered significant media attention internationally. The polar expedition had already become a media sensation long before Wellman embarked on his journey to the northern regions. Diesen and Fulton also argue that Wellman's inadequate preparations affected all of his several attempts to reach the North Pole through various means: on foot (1894), by dogsled (1898–1899), and using airships and motorized sledges based in Virgohamna (1906–1909) [120].

The interest surrounding the expedition in its time was significant, and tourist cruises were organized to witness the departure [4,42,109,120,121]. Wellman had made sure to have photographers present to document his every step [120]. Andrée's story was greatly influenced by the fact that the expedition disappeared. Swedish author, illustrator, and medical doctor Bea Uusma, who has devoted her life to unraveling the fate of the Swedish expedition members, succinctly captures the mystery surrounding the departure of the Eagle from Virgohamna. She describes how the Eagle vanished in the mist over Norskøyene (the Norwegian islands) northeast of Virgohamna and disappeared [122]. A carrier pigeon with a message and a buoy were the only signs of life. Despite extensive search expeditions, no trace of the Eagle with its three men was found [120,122]. It was not until 33 years later that their tent and human remains were discovered on Kvitøya [42]. This mysterious "disappearance" and the wonderment about what had happened may have added to the mystique surrounding the expedition. And when unexposed film that could be developed and diaries that could be deciphered were found after 33 years, the interest was revived.

Through his archaeological research, Capelotti has shown that there is more than just the homeland of Andrée and Wellman that makes the cultural environment in Virgohamna an international cultural setting [46]. Capelotti has demonstrated that both Andrée and Wellman utilized French technology for generating hydrogen, and even Andrée's hot air balloon was of French origin. Based on these findings and subsequent expeditions that also utilized international technology, Capelotti suggests that despite the national historical portrayal of these expeditions, they were technologically very international [46]. Capelotti has contributed to elevating Wellman's "reputation"—emphasizing that both Andrée and Wellman have contributed to "testing the frontiers of technology, geography, and personal and national ambitions" [11] (p. 85), also stating that these flights, especially Andrée's, "remain fixed in the collective human consciousness" [11] (p. 85). The discovery of the diaries and the developed films significantly contributed to the fascination surrounding Andrée's expedition [42,122,123]. The insight into intimate details and firsthand descriptions of what actually happened, right up until the men landed on Kvitøya, added to the intrigue.

According to Bea Uusma, the Andrée expedition is Sweden's most extensively covered polar expedition, with over 50 books published about it [122]. The narratives are perceived and shaped based on individuals' own experiences, knowledge, and observations. This seemed evident in the observations of Swedish tourists visiting Virgohamna. In more recent times, doubts have been raised about Andrée's heroic motives [42]. Nevertheless, Andrée remains important to the Swedish people. Observations in Virgohamna showed that Swedish tourists almost always went straight to the Andrée monument where the balloon house once stood. After spending some time by the remains of the balloon house and listening to the guide, they walked back to the beach and proceeded to Wellman's base.

Another narrative can shed light on how difficult it is to understand, even for an official person, that what may seem like garbage can be worthy of preservation. In 1995, the current Minister of Environment, who in Norway is responsible for nature and cultural heritage conservation, visited Virgohamna. When he saw the historical remains of Wellman's expedition in the form of rusty barrels, scrap iron, broken ceramic pipes, and scattered wood, he exclaimed that it needed to be cleaned up, and he would ensure funding for it [124]. However, at that time, the entire area was automatically protected under the new regulation of 1992, stating that historical remains predating 1 January 1946 are automatically protected [73]. No cleanup took place.

3.4.2. Cultural Heritage Value

Beyond their historical significance, the historical remains in Virgohamna possess cultural, scientific, educational, and experiential value, reflecting the tenacity of human endeavors, the spirit of scientific curiosity, and the desire to conquer and comprehend the natural world. This assessment examines the multifaceted heritage value of the Virgohamna historical remains, highlighting their contributions to research, education, public awareness, and the preservation of Svalbard's cultural and historical heritage.

Historical Value: The historical remains in Virgohamna hold significant historical value as they represent a large time span in Svalbard's cultural history. The heritage site is associated with Wellman's and Andrée's spectacular polar expeditions, representing pioneering efforts in exploring the Arctic and pushing the boundaries of human achievement. Despite the expeditions' ultimate lack of success, they encapsulate crucial historical trends and events, such as the race for the North Pole and the emergence of aviation in Arctic regions. Andrée and Wellman's experiences were significant for Amundsen when he planned his expeditions together with Ellsworth using the Dornier Wal seaplanes N 24 and N 25 in 1925 and with Ellsworth and Nobile using the airship Norge in 1926 [112]. Linked to the remarkable polar expeditions of Wellman and Andrée, these remains showcase pioneering efforts in Arctic exploration and the pursuit of extraordinary human accomplishments.

Cultural Value: The cultural significance of the historical remains lies in their representation of human endeavors, challenges, and achievements in the harsh Arctic environment. They reflect the spirit of exploration, scientific curiosity, and the human desire to conquer and understand the natural world. Adding to the cultural value is the possibility for visitors including tourists and researchers to experience the site.

Scientific Value: The historical remains in Virgohamna provide valuable insights into the technological advancements, survival strategies, and cultural practices of past polar expeditions. They offer opportunities for scientific research, including studies on material deterioration, climate change impact, and historical documentation. The landscape on the north-western part of Spitsbergen surrounding Virgohamna including the glaciers, the fiords, and the valleys provides information on past climate and might offer valuable data for understanding current and future climate. Ecosystems are particularly vulnerable in Arctic regions, and climate change is considered the most important influencing factor. The environment surrounding Virgohamna offers extreme living conditions for flora and fauna, adding to the scientific value. Key species such as polar bears and most mountain bird species are found in the area, and the sparse vegetation belongs to the northern Arctic tundra zone.

Educational Value: The historical remains serve as educational resources, offering opportunities for learning about Arctic exploration, the history of scientific advancements, and the challenges faced by early explorers. They contribute to the public awareness and understanding of the region's cultural and historical heritage.

Experiential value: The narrative of the Andrée expedition holds a strong presence in the experience of the cultural environment in Virgohamna. The powerful story of the three men who disappeared in the fog and were later rediscovered after 33 years is fascinating. The fact that the expedition took place in such a remote and harsh landscape further enhances the experiential value and sense of awe. The heritage value of the historic remains in Virgohamna is enhanced by the breathtaking landscape, which further contributes to the overall experience. The location within the Arctic environment adds an additional layer of significance to the site. The remains are situated amidst stunning natural landscapes, including coastal areas, mountains, and glaciers, enhancing their aesthetic and ecological significance. Although the expeditions were unsuccessful, they reflect important historical trends and events: the race for the North Pole and the relatively new field of aviation, both in general and specifically in Arctic regions. Andrée and Wellman's experiences were significant for Amundsen when he planned his expeditions together with Ellsworth using the Dornier Wal seaplanes N 24 and N 25 in 1925 and with Ellsworth and Nobile using the airship Norge in 1926.

The historical remains in Virgohamna hold significant cultural heritage value. They represent a rich historical timeline in Svalbard's cultural history, associated with the pioneering polar expeditions of Wellman and Andrée. The heritage value assessment of the historical remains highlights their multidimensional significance. With their historical, cultural, scientific, educational, and experiential value, they contribute to preserving and promoting Svalbard's cultural and historical heritage. The remains not only reflect important historical events and trends but also offer opportunities for scientific research and educational exploration. The combination of their captivating narrative, the harsh Arctic environment, and the breathtaking natural landscapes further enhances their overall value. As a result, preserving and understanding the historical remains in Virgohamna is crucial for maintaining the cultural heritage of the region and facilitating a deeper appreciation for Arctic exploration and human achievements in extreme environments.

3.5. Impacts of Tourism

Tourism in Svalbard is highly regulated [87,88,125–127]. While the overarching guidelines for the preservation of natural and cultural heritage values in Svalbard are strict, tourism is also a desired activity [80]. Therefore, arrangements are made for cruise ships with tourists to visit large parts of the archipelago and disembark at numerous visitation sites. The medium-sized coastal cruise ships and small, private boats explore a greater number of locations, while larger cruise ships primarily focus on one or two sites outside of the settlements [128]. Cruise traffic in Svalbard is typically divided into three main groups, as shown in Table 1. Only cruise ships and vessels belonging to categories 2 and 3 visit Virgohamna.

Table 1. Type of cruise ships visiting Svalbard [104] (p. 5) (translated by the author).

Category 1: Conventional Cruise Ships	Category 2: Expedition Cruise Ships	Category 3: Small Vessels
Also referred to as overseas cruise ships: large vessels often accommodating thousands of people on board. Cruises usually start in Europe, sail along the Norwegian coast with stops at several ports and include a couple of days in Svalbard before returning south. Visiting Longyearbyen, Ny-Ålesund and some also sail to Gravneset in Magdalenefjorden.	Also known as coastal cruise ships: come in various sizes, typically accommodating 12–500 passengers. Those vessels commonly exchange passengers in Longyearbyen (fly to/from Svalbard). Cruises last 4–20 days, with an average of 7–10 days. In normal operating years, many sail around Spitsbergen, some navigate the entire archipelago. Almost all operators are members of AECO [1]	Boats less than 24 m in length carrying 12 or fewer passengers. These are often sailboats, some are operated purely for commercial purposes, others are privately owned vessels with commercial operations, and some are chartered with a skipper/crew. During the summer, there can be a relatively large number of such vessels in Svalbard (around 40 in 2019).

[1] Association of Arctic Expedition Cruise Operators.

According to the Governor, cruise tourism in Svalbard has significantly increased in the last 10–15 years [128]. The Governor and the tourism industry in Svalbard consider the 2019 season as an "all-time high" (Governor's Annual Report 2019, 2020, 2021, and 2022) and realize increasing tourism may have negative effects on the environment and cultural heritage [128].

The Governor's office monitors the tourism industry, which must annually report the locations they wish to visit and the number of tourists they intend to disembark. The number of people disembarking each year remained stable from 1996 to 2000. 2001 saw an increase of approximately 72%. After a period of steady growth, 2010 and 2011 witnessed a significant reduction. The largest increase occurred in 2015, with an approximately 40% rise compared to the previous year. Expedition cruise ships have shown a steady increase over the years and contributed the most to the 2015 increase. In the subsequent years, they also had the highest number of disembarkations and utilized the most sites outside the settlements. During the period of 2015–2019, the number of foreign ships decreased, but they became larger with more passengers on board. The low figures in 2020 and 2021 were due to the COVID-19 situation (see Figure 14). Figures from 2022 indicate that the number of people disembarking is approximately 90% of the level from the previous normal year (2019). According to the Governor [128], cruise traffic is continuously spreading to new areas in Svalbard. The Governor also indicates that if there are many vessels in one area, tour operators often seek new areas to disembark guests and go on excursions. The number of landing sites has steadily increased along with the number of people disembarking. In 2020 and 2021, there were hardly any tourists disembarking at the visitation sites due to the COVID-19 situation.

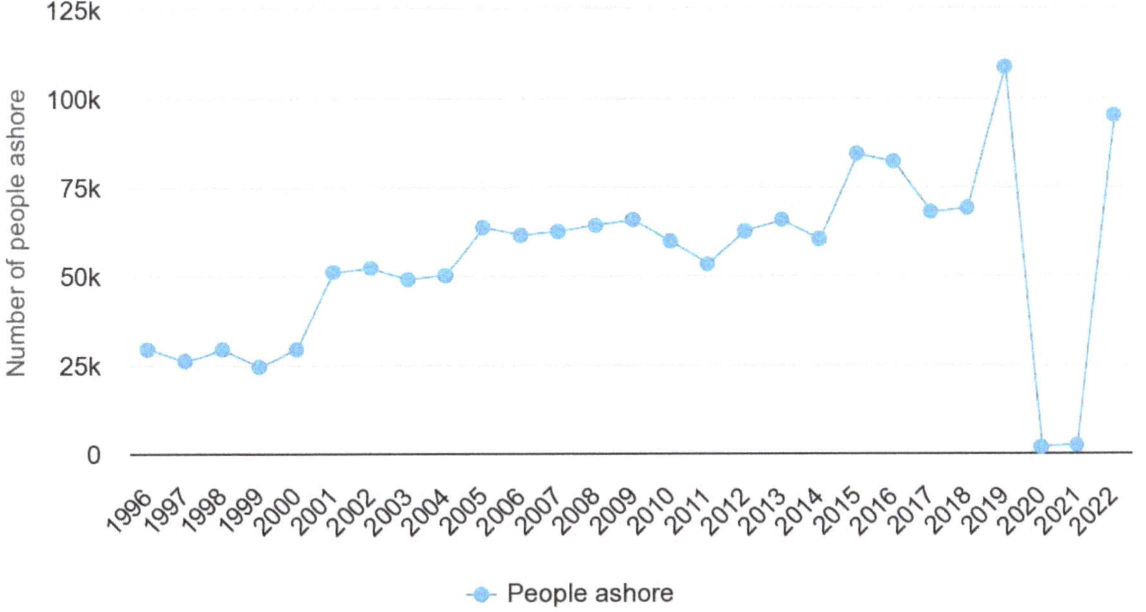

Figure 14. Number of visitors going ashore on sites in Svalbard outside the settlements and the Isfjord area [128].

The guides on the expedition cruise ships are hired by the cruise ship operators and are mostly based in Longyearbyen. Most guides only stay for a few seasons, and the turnover is high. Currently, there are no legal regulations or formal requirements demanding certification or training of guides. A course plan for guides exists; however, it is voluntary. Accordingly, the knowledge on the sites is varying. The sites outside the settlements are managed by the Governor of Svalbard. However, due to vast distances and a lack of infrastructure, the sites are seldom visited by the Governor. Some listed buildings are conserved and maintained, although most historic structures are left to deteriorate at their own pace.

In Virgohamna, the necessity of applying for permission for landing regulates the traffic. However, the Governor's office states that everyone who applies is granted permission but must adhere to the guidelines for travel. The Governor's office keeps statistics on the number of permits issued, and in 2022, 73 ships applied for and were granted permission for landing [129], as shown in Table 2. Each cruise ship is normally visiting Virgohamna several times during one season. There are no statistics available on the number of visitors.

Table 2. Number of permissions to land in Virgohamna issued by the Governor's office [129].

Statistics	Number of Permits 2021	Number of Permits 2022
Permits for sensitive natural and cultural environments in Virgohamna	32 permits	69 permits for 73 vessels

Tourism in Virgohamna has both negative and positive consequences. Cultural environments do not possess inherent value like living organisms. Cultural heritage and cultural environments derive their most significant value from being experienced by people. In Svalbard, cultural environments constitute important elements of the archipelago's history, representing visible traces of international activities. Therefore, cultural environments are also important in the tourism industry, even though studies initially show that tourists come to Svalbard to experience pristine nature [64]. Nevertheless, many tourists report that cultural environments have made a significant impression [64].

Norway has an international obligation to preserve the cultural environments in Svalbard [80] and has declared that tourism should be a vital industry in the archipelago. Therefore, it is natural for cultural environments to be facilitated for tourism. In Virgohamna, Norwegian authorities have supported careful behavior through the established information system. No other measures to preserve the historic remains have been approved or planned. So, how has tourism influenced the perception and interpretation of the cultural environment?

In terms of positive impacts, tourism has brought increased awareness and greater attention to the cultural environment in Virgohamna, making it more widely known and recognized among visitors. The presence of tourists has reinforced the historical significance of the cultural environment in Virgohamna, highlighting its importance as a site of exploration, scientific endeavors, and human activities in the past. Tourism has provided a platform for interpretation and storytelling, allowing guides and experts to share knowledge and narratives about the cultural heritage of Virgohamna with visitors. The influx of tourists has also led to increased awareness of the need for preservation and conservation of the cultural environment. Efforts have been undertaken to protect the site and its artifacts from potential damage caused by tourism activities. Tourists' behavior and interaction with the cultural environment play a role in shaping its interpretation. Their respect for preservation guidelines and adherence to responsible tourism practices can contribute positively to the understanding and conservation of the site. However, it is important to note that while tourism can have positive impacts on the perception and interpretation of the cultural environment, there are also potential negative effects such as overcrowding, degradation of sensitive areas, and the commodification of cultural heritage.

Most Svalbard guides do not stay for an extended period on the island [36,130]. Observations on-site indicate that most guides lack proper knowledge and control over the vulnerable cultural environments in Virgohamna. They often positioned themselves on top of the delicate whaling ovens to gain a higher vantage point when providing information about the visitation sites. Furthermore, they allowed tourists to trespass into restricted areas, disregarding the regulations. While there was no direct evidence of souvenir taking, many tourists handled and examined historical artifacts and remnants before returning them.

Consequently, this behavior contributes to the deterioration of the cultural environment, and the guides themselves bear a significant responsibility in this process.

The observations provided a comprehensive overview of how guides and tourists navigate through the cultural landscape. Particularly within the section encompassing Wellman's base, there are abundant historical remains. Paths meander both within and atop these historical remnants. The information provided by the Governor's Office indicates the designated paths to follow, but the older paths from previous visits are more prominent than the new ones to be followed. As a result, tourists and guides tend to walk on the most visible paths, mistakenly assuming they are the correct ones.

The guides appeared to focus primarily on Andrée's and Wellman's bases, placing less emphasis on presenting the complete historical picture of Virgohamna. The history of whaling was showcased to a much lesser extent. The path from the beach near the landing site up to the remains of Andrée's balloon house partially traverses several whaling graves. Although these graves are not highly visible, they remain vulnerable to inadvertent footsteps. The guides did not try to indicate to tourists where they should avoid stepping near the graves. The remains of Andrée are less prominent, yet his story appears to resonate the strongest. The remains of Wellman's expedition base are numerous and detailed, and even though tourists also explored among them, many tourist groups seemed more captivated by Andrée's base and history.

4. Discussion

Analyzing the heritagization process of Virgohamna indicates that the cultural heritage landscape has been shaped by multiple complex processes. The formal process of selecting, identifying, and designating the historic elements and what unfolded there, forming the historic landscape in Virgohamna, was initiated by the authorities. However, the cultural environment itself was not defined as worthy of protection through a separate process but became part of a larger process where all traces of human activity on the archipelago before 1900 were automatically protected under a specific regulation for cultural heritage on Svalbard. Later, this boundary was extended to include all cultural environments before 1946. This preservation approach has likely saved many cultural environments from being cleared, including Virgohamna, and has influenced the development of the cultural landscape there. It is, however, important to note that the preservation measures were implemented almost 90 years after Andrée and his men left Virgohamna in the balloon Ørnen and 65 years after Wellman packed up and departed. Much had happened during this time, leaving only ruins and remnants to be protected. On paper, Norwegian authorities expressed high ambitions for the preservation of cultural environments in Svalbard, and since the first cultural heritage plan was launched for Svalbard's cultural environments, Virgohamna has been a prioritized location for conservation and follow-up due to its high heritage value. However, restoring ruins is not an easy task, and little was done to preserve them. Under different circumstances, the ruins would have been defined as rubbish and discarded, as described by a visiting Norwegian environmental minister. Fortunately, it was clarified that he was referring to valuable cultural heritage sites. Nevertheless, this incident can be seen as a symbol of how natural values seem to override cultural heritage values in Svalbard, despite the high and declared ambitions of the authorities to preserve both. In Norway, natural values are highly regarded, and in Svalbard, the interest in and awareness of natural values are far higher than cultural heritage values. When natural decay and wear from visitors became too pronounced, restrictions on visits and an information program were implemented. According to the authorities, this has proven effective, but in practice, wear and tear continue. In this way, it can be said that the formal heritagization of the Virgohamna cultural environment has happened somewhat randomly and was unplanned, despite the authorities' high ambitions. In its own way, this has also influenced the heritagization processes in Virgohamna.

The narrative of Virgohamna and what transpired there is strong and alive. The story of Andrée and his men, their bravery, and their fate is particularly impactful. Historic remains from Wellman's base are still numerous and visible, and the narratives surrounding his attempts to reach the North Pole are flourishing. Tourism has greatly contributed to maintaining the narrative of Virgohamna and its expeditions, both through visiting the site and retelling the stories. Although few visible traces of Andrée's base remain, it is still an obvious and significant experience to visit the place. This offers hope that the stories of what happened there can be carried on even when the physical traces have disappeared in Virgohamna. Perhaps it is precisely the tourists who have played the most significant role in preserving Virgohamna as an important cultural environment through their eagerness to visit the site and their awareness of the stories, despite the impact of tourism wearing down the place.

Further Development of the Cultural Environment in Virgohamna

The delicate cultural heritage of Svalbard faces a growing threat from the combined effects of climate change and increasing tourism. The impact of human activities, along with natural decay processes, creates a complex web of degradation, highlighting notable gaps in our understanding. To tackle these challenges, the Cultural Heritage Administration is advocating for an evidence-based management approach that aligns with the United Nations Sustainable Development Goals, placing a strong emphasis on the preservation of both global cultural and natural heritage. The analysis performed in this article can inform the development of effective management strategies for Virgohamna. It can provide insights into sustainable tourism practices, conservation measures, interpretation methods, and visitor management, ensuring the site's preservation while promoting educational and recreational opportunities. Involving the visitors more consciously and firmly in the preservation of the historic environment, for example, through monitoring activities and citizen science, can make visitors aware of heritage values and preserve the site. Tourism revenue, which has previously been introduced in Svalbard, might also contribute to the preservation of Virgohamna.

Sustainable tourism practices and responsible visitor behavior are crucial for balancing the benefits and impacts of tourism on the cultural environment. Visit Svalbard states in the Master Plan for Svalbard following the "White Paper for Destination Development" that while the visits of large and medium-sized cruise ships are decreasing due to regulations, there is growth potential in smaller expedition cruises [131]. These are the types of cruises that visit Virgohamna, and therefore, increased visits to the site can be expected. In addition, the Governor has announced that they will remove the visitation restrictions in Virgohamna. This will likely lead to the loss of the restrictive effect on visitation activities. Therefore, it is likely that the deterioration will accelerate. Alongside the expected natural deterioration, it will probably not take many years before this increased burden becomes evident on the historical remnants in Virgohamna.

Despite apparent high goals set by the authorities for the management of cultural environments in Svalbard, there are indications that these goals are not effectively implemented in practice regarding cultural environments. Environmental monitoring organized by the Governor heavily favors wildlife and nature, with hardly any programs dedicated to monitoring cultural environments. MOSJ (Environmental Monitoring of Svalbard and Jan Mayen) is an environmental monitoring system that is part of the national environmental monitoring in Norway [132]. One of its important functions is to provide a basis for assessing whether the political goals set for the environment in the northern regions are being achieved. This system has been in place since 1996, and monitoring reporting is conducted through numerous programs focused on wildlife and the natural side [132], but nothing has been initiated yet regarding cultural environments [132]. The Destination Svalbard Towards 2025 discussion, for example, only addresses nature and nature experiences, with no mention of cultural environments [131].

The narratives surrounding the expeditions originating from Virgohamna are powerful, strong, and alive. The story of Andrée and his men, their bravery, and their fate is particularly impactful. Tourism has greatly contributed to maintaining the narrative of

Virgohamna and its expeditions, both through visiting the site and retelling the stories. The physical remnants contribute to telling the story and strengthening the narrative. Although few visible traces of Andrée's base remain, it is still an obvious and significant experience to visit the place. This offers hope that the stories of what happened there can be carried on even when the physical traces have disappeared in Virgohamna. Perhaps it is precisely the tourists who have played the most significant role in preserving Virgohamna as an important cultural environment through their eagerness to visit the site and their awareness of the stories, despite the impact of tourism wearing down the place. It is, nevertheless, an open question how the narrative will be affected when the physical remnants disappear. Archaeological investigations and documentation work have also been conducted in Virgohamna. However, the physical traces of the expeditions are the most tangible historical evidence—and much remains hidden in the remnants of the expeditions in Virgohamna.

The authorities characterize Virgohamna as one of Svalbard's most valuable heritage sites. However, even the most valuable environments will eventually disappear.

Future research directions in Virgohamna may include exploring alternative methods of preservation or documentation to ensure the continuation of the narrative and historical significance even after the physical site has deteriorated. Additionally, examining the impact of tourism and visitor management on the preservation of the cultural landscape could provide insights into sustainable strategies for protecting similar heritage sites.

Overall, the findings highlight the complex processes that have influenced the cultural heritage landscape in Virgohamna, the challenges of preserving ruins, the enduring narrative associated with the site, and the need for continued efforts to ensure the preservation and dissemination of its historical significance.

5. Concluding Remarks

This analysis of the evolution of the heritage site, including the heritagization process of Virgohamna, can contribute to a deeper understanding of its historical significance, identifying heritage values, engaging stakeholders, developing management strategies, and informing policy recommendations. This comprehensive examination serves as a foundation for responsible and sustainable heritage management, ensuring the preservation and promotion of Virgohamna's cultural heritage for present and future generations.

All four key areas introduced in the Introduction have influenced the development in Virgohamna. It is primarily the expedition bases of Andrée and Wellman that best illustrate this development. The development has gone through several main stages, from initially being built as main bases for expeditions to the North Pole, then functioning as storage facilities for trappers, to being seen as almost a dump, and finally emerging as a valuable protected cultural environment where no parts can be touched or moved.

1. Policy documents and legal framework. The overarching political framework.

The overarching political framework has shaped this development, but the political guidelines were not put in place until Norway assumed sovereignty over Svalbard in 1925. Even then, the guidelines for cultural heritage management were inadequate. Trappers searching for materials for their cabins and tourists on souvenir hunts helped themselves for many years before legislation and overarching political guidelines put a stop to it. Originating from before 1900, the cultural heritage sites in Virgohamna, with the exception of Wellman's base, were automatically protected in 1974 when the first cultural heritage regulations were implemented. Wellman's base was included in the protection when the boundary was moved to 1 January 1946, with new regulations in 1992. That was also when the first cultural heritage plan for Svalbard was introduced, and Virgohamna was among the 35 initial cultural environments prioritized for follow-up.

2. Physical transformations of the region.

The deterioration is amplified by the natural processes taking place in Virgohamna, where strong winds, erosion, solifluction, decay, and rust are among the most evident.

Combined with the gradual dismantling of cultural heritage by trappers and tourists, degradation occurs.

3. Narratives and cultural heritage value.

The focus of cultural heritage management on the understanding that what may appear as garbage can also be valuable for preservation has likely saved the historical remains in Virgohamna. Over time, this understanding has become more widespread. However, if it were not for the strong stories and the fate of the participants in the expeditions, the interest would probably not be as strong. Perhaps there would have been less wear and tear from tourism, while the nation of Norway's strong commitment to environmental protection and pristine nature could have led to the "garbage" being cleaned up in Virgohamna. In addition to this, Capelotti's archaeological surveys and documentation work have revealed a wealth of information hidden in the "garbage".

4. Impact of tourism.

The strict overarching goals set by the Parliament and Government from the 1980s onward contributed to focusing on the preservation of cultural environments, even though the emphasis was on the natural environment. The strict goals of managing Svalbard's natural and cultural values as the best-managed in the world likely played a part in putting an end to souvenir picking and scavenging for building parts. However, wear and tear from tourism have increased significantly in recent years, and the goal of "best-managed cultural and natural environments" is difficult to reconcile with the clearly stated goal of increasing tourism as part of economic activity in Svalbard. The Governor's plan for visitor behavior in Virgohamna has many weaknesses and contributes partly to confusion and increased deterioration.

Today, Virgohamna is a heavily visited and highly sought-after cultural environment. If it is to continue being a source of experiences and wonder, providing information about technological development and insights into European history, some action should be taken. The impact of climate change is accelerating, and increasing tourism is exerting pressure. Furthermore, there is still more information hidden in the historical remains.

Three keywords can contribute to the future preservation of cultural heritage values: comprehensive documentation, improved visitor management, and knowledge dissemination/training of guides and tourists. Digital documentation of the entire vast cultural landscape can contribute to an increased understanding of cultural heritage values and ensure that at least a digital model exists when the cultural environment eventually breaks down completely. Better prepared and clearer information provision can make it easier for visitors to contribute to the preservation of the most vulnerable parts of the cultural environment. Training guides and visitors, emphasizing how to behave and move in the field, and not just focusing on the history itself, can spare the cultural heritage from excessive impact by visitors.

This way, the fascinating cultural environment with all its incredibly exciting stories and myths can live on for a long time, despite the physical decay that nothing can resist.

Funding: This research was funded by the Research Council of Norway, grant numbers 294314, 320507, and 437035, and the Svalbard Environmental Protection Fund, grant number 08/11.

Institutional Review Board Statement: Ethical review and approval were waived for this study as no sensitive information was collected in this project. Data from the interviews do not contain any sensitive information and names were not collected. The interviews will be reported to Norwegian Agency for Shared Services in Education and Research, a public administrative body under the Ministry of Education and Research.

Informed Consent Statement: Informed consent was obtained from all subjects involved in the study concerning interviews.

Data Availability Statement: Not applicable.

Acknowledgments: The author would like to thank the two anonymous reviewers for valuable comments and input and the co-researchers for fruitful discussions and smooth collaboration during simultaneous fieldwork. The author would also like to thank Kari Flyen for valuable assistance in proofreading the manuscript.

Conflicts of Interest: The author declares no conflict of interest. The funders had no role in the design of the study; in the collection, analyses, or interpretation of data; in the writing of the manuscript; or in the decision to publish the results.

References

1. Jones, S. Wrestling with the Social Value of Heritage: Problems, Dilemmas and Opportunities. *J. Community Archaeol. Herit.* **2017**, *4*, 21–37. [CrossRef]
2. Smith, L. *Uses of Heritage*; Routledge: London, UK, 2006; 351p.
3. Gong, Z.; Zang, Z.; Zhou, J.; Zhou, J.; Wang, W. The Evolutionary Process and Mechanism of Cultural Landscapes: An Integrated Perspective of Landscape Ecology and Evolutionary Economic Geography. *Land* **2022**, *11*, 2062. [CrossRef]
4. Nansen, F. *En Ferd til Spitsbergen*; Dybvad: Oslo, Norway, 1941; 227p.
5. Arlov, T.B. Forsknings- og ekspedisjonshistorie (ca. 1800–i dag). In *Katalog Prioriterte Kulturminner og Kulturmiljøer på Svalbard*; Sandodden, I.S., Ed.; Sysselmannen på Svalbard: Longyearbyen, Norway, 2013; pp. 19–22.
6. Harrison, R. *Heritage. Critical Approaches*; Routledge: London, UK; New York, NY, USA, 2013.
7. Løkka, N.; Hjemdahl, A.S.; Kleppe, B.R. *Frivillig kulturvern. Et kulturvern Innenfra*; Fagbokforlaget: Bergen, Norway, 2021; 286p.
8. Swensen, G.; Guttormsen, T.S. Introduksjon. Å lage kulturminner. In *Å Lage Kulturminner—Hvordan Kulturarv Forstås, Formes og Forvaltes*; Swensen, G., Ed.; Novus Forlag: Oslo, Norway, 2013; 370p.
9. Sandodden, I.S.; Reymert, P.K.; Hultgreen, T.; Hauan, M.A.; Arlov, T.B. *Katalog Prioriterte Kulturminner og Kulturmiljøer på Svalbard*; Sysselmannen på Svalbard: Longyearbyen, Norway, 2013; 220p.
10. Arlov, T.B. *A Short History of Svalbard*; Norsk Polarinstitutt: Oslo, Norway, 1994; 95p.
11. Capelotti, P.J. *The Wellman Polar Airship Expeditions at Virgohamna, Danskøya, Svalbard: A Study in Aerospace Archaeology*; Norsk Polarinstitutt: Oslo, Norway, 1997; 101p.
12. Labadi, S. Representations of the nation and cultural diversity in discourses on World Heritage. *J. Soc. Archaeol.* **2007**, *7*, 147–170. [CrossRef]
13. Waterton, E.; Smith, L.; Campbell, G. The Utility of Discourse Analysis to Heritage Studies: The Burra Charter and Social Inclusion. *Int. J. Herit. Stud.* **2006**, *12*, 339–355. [CrossRef]
14. Andersson, F.K.; Frihammar, M. As above, so below? On AHD critique, identity, essence and Cold War heritagizations in Sweden. *J. War Cult. Stud.* **2022**, 1–20. [CrossRef]
15. Gentry, K.; Smith, L. Critical heritage studies and the legacies of the late-twentieth century heritage canon. *Int. J. Herit. Stud.* **2019**, *25*, 1148–1168. [CrossRef]
16. Paradiso, C. A vineyard landscape, a UNESCO inscription and a National Park. In *Landscape as Heritage International Critical Perspectives*; Pettenati, G., Ed.; Routledge: London, UK, 2023; pp. 56–66.
17. Simensen, T.; Halvorsen, R.; Erikstad, L. Methods for landscape characterisation and mapping: A systematic review. *Land Use Policy* **2018**, *25*, 557–569. [CrossRef]
18. Sánchez-Carretero, C. Heritagization of the Camino to Finisterre. In *Heritage, Pilgrimage and the Camino to Finisterre*; Sánchez-Carretero, C., Ed.; Springer: Cham, Germany, 2015; Volume 117, pp. 95–119.
19. Meyer, B. Recycling the Christian Past. The Heritagization of Christianity and National Identity in the Netherlands. In *Cultures, Citizenship and Human Rights*; Buikema, R., Buyse, A., Robben, A.C.G.M., Eds.; Routledge: London, UK; New York, NY, USA, 2020; pp. 64–88.
20. Zhang, Q. Intangible Cultural Heritage Safeguarding in Times of Crisis. *Asian Ethnol.* **2020**, *7*, 91–113.
21. O'Hare, D. Interpreting the cultural landscape for tourism development. *Urban Des. Int.* **1997**, *2*, 33–54. [CrossRef]
22. Pastor Pérez, A.; Barreiro Martínez, D.; Parga-Dans, E.; Alonso González, P. Democratising Heritage Values: A Methodological Review. *Sustainability* **2021**, *13*, 12492. [CrossRef]
23. Bachi, L.; Ribeiro, S.C.; Hermes, J.; Saadi, A. Cultural Ecosystem Services (CES) in landscapes with a tourist vocation: Mapping and modeling the physical landscape components that bring benefits to people in a mountain tourist destination in southeastern Brazil. *Tour. Manag.* **2020**, *77*, 104017. [CrossRef]
24. Plieninger, T.; Kizos, T.; Bieling, C.; Le Dû-Blayo, L.; Budniok, M.A.; Bürgi, M.; Crumley, C.L.; Girod, G.; Howard, P.; Kolen, J.; et al. Exploring ecosystem-change and society through a landscape lens: Recent progress in European landscape research. *Ecol. Soc.* **2015**, *20*, 5. [CrossRef]
25. Labadi, S. *UNESCO, Cultural Heritage, and Outstanding Universal Value: Value-Based Analyses of the World Heritage and Intangible Cultural Heritage Conventions*; AltaMira Press: Plymuth, UK, 2013.
26. Alberts, H.C.; Hazen, H.D. Maintaining Authenticity and Integrity at Cultural World Heritage Sites. *Geogr. Rev.* **2010**, *100*, 56–73. [CrossRef]
27. Nordberg-Schulz, C. *Genius Loci: Towards a Phenomenology of Architecture*; Academy Editions: London, UK, 1980.

28. Harvey, D.C. Heritage Pasts and Heritage Presents: Temporality, meaning and the scope of heritage studies. *Int. J. Herit. Stud.* **2001**, *7*, 319–338. [CrossRef]
29. Grydehøj, A. Uninherited heritage: Tradition and heritage production in Shetland, Åland and Svalbard. *Int. J. Herit. Stud.* **2010**, *16*, 77–89. [CrossRef]
30. Graham, B.J.; Ashworth, G.J.; Tunbridge, J.E. *A Geography of Heritage: Power, Culture and Economy*; Routledge: London, UK; New York, UY, USA, 2016; pp. 29–53.
31. Altaba Tena, P.; García-Esparza, J.A. The Heritagization of a Mediterranean Vernacular Mountain Landscape: Concepts, Problems and Processes. *Herit. Soc.* **2019**, *11*, 189–210. [CrossRef]
32. Yang, H.; Qiu, L.; Fu, X. Toward Cultural Heritage Sustainability through Participatory Planning Based on Investigation of the Value Perceptions and Preservation Attitudes: Qing Mu Chuan, China. *Sustainability* **2021**, *13*, 1171. [CrossRef]
33. Megeirhi, H.A.; Woosnam, K.M.; Ribeiro, M.A.; Ramkissoon, H.; Denley, T.J. Employing a value-belief-norm framework to gauge Carthage residents' intentions to support sustainable cultural heritage tourism. *J. Sustain. Tour.* **2020**, *28*, 1351–1370. [CrossRef]
34. De Luca, G.; Shirvani Dastgerdi, A.; Francini, C.; Liberatore, G. Sustainable cultural heritage planning and management of overtourism in art cities: Lessons from atlas world heritage. *Sustainability* **2020**, *12*, 3929. [CrossRef]
35. Landorf, C. A framework for sustainable heritage management: A study of UK industrial heritage sites. *Int. J. Herit. Stud.* **2009**, *15*, 494–510. [CrossRef]
36. Hagen, D.; Vistad, O.I.; Eide, N.E.; Flyen, A.C.; Fangel, K. Managing visitor sites in Svalbard: From a precautionary approach towards knowledge-based management. *Polar Res.* **2012**, *31*, 18432. [CrossRef]
37. Kruse, F. Catching up: The state and potential of historical catch data from Svalbard in the European Arctic. *Polar Rec.* **2017**, *53*, 520–533. [CrossRef]
38. Hacquebord, L.; Avango, D. Industrial heritage sites in Spitsbergen (Svalbard), South Georgia and the Antarctic Peninsula: Sources of historical information. *Polar Sci.* **2016**, *10*, 433–440. [CrossRef]
39. Capelotti, P.J. Extreme archaeological sites and their tourism: A conceptual model from historic American polar expeditions in Svalbard, Franz Josef Land and Northeast Greenland. *Polar J.* **2012**, *2*, 236–255. [CrossRef]
40. Arlov, T.B. The discovery and early exploitation of Svalbard. Some historiographical notes. *Acta Boreal.* **2005**, *22*, 3–19. [CrossRef]
41. Jørgensen, R. Archaeology on Svalbard: Past, present and future. *Acta Boreal.* **2005**, *22*, 49–61. [CrossRef]
42. Arlov, T.B. *Svalbards Historie*, 2nd ed.; Tapir Akademisk Forlag: Trondheim, Norway, 2003; 499p.
43. Tora Hultgreen, T. When Did the Pomors Come to Svalbard? *Acta Boreal.* **2002**, *19*, 125–145. [CrossRef]
44. Jasinski, M.E. Russian hunters on Svalbard and the polar winter. *Arctic* **1991**, *44*, 156–162. [CrossRef]
45. Capelotti, P.J. A preliminary archaeological survey of Camp Wellman at Virgohamn, Danskøya, Svalbard. *Polar Rec.* **1994**, *30*, 265–276. [CrossRef]
46. Capelotti, P.J. Virgohamna and the archaeology of failure. In *Centennial of SA Andrée's North Pole Expedition, Proceedings of the Conference on SA Andrée and the Agenda for Social Science Research of the Polar Regions, Longyearbyen, Svalbard, 16–17 August 1999*; Wråkberg, U., Ed.; Royal Swedish Academy of Sciences: Stockholm, Sweden, 1999; Volume 29, pp. 37–52.
47. Capelotti, P.J.; Van Dyk, H.; Cailliez, J.C. Strange interlude at Virgohamna, Danskøya, Svalbard, 1906: The merkelig mann, the engineer and the spy. *Polar Res.* **2007**, *26*, 64–75. [CrossRef]
48. Johansen, B.F.; Prestvold, K.; Overrein, Ø. *The Cruise Handbook for Svalbard*; Norwegian Polar Institute: Tromsø, Norway, 2015. Available online: https://cruise-handbook.npolar.no/en/nordvesthjornet/virgohamna.html (accessed on 1 March 2023).
49. Roll, L. *Kulturminneplan for Svalbard 1994–1998*; Sysselmannen på Svalbard: Longyearbyen, Svalbard, 1994; Unpublished manuscript, last modified April 1994.
50. Dahle, K.; Bjerck, H.B.; Prestvold, K. *Kulturminneplan for Svalbard 2000–2010*; Sysselmannen på Svalbard: Longyearbyen, Svalbard, 2010; 125p.
51. Norwegian Polar Institute. *The Place Names of Svalbard*; Norwegian Polar Institute: Tromsø, Norway, 2003; 537p.
52. Wilse, A.B. Spitsbergen—Walter Wellmanns Balong Station fra Heien 18/8 1906. 1906. Photograph. DigitaltMuseum. Available online: https://digitaltmuseum.no/0210111916339/prot-spitsbergen-walter-wellmanns-balong-station-fra-heien-18-8-1906 (accessed on 24 April 2023).
53. Directorate for Cultural Heritage. Kulturminnesøk. Available online: https://www.kulturminnesok.no/kart/?q=virgohamna,svalbard&am-county=&lokenk=location&am-lok=&am-lokdating=&am-lokconservation=&am-enk=&am-enkdating=&am-enkconservation=&bm-county=&cp=1&bounds=79.72408832956863,10.85423469543457,79.71575568971976,10.93663215637207&zoom=16&id= (accessed on 10 March 2023).
54. Svalbardmiljøloven. Lov om Miljøvern på Svalbard (LOV-2021-06-18-122). 2001. Available online: https://lovdata.no/dokument/NL/lov/2001-06-15-79 (accessed on 1 March 2023).
55. Hoel, A. *Svalbard: Svalbards Historie 1596–1965*; Bind 1; S. Kildahls Boktrykkeri: Oslo, Norway, 1966; 487p.
56. Reilly, J.T. *Greetings from Spitsbergen: Tourists at the Eternal Ice 1827–1914*; Tapir Academic Press: Trondheim, Norway, 2009; 227p.
57. Chapman, A. *Wild Norway: With Chapters on Spitzbergen, Denmark*; Etc. E. Arnold: London, UK; New York, NY, USA, 1897; 351p.
58. Bjerck, H.B. *Overvåkning av Kulturmiljø på Svalbard. Målsetting, Metode, Lokaliteter og Overvåkning*; Sysselmannen på Svalbard: Longyearbyen, Svalbard, 1999; 63p.
59. Bjerck, H.B.; Johannessen, L.J. *Virgohamna. I lufta mot Nordpolen*; Sysselmannen på Svalbard: Longyearbyen, Svalbard, 1999; 35p.
60. Kræmer, W. *Fangstliv i 45 år*; Vågemot Miniforlag: Skien, Norway, 2007; 44p.

61. Wilse, A.B. Wellmans Station—Pikes hus og Andrees Støtte 19/8 1908. 1908. Photograph. DigitaltMuseum. Available online: https://digitaltmuseum.no/021015795775/wellmans-station-pikes-hus-og-andrees-stotte-19-8-1908 (accessed on 1 June 2023).
62. Grenna Museum–Andréexpeditionen Polarcenter. Danskön Med Ballonghuset i Centrum. *Till Vänster Skymtar så Kallade Pikes hus Intill Vätgasapparaten Med Ett Stort Antal kärl och Trälådor för Svavelsyra och Järnfilspån. I Förgrunden en upp och Nedvänd Mindre Båt. 1896–1897. Photograph. DigitaltMuseum.* Available online: https://digitaltmuseum.se/021019241162/danskon-med-ballonghuset-i-centrum-till-vanster-skymtar-sa-kallade-pikes (accessed on 24 April 2023).
63. Wilse, A.B. Spitsbergen, Wellmans Motorer, Gondol og Balonghus. 1906. Photograph. DigitaltMuseum. Available online: https://digitaltmuseum.no/0210111916342/prot-spitsbergen-wellmans-motorer-gondol-og-balonghus (accessed on 1 June 2023).
64. Hagen, D.; Eide, N.E.; Flyen, A.C.; Fangel, K.; Vistad, O.I. *Håndbok for Sårbarhetsvurdering av Ilandstigningslokaliteter på Svalbard*; Norsk Institutt for Naturforskning: Trondheim/Lillehammer/Oslo, Norway, 2014; 65p.
65. Hagen, D.; Eide, N.E.; Fangel, K.; Flyen, A.C.; Vistad, O.I. *Sårbarhetsvurdering og bruk av Lokaliteter på Svalbard. Sluttrapport fra Forskningsprosjektet "Miljøeffekter av ferdsel?"*; Norsk Institutt for Naturforskning: Trondheim/Lillehammer/Oslo, Norway, 2012; 123p.
66. "TopoSvalbard". *Norsk Polarinstitutt*. Available online: https://toposvalbard.npolar.no/ (accessed on 2 March 2023).
67. "Photo Library". *Norsk Polarinstitutt*. Available online: https://www.npolar.no/en/photo-library/ (accessed on 1 February 2023).
68. "Kulturminnebilder". *Riksantikvaren*. Available online: https://kulturminnebilder.ra.no/fotoweb/ (accessed on 1 February 2023).
69. "DigitaltMuseum". *KulturIT AS*. Available online: https://digitaltmuseum.org/ (accessed on 1 February 2023).
70. Norway. Ministry of Justice and Public Security. Svalbardloven (1925) Lov om Svalbard. Available online: https://lovdata.no/dokument/NL/lov/1925-07-17-11 (accessed on 10 February 2023).
71. Sandodden, I.S.; Yri, H.T.; Solli, H. *Kulturminneplan for Svalbard 2013–2023*; Sysselmannen på Svalbard: Longyearbyen, Svalbard, 2013; 110p.
72. Norway. Ministry of Climate and Environment. Forskr. om Fredning av Kulturminner, Jan Mayen. Forskrift om Fredning av Kulturminner på Jan Mayen. Repealed. Available online: https://lovdata.no/dokument/SFO/forskrift/1974-06-21-8792?q=forskrift%20om%20kulturminner%20p%C3%A5%20svalbard (accessed on 10 February 2023).
73. Norway. Ministry of Climate and Environment. Forskrift om Kulturminner på Svalbard. Forskrift om Kulturminner på Svalbard–Repealed. Available online: https://lovdata.no/dokument/SFO/forskrift/1992-01-24-34 (accessed on 10 February 2023).
74. Norway. Ministry of Climate and Environment. The Svalbard Environmental Protection Act. Available online: https://www.regjeringen.no/en/dokumenter/svalbard-environmental-protection-act/id173945/ (accessed on 10 February 2023).
75. Jensen, Ø. The Svalbard Treaty and Norwegian Sovereignty. *Arct. Rev. Law Politics* **2020**, *11*, 82–107. [CrossRef]
76. Ministry of Justice and Public Security. *White Paper No. 39 (1974–75) Vedrørende Svalbard*; Ministry of Justice and Public Security: Oslo, Norway, 1973. Available online: https://www.stortinget.no/no/Saker-og-publikasjoner/Stortingsforhandlinger/Lesevisning/?p=1974-75&paid=3&wid=c&psid=DIVL814&s=False&pgid=c_0835 (accessed on 25 June 2023).
77. Ministry of Justice and Public Security. *White Paper No. 40 (1985–86) Svalbard*; Ministry of Justice and Public Security: Oslo, Norway, 1985. Available online: https://www.stortinget.no/no/Saker-og-publikasjoner/Stortingsforhandlinger/Lesevisning/?p=1985-86&paid=3&wid=c&psid=DIVL1260&s=True (accessed on 25 June 2023).
78. Ministry of Justice and Public Security. *White Paper No. 9 (1999–2000) Svalbard*; Ministry of Justice and Public Security: Oslo, Norway, 1999. Available online: https://www.regjeringen.no/contentassets/44611dfd09664fcd9beac74a91016fb0/no/pdfa/stm199920000009000dddpdfa.pdf (accessed on 25 June 2023).
79. Ministry of Justice and the Police. *White Paper No. 22 (2008–2009) Svalbard*; Ministry of Justice and the Police: Oslo, Norway, 2009. Available online: https://www.regjeringen.no/contentassets/e70b04df32ad45f483f2619939c5636d/en-gb/pdfs/stm200820090022000en_pdfs.pdf (accessed on 25 June 2023).
80. Ministry of Justice and Public Security. *White Paper 32 (2015–2016) Svalbard*; Ministry of Justice and Public Security: Oslo, Norway, 2016. Available online: https://www.regjeringen.no/contentassets/379f96b0ed574503b47765f0a15622ce/en-gb/pdfs/stm201520160032000engpdfs.pdf (accessed on 25 June 2023).
81. Moe, A.; Jensen, Ø.H. Svalbard og havområdene–nye utenrikspolitiske utfordringer for Norge? *Int. Polit.* **2020**, *78*, 511–522. [CrossRef]
82. Ministry of Climate and Environment. *White Paper No. 22 (1994–95) Om miljøvern på Svalbard*; Ministry of Climate and Environment: Oslo, Norway, 1995; Available online: https://www.stortinget.no/no/Saker-og-publikasjoner/Stortingsforhandlinger/Lesevisning/?p=1994-95&paid=3&wid=b&psid=DIVL618 (accessed on 25 June 2023).
83. Ministry of Climate and Environment. *White Paper No. 16 (2004–2005) Living with Our Cultural Heritage*; Ministry of Climate and Environment: Oslo, Norway, 2005. Available online: https://www.regjeringen.no/contentassets/c6311a10138f40e590c81f3891c96a87/no/pdfs/stm200420050016000dddpdfs.pdf (accessed on 25 June 2023).
84. Ministry of Climate and Environment. *White Paper No. 21 (2004–2005) the Government's Environmental Policy and the State of the Environment in Norway*; Ministry of Climate and Environment: Oslo, Norway, 2005. Available online: https://www.regjeringen.no/contentassets/b6d22b1df80143279e3d74601ec7f5d7/en-gb/pdfs/stm200420050021000en_pdfs.pdf (accessed on 25 June 2023).
85. The Governor of Svalbard. Roles in Cultural Heritage Management. Available online: https://www.sysselmesteren.no/en/the-governor-of-svalbard/environmental-protection/cultural-heritage-management/roles-in-cultural-heritage-management/ (accessed on 31 May 2023).

86. Ministry of Justice and Public Security; Ministry of Climate and Environment. *Tildelingsbrev 2023. Sysselmesteren på Svalbard*; Ministry of Justice and Public Security: Oslo, Norway; Ministry of Climate and Environment: Oslo, Norway, 2022. Available online: https://www.regjeringen.no/contentassets/cd6eaa9166364869b9bb95bd738a5953/tildelingsbrev-sysselmesteren-2023.pdf (accessed on 25 June 2023).
87. Norway. Ministry of Climate and Environment. Forskrift om Områdefredning og Ferdselsregulering i Virgohamna på Svalbard. Available online: https://lovdata.no/dokument/SF/forskrift/2000-05-03-526 (accessed on 10 February 2023).
88. The Governor of Svalbard. VIRGOHAMN. Map of Area with Entrance Prohibition and Entrance Regulations. Available online: https://www.sysselmesteren.no/siteassets/kart/temakart/naturvernomrader/kulturminneomrader/kart-virgohamna.pdf (accessed on 31 May 2023).
89. Sesana, E.; Gagnon, A.S.; Ciantelli, C.; Cassar, J.A.; Hughes, J.J. Climate change impacts on cultural heritage: A literature review. *Wiley Interdiscip. Rev. Clim. Chang.* **2021**, *12*, e710. [CrossRef]
90. Flyen, A.C.; Thuestad, A.E. A Review of Fungal Decay in Historic Wooden Structures in Polar Regions. *Conserv. Manag. Archaeol. Sites* **2023**, 1–33. [CrossRef]
91. Flyen, A.C.; Flyen, C.; Mattsson, J. Climate change impacts and fungal decay in vulnerable historic structures at Svalbard. *E3SWeb Conf.* **2020**, *172*, 20006. [CrossRef]
92. Nicu, I.C.; Stalsberg, K.; Rubensdotter, L.; Martens, V.V.; Flyen, A.C. Coastal erosion affecting cultural heritage in Svalbard. A case study in Hiorthhamn (Adventfjorden)—An abandoned mining settlement. *Sustainability* **2020**, *12*, 2306. [CrossRef]
93. Nicu, I.C.; Fatorić, S. Climate change impacts on immovable cultural heritage in polar regions: A systematic bibliometric review. *Wiley Interdiscip. Rev. Clim. Chang.* **2023**, *14*, e822. [CrossRef]
94. Øian, H.; Kaltenborn, B. *Turisme på Svalbard og i Arktis. Effekter på Naturmiljø, Kulturminner og Samfunn med Hovedvekt på Cruiseturisme. Turisme på Svalbard og i Arktis*; NINA Rapp. 1745; Norsk Institutt for Naturforskning: Lillehammer, Norway, 2020; 56p.
95. Jørgen Hollesen, J.; Callanan, M.; Dawson, T.; Fenger-Nielsen, R.; Friesen, T.M.; Jensen, A.M.; Markham, A.; Martens, V.V.; Pitulko, V.V.; Rockman, M. Climate change and the deteriorating archaeological and environmental archives of the Arctic. *Antiquity* **2018**, *92*, 573–586. [CrossRef]
96. Hall, C.M.; Baird, T.; James, M.; Ram, Y. Climate change and cultural heritage: Conservation and heritage tourism in the Anthropocene. *J. Herit. Tour.* **2016**, *11*, 10–24. [CrossRef]
97. Barr, S. The effects of climate change on cultural heritage in the polar regions. *Herit. Risk* **2009**, *2006/2007*, 203–205.
98. Masson-Delmotte, V.; Zhai, P.; Pirani, A.; Connors, S.L.; Péan, C.; Berger, S.; Caud, N.; Chen, Y.; Goldfarb, L.; Gomis, M.I. (Eds.) *Climate Change 2021: The Physical Science Basis. Contribution of Working Group I to the Sixth Assessment Report of the Intergovernmental Panel on Climate Change*; Intergovernmental Panel on Climate Change: Geneva, Switzerland, 2021; pp. 1–250.
99. Miller, L.B.; Miller, L.B.; Hallo, J.C.; Dvorak, R.G.; Fefer, J.P.; Peterson, B.A.; Brownlee, M.T.J. On the edge of the world: Examining proenvironmental outcomes of last chance tourism in Kaktovik, Alaska. *J. Sustain. Tour.* **2020**, *28*, 1703–1722. [CrossRef]
100. Hindley, A.; Font, X. Values and motivations in tourist perceptions of last-chance tourism. *Tour. Hosp. Res.* **2018**, *18*, 3–14. [CrossRef]
101. Dawson, J.; Johnston, M.J.; Stewart, E.J.; Lemieux, C.J.; Lemelin, R.H.; Maher, P.T.; Grimwood, B.S.R. Ethical considerations of last chance tourism. *J. Ecotour.* **2011**, *10*, 250–265. [CrossRef]
102. Harvey Lemelin, H.; Dawson, J.; Stewart, E.J.; Maher, P.; Lueck, M. Last-chance tourism: The boom, doom, and gloom of visiting vanishing destinations. *Curr. Issues Tour.* **2010**, *13*, 477–493. [CrossRef]
103. Norwegian Polar Institute. MOSJ. Environmental Monitoring of Svalbard and Jan Mayen. Climate. Available online: https://mosj.no/en/indikator/climate/ (accessed on 7 June 2023).
104. Sysselmesteren på Svalbard. *Veileder for Ekspedisjonscruise (Kystcruise) på og Rundt Svalbard under COVID-19 Utbruddet 2020*, 2nd ed.; Sysselmesteren på Svalbard: Longyearbyen, Svalbard, 2020; pp. 1–68. Available online: https://www.sysselmesteren.no/contentassets/6e10fd653e404de99941e967de5d1dd7/2020-07-14_veileder-for-ekspedisjonscruise-kystcruise-pa-og-rundt-svalbard-under-utbruddet-av-covid-19_versjon-2.0.pdf (accessed on 5 May 2023).
105. Blankholm, H.P. Long-Term Research and Cultural Resource Management Strategies in Light of Climate Change and Human Impact. *Arct. Anthropol.* **2009**, *46*, 17–24. [CrossRef]
106. Nicu, I.C.; Rubensdotter, L.; Stalsberg, K.; Nau, E. Coastal Erosion of Arctic Cultural Heritage in Danger: A Case Study from Svalbard, Norway. *Water* **2021**, *13*, 784. [CrossRef]
107. Mattsson, J.; Flyen, A.C.; Nunez, M. Wood-decaying fungi in protected buildings and structures on Svalbard. *Agarica* **2010**, *29*, 5–14.
108. Bjørvik, P. *Hardhausen. Dagboksnotater frå Frans Josefs Land og Danskøya*; Vågemot Miniforlag: Skien, Norway, 2009; 60p.
109. Schibsted, A.; Resvoll-Dieset, H. *Sommerliv på Spitsbergen 1908*; Vågemot Miniforlag: Skien, Norway, 1998; 64p.
110. Bjørvik 1908: Bjørviks Dagbok «Stile Tanker om Vinteren 1908-9 paa Spidsbergen», Tilgjengelig på. Available online: https://www.polarhistorie.no/filearchive/010_Bjorvig.pdf (accessed on 1 June 2023).
111. Alme, H.H. *Om Spitsbergen og den Wellmanske Polarekspedisjonen*; Vågemot Miniforlag: Skien, Norway, 1998; 48p.
112. Cameron, G.J. *From Pole to Pole: Roald Amundsen's Journey in Flight*; Pen & Sword Books Limited: Barnsley, UK, 2013; 193p.
113. Robinson, M. Manliness and Exploration: The Discovery of the North Pole. *Sci. Masculinities* **2015**, *30*, 89–109. Available online: https://www.jstor.org/stable/10.1086/682968 (accessed on 25 June 2023). [CrossRef]

114. Basberg, B.L. Commercial and economic aspects of Antarctic exploration—From the earliest discoveries into the "Heroic Age". *Polar J.* **2017**, *7*, 205–226. [CrossRef]
115. Headland, R.K. *A Chronology of Antarctic Exploration*; New Quaritch Publ.: London, UK, 2009.
116. Headland, R.K. Antarctic Odyssey. Historical Stages in Development of Knowledge of the Antarctic. In *Antarctic Challenges: Historical and Current Perspectives on Otto Nordenskjöld's Antarctic Expedition 1901–1903*; Elzinga, A., Nordin, T., Turner, D., Wråkberg, U., Eds.; Royal Society of Arts and Sciences: Göteborg, Sweden, 2004; pp. 15–24.
117. Headland, R.K. *The Island of South Georgia*; Cambridge University Press: Cambridge, UK, 1984.
118. Rawlins, D. Byrd's heroic 1926 North Pole failure. *Polar Rec.* **2000**, *36*, 25–50. [CrossRef]
119. Skattum, O.J. Roald Amundsen som geografisk opdager. *Nor. J. Geogr.* **1928**, *2*, 147–175. [CrossRef]
120. Diesen, J.A.; Fulton, N. Wellman's 1906 polar expedition: The subject of numerous newspaper stories and one obscure film. *Polar Res.* **2007**, *26*, 76–85. [CrossRef]
121. Barr, S. 2019 Cultural heritage, or how bad news can also be good. In *Arctic Triumph: Northern Innovation and Persistence*; Sellheim, N., Zaika, Y.V., Kelman, I., Eds.; Springer Polar Sciences: Cham, Switzerland, 2019; pp. 43–57. [CrossRef]
122. Uusma, B. *Expeditionen. Min Karlekshistoria*; Norstedts: Stockholm, Sweden, 2013; 315p.
123. Martinson, T. Recovering the visual history of the Andrée Expedition: A case study in photographic research Tyrone Martinsson. *Res. Issues Art Des. Media* **2004**, *6*, 1–13.
124. Chaplin, B.; Barr, S. Polar Heritage-Rubbish or Relics? *Herit. Risk* **2003**, *2002/2003*, 233–235.
125. Hovelsrud, G.K.; Olsen, J.; Nilsson, A.E.; Kaltenborn, B.; Lebel, J. Managing Svalbard Tourism: Inconsistencies and Conflicts of Interest. *Arct. Rev. Law Politics* **2023**, *14*, 86–106. [CrossRef]
126. Norway. Ministry of Climate and Environment. Forskrift om Leiropphold. 2002. Available online: https://lovdata.no/dokument/SF/forskrift/2002-06-27-731?q=leirforskriften (accessed on 10 February 2023).
127. Norway. The Ministry of Justice and Public Security. Forskrift om Turisme mv. på Svalbard. Forskrift om Turisme, Feltopplegg og Annen Reisevirksomhet på Svalbard. 1992. Available online: https://lovdata.no/dokument/SF/forskrift/1991-10-18-671 (accessed on 10 February 2023).
128. Norwegian Polar Institute. MOSJ. Environmental Monitoring of Svalbard and Jan Mayen. Influence. Available online: https://mosj.no/en/indikator/influence/traffic/cruise-tourism/ (accessed on 7 June 2023).
129. Sysselmesterns Årsrapport 2022. Available online: https://www.sysselmesteren.no/contentassets/cc1280ad5c8841638bf6c58d9172ddca/arsrapport-2022.pdf (accessed on 5 May 2023).
130. Dannevig, H.; Søreide, J.E.; Sveinsdóttir, A.G.; Olsen, J.; Hovelsrud, G.K.; Rusdal, T.; Dale, R.F. Coping with rapid and cascading changes in Svalbard: The case of nature-based tourism in Svalbard. *Front. Hum. Dyn.* **2023**, *5*, 1178264. [CrossRef]
131. Brunvoll, R.; Jervan, B.; Pedersen, A.-J. *Destinasjon Svalbard mot 2025. Masterplan for Svalbard, 2015–2025*; Visit Svalbard: Longyearbyen, Svalbard, 2015; 52p. Available online: https://www.visitsvalbard.com/dbimgs/Masterplan%20Destinasjon%20Svalbard%20mot%202025.pdf (accessed on 25 June 2023).
132. Norwegian Polar Institute. MOSJ. Environmental Monitoring of Svalbard and Jan Mayen. About MOSJ. Available online: https://mosj.no/en/about-mosj/ (accessed on 7 June 2023).

Article

The Evolution of Historic Agroforestry Landscape in the Northern Apennines (Italy) and Its Consequences for Slope Geomorphic Processes

Filippo Brandolini [1,*], Chiara Compostella [2], Manuela Pelfini [2] and Sam Turner [1]

[1] McCord Centre for Landscape, School of History, Classics and Archaeology, Newcastle University, Armstrong Building, Newcastle upon Tyne NE17RU, UK
[2] Dipartimento di Scienze della Terra "Ardito Desio", Università degli Studi di Milano, 20133 Milano, Italy
* Correspondence: filippo.brandolini@newcastle.ac.uk

Abstract: Historic agricultural practices have played a dominant role in shaping landscapes, creating a heritage which must be understood and conserved from the perspective of sustainable development. Agroforestry (i.e., the practice of combining trees with agriculture or livestock) has existed since ancient times in European countries, and it has been recognised as one of the most resilient and multifunctional cultural landscapes, providing a wide range of economic, sociocultural, and environmental benefits. This research explores aspects of the history, physical characteristics, decline, and current state of conservation of historic agroforestry systems on the Northern Apennines in Italy, using an interdisciplinary approach combining archival sources, landscape archaeology, dendrochronology, and GIS analysis. Furthermore, through computer-based modelling, this research aims to evaluate how the abandonment of this historic rural land-use strategy impacted slope geomorphic processes over the long term. The importance of environmental values attached to traditional rural landscapes has received much attention even beyond the heritage sector, justifying the definition of transdisciplinary approaches necessary to ensure the holistic management of landscapes. Through the integration of the Unit Stream Power-Based Erosion Deposition (USPED) equation with landscape archaeological data, the paper shows how restoring the historic agroforestry landscape could significantly mitigate soil mass movements in the area. Thus, the interdisciplinary workflow proposed in this study enables a deep understanding of both the historical evolution of agroforestry systems and its resulting effects for cumulative soil erosion and deposition in the face of climate change.

Keywords: remote sensing and GIS; historic landscape characterisation; slope processes; landscape archaeology; landscape modelling; transdisciplinary landscape studies; geomorphometry; alberata emiliana

Citation: Brandolini, F.; Compostella, C.; Pelfini, M.; Turner, S. The Evolution of Historic Agroforestry Landscape in the Northern Apennines (Italy) and Its Consequences for Slope Geomorphic Processes. *Land* **2023**, *12*, 1054. https://doi.org/10.3390/land12051054

Academic Editor: Rui Alexandre Castanho

Received: 3 April 2023
Revised: 1 May 2023
Accepted: 9 May 2023
Published: 12 May 2023
Corrected: 3 November 2023

1. Introduction

Agriculture represents the largest land-use type worldwide, and deciphering the processes that created today's rural landscapes is fundamental to understanding how human activities have altered natural resources and geomorphic processes in the past [1–5]. Recent environmental studies and policies have recommended maintaining traditional rural landscape features such as intercropping, agroforestry, and cross-slope barriers (e.g., hedgerows, stone walls, earth banks, ridges, and furrows) for their potential benefits to ecosystems [6,7]. Over the long term, agroforestry systems (i.e., the practice of combining trees with agriculture or livestock) are among the most resilient types of rural land use [8].

Agroforestry has existed since ancient times in European countries, and it has given rise to a wide variety of multifunctional historic landscapes [9], such as *dehesa* in Spain [10], *montado* in Portugal [11], *plužiny* in the Czech Republic [12], and *streuobst* in Germany [13]. Agroforestry systems are still widely implemented [14] in tropical areas where geoarchaeological studies have also demonstrated their central economic, cultural, and ecological role even in past societies [15–17]. During the 20th century, as climate change emerged as a pressing global issue, agroforestry as a polyculture strategy garnered significant attention as a promising way to manage land, providing a wide range of economic, sociocultural, and environmental benefits [18,19]. Some of those include sustaining biodiversity [14], promoting carbon sequestration [20], improving the water balance, lowering risk of wildfire, and preserving traditional agricultural landscapes and rural knowledge [21]. Moreover, numerous studies [21–24] have demonstrated the significant impact of agroforestry systems on slope stability. Furthermore, the practice of integrating trees, shrubs, and other perennial plants with crops plays a crucial role in holding soil together and resisting mass movement in landslides, making them an effective measure to prevent slope instability and promote soil conservation.

Soil erosion caused by water is a complex process that occurs in three stages. Firstly, the soil particles become detached from the soil mass due to the force of rainfall or runoff. Secondly, the detached particles are transported by the moving water, as either dissolved or suspended solids. Finally, the transported soil is deposited somewhere away from its original location [25]. The extent and severity of soil erosion can vary widely depending on a range of site-specific and regional factors. These include the slope gradient, soil type, vegetation cover, and rainfall intensity and frequency [26,27]. Intensive storm events can trigger rainfall-induced landslides [28] with severe consequences on the environment (e.g., loss of topsoil and destruction of vegetation and habitat for wildlife) [29], human settlements (e.g., damage to infrastructure), economy (e.g., increased costs for disaster response and recovery efforts) [30], and cultural heritage [31,32]. Since land-use and land-cover dynamics are the major anthropogenic drivers of soil erosion and degradation [33], the development of sustainable rural strategies is fundamental to cope with this environmental hazard.

Italy is among the European countries most affected by the natural hazard of slope instability, which leads to an increased risk of landslides [34]. In Northern and Central Italy, agroforestry was widespread in the past [35] but survives only in a few areas in the form of relics. The Italian term *coltura promiscua* indicates the typical association of trees, vines and arable crops. It was practised widely in the Po-Venetian Plain and on the Tuscan–Emilian Apennines under different names corresponding to regional agroforestry subtypes with their own technical characteristics (*alteno* in Piedmont, *piantata* in Lombardy, Emilia Romagna, and Veneto, and *alberata* on the Apennines) [36–38].

Through an interdisciplinary archaeo-historical approach, this research aims to reconstruct the origin and physical characteristics of the typical *coltura promiscua* of the Northern Apennines (aka "*alberata emiliana*") and to evaluate its current state of conservation. Furthermore, in the last two decades, intensive rainfall events have triggered dozens of soil slips in the Northern Apennines [39], and recent climate change projections indicate that increasingly severe storm intensity will induce greater soil mass movements via water erosion in the future than in the past [40,41].

To address these challenges, GIS (Geographic Information Systems) modelling has been employed to simulate the effect of historic rural landscape change for slope geomorphic processes. Computer-based modelling can provide a quantitative and consistent approach to estimate soil erosion under a wide range of conditions, representing one of the most versatile tools for planning suitable landscape protection measures. In the Central and Northern Apennines, researchers have employed several computational methods to measure, estimate, and monitor soil erosion rates [42,43]. Of these methods, the Revisited Universal Soil Loss Equation (RUSLE) is the most widely applied model for identifying areas susceptible to soil erosion in a region of interest. This empirical model predicts annual soil loss due to sheet and rill water erosion [44]. The results of RUSLE modelling often identify human activities such as grazing, forestry, and agriculture as the most responsible factors for land degradation [45].

However, one limitation of the RUSLE equation is its inability to simulate deposition processes. Conversely, the GIS modelling approach adopted in this study provides a comprehensive understanding of soil mass movements, highlighting areas where soil is removed and deposited using the Unit Stream Power-Based Erosion Deposition (USPED) equation. A similar approach was successfully employed in the Central Apennines to assess human-induced soil erosion processes resulting from forest harvesting [46]. This previous study revealed that forestry activities led to a noticeable increase in soil mass movements, although it did not account for the effects of historic landscape transformation in the model. On the other hand, a recent paper used a GIS modelling approach to estimate soil loss variation in the Northern Apennines in accordance with the level of conservation of historic landscape features (e.g., terrace farming and field boundaries) [47], but it did not explore the effect of historic land-use change on soil deposition processes.

The main innovation of this study is the integration of the USPED equation with information regarding changes in the historic rural landscape. This could be used to develop more effective landscape conservation strategies in the region. Therefore, this research not only aims (1) to understand the historic background of polyculture strategies in the study area (see Section 4.1), but also seeks (2) to explore its potential for mitigating downslope soil erosion and deposition in the face of climate change (see Section 4.2).

2. Study Area

This research focused on a portion of the Tuscan–Emilian Apennines coinciding with the municipality of Vetto d'Enza (Emilia Romagna Region, northern Italy). The main characteristics of this historic rural landscape trace their origin back to the Middle Ages in the period of the Great Countess Matilda of Canossa (10th–11th century CE) and the area's land management system appears to have remained largely unaltered until the end of the 19th century CE [47] (Figure 1).

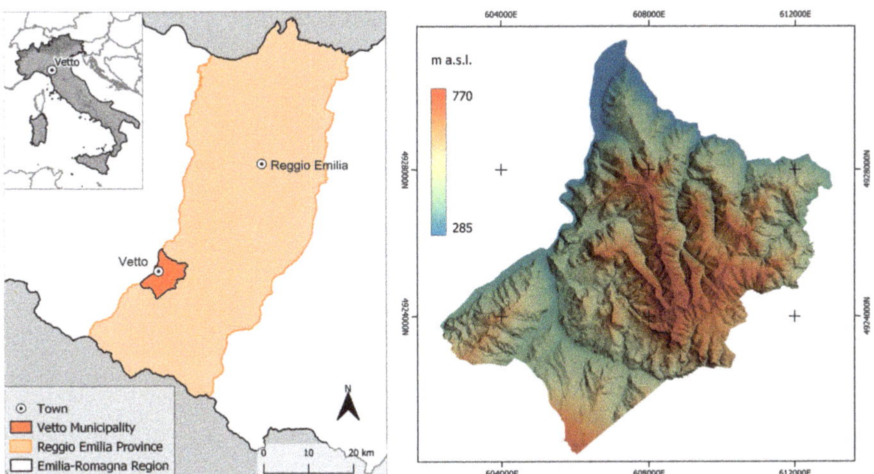

Figure 1. Location of the study area and its topographical setting.

Among the most distinctive characteristics of this historic landscape are relics of traditional *alberata emiliana* and well-preserved stone walls and earth banks that have been used extensively between steeply sloping fields to delimit tenurial boundaries and to face agricultural terraces [48].

The environmental setting of the study area, including the lithological composition, fault systems, soil properties, and climatic factors, contribute to the prevalence of geomorphological slope processes [49]. According to the Köppen–Geiger Climate Classification, the prevailing climate in the study area is warm-temperate, with warm summers and no distinct dry season (Cf—subcontinental/continental temperate) [50]. The faults are mainly located close to the surface, and, when combined with the active uplift of the external rim of the outer chain, they significantly amplify the surface effects of seismic events [49]. The predominant rock type in the study area consists of sedimentary rocks that exhibit a high proportion of clay, such as sandstone and marl [51]. Soils in the area exhibit varying levels of development, ranging from scarce to moderately developed, and are characterised by moderate alkalinity, considerable depth (ranging from 3 to 4 m), and high fertility. Slope geomorphic processes are particularly evident in soils that originate from silty–clayey flysch formations which are particularly vulnerable to erosion and degradation [52,53] (Figure 2).

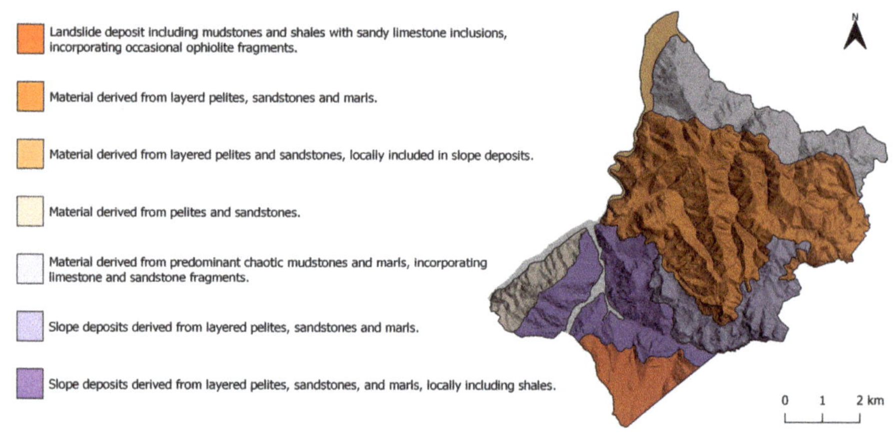

Figure 2. Geolithological setting of the study area [54].

3. Materials and Methods

The interdisciplinary approach proposed combines different disciplines and tools to study the transformation of the historic landscape in the area and the resulting implications for slope geomorphic processes. The first part focuses on the development of the historic AFS landscape. Historical sources (see Section 3.1) were employed to reconstruct the genesis and development of the historic AFS. Using historic cartography, as well as aerial and satellite images, Historic Landscape Characterisation (HLC, see Section 3.2) was used to analyse the landscape transformation of the area from the late 19th century CE to the present day. Dendrochronological analysis (see Section 3.3) on relics of AFS completed the HLC retrogressive analysis, marking the last possible phase of use of AFS in the region. Secondly, the GIS HLC data were employed to model the effect of historic landscape transformation for slope geomorphic processes (see Section 3.4) in order to provide insights for potential future holistic landscape management strategies.

3.1. Historical Sources

For a general overview of the origin of the *alberata emiliana*, the book published by Emilio Sereni in the mid-20th century CE [55] provides a helpful starting point. The oldest agronomical documentation available about *coltura promiscua* dates back to 1674 [56], while the most exhaustive historical report about the rural landscape of the study area was carried out in the 19th century CE [57].

3.2. GIS—Historic Landscape Characterisation

The application of GIS and remote sensing technologies is becoming increasingly acknowledged as a potent tool in landscape archaeology [58,59], as well as in geomorphological studies [60]. Furthermore, the advent of free and open-source software (FOSS) geospatial tools has further widened the user base and improved accessibility to these powerful technologies [61].

HLC is a specific landscape archaeological GIS tool to map the chronological and spatial complexity of historic landscapes with particular reference to their historical development through a systematic interpretation of rural landscape components [62,63]. In each

HLC study, GIS is used to map the "Historic Landscape Character types" (HLC types) on the basis of unique features resulting from known historical processes. The HLC method employs a qualitative but formalised technique to map the chronological and spatial intricacies of historical landscapes [64]. The mapping procedure involves identifying the smallest "uniform diachronic unit" (UDU), represented by a polygon whose size and shape depend on the variability of the HLC type over time [63]. In the study area, the GIS HLC dataset was developed using various sources including historical maps, 19th century cadastral records, declassified satellite images, and aerial photography (Table 1). Further data about historic land use were recovered in the regional geodatabase [65]. The resulting geodataset consisted of a GeoPackage (.gpkg) [66] vectorial layer of the historic landscape changes that occurred since the 19th century CE [48].

Table 1. List of the historic sources employed in this study.

Name	Publication	Type	Scale	Source
Google© Satellite	2022	Satellite images	-	QuickMapServices plugin [67] in QGIS 3.22 [68]
Bing© Satellite	2022	Satellite images	-	QuickMapServices plugin [67] in QGIS 3.22 [68]
Carta Tecnica Regionale (CTR)	2018	Cadastral map	1:5000	WMS service [69]
Compagnia Generale Riprese (CGR) Aeree	2018	Aerial photos	-	WMS service [70]
AGEA (Agenzia per le Erogazioni in Agricoltura) 11	2011	Aerial photos	-	WMS service [71]
AGEA (Agenzia per le Erogazioni in Agricoltura) 08	2008	Aerial photos	-	WMS service [72]
Volo Compagnia Generale Riprese Aeree (CGRA)	1976–1978	Aerial photos	1:13,500	Photos retrieved at the Ufficio cartografico della Provincia di Reggio Emilia [73]
KH-9 (Hexagon)	1974	Satellite images	-	Declassified image retrieved at the US Geological Survey website [74]
Volo GAI (Gruppo Aereo Italiano)	1954–1955	Aerial photos	1:33,000	Photos retrieved at the Istituto Geografico Militare (IGM) website [75]
Nuovo Catasto Terreni	1886–1900	Cadastral map	1:2000	Map retrieved at the Ufficio cartografico della Provincia di Reggio Emilia [76]
Carta Storica Regionale Emilia Romagna	1853	Historical map	1:50,000	WMS service [77]
Second military survey of the Habsburg Empire	1818–1829	Historical map	1:28,800	Map retrieved at the Mapire website [78,79]
Viaggio Agronomico per La Montagna Reggiana E Dei Mezzi Di Migliorare L'agricoltura Delle Montagne Reggiane	1800	Historic document	-	[57]
L'economia del cittadino in villa, del signor Vincenzo Tanara libri 7. Riueduta, ed accresciuta in molto luoghi dal medesimo auttore, con l'aggiunta delle qualita del cacciatore	1713	Historic document	-	[56]

3.3. *Dendrochronology*

Dendrochronology as a method for scientific dating provides accurate chronologies because, in principle, each ring represents a year in a tree's life. In geomorphological studies this technique has been employed in exploring the temporal variation of slope geomorphic processes and the resulting impact on slope instability in recent decades [80,81]. Furthermore, tree-ring analyses have been applied to the study of agroforestry, especially to assess which species return the best benefit in terms of climate change mitigation (e.g., carbon storage) [82,83].

In this study, dendrochronological analysis was employed to assess the last phase of *alberata emiliana* in relics of agroforestry systems detected during the GIS HLC mapping process. An increment borer (Pressler; inner diameter: 5 mm) was used to extract core samples from the trunks at "breast height" (about 1.30 m above the ground) [84]. All extracted samples were dried at room temperature, glued on wood profiles, and sanded, to clearly expose all tree rings. The number of annual rings was used to estimate the *terminus post quem* since each tree was established.

3.4. GIS Geomorphic Modelling

The Geographic Resources Analysis Support System (GRASS) GIS software [85] was employed to simulate how historic land-use changes affected downslope soil erosion and deposition. The module `r.landscape.evol` was specifically designed to simulate the cumulative effect of erosion and deposition on a landscape over time [86]. It takes as input a raster digital elevation model (DEM) of surface topography and an input raster DEM of bedrock elevations, as well as several environmental variables, such as the rainfall erosivity factor (R factor, measured in $MJ \cdot mm \cdot ha^{-1} \cdot h^{-1} \cdot year^{-1}$), the soil erodibility factor (K factor, $Mg \cdot h \cdot MJ^{-1} \cdot mm^{-1}$), the land-cover management factor (C factor, dimensionless value ranging between 0 and 1), and the support practices factor (P factor, dimensionless value ranging between 0 and 1). The R factor represents the erosive power of raindrops on the soil surface, while the K factor represents the susceptibility of the soil to erosion. The C factor represents the effect of vegetation and land management practices on erosion. Vegetation cover can protect the soil from erosion by intercepting raindrops and reducing runoff [87]. Lastly, the P factor represents the effect of erosion control practices to reduce erosion by slowing down water flow and reducing the length of slope (e.g., contour farming, terracing, and hedgerows) [44].

In the module `r.landscape.evol`, three different equations can be used to compute the net change in elevation due to erosion and deposition: the stream power equation, the shear stress equation, and the USPED equation [88]. All three equations estimate transport capacity as the force required to move a unit area of fluid at a unit velocity ($kg/m \cdot s$), thus eventually yielding the erosion/deposition (ED) rate as mass flux per unit area per unit time ($kg/m^2 \cdot s$). The unit $kg/m^2 \cdot s$ represents a measure of mass flux density or mass transfer rate per unit area per unit time. It is commonly used in chemical engineering and related fields to express the rate of mass transfer between two phases or the rate of mass flow through a surface. It is also the standard unit of *momentum* (i.e., mass in motion) [89].

The USPED equation was employed in this study because it is best suited for modelling erosion and deposition on hillslopes and relies on the rainfall intensity factor during the simulation process.

The DEM (5 m resolution), R, and K environmental parameters employed in this study were supplied by the Emilia–Romagna region geological service [90], as well as the regional soil maps used to generate the bedrock elevation raster map. The first modelling step focused on estimating the soil depth. This parameter is crucial as it provides a depth-based limitation on the amount of erosion that can occur at any particular cell. In this research, these data were retrieved from the regional soil map vector file [65] and transformed into a raster file using the GRASS module `v.to.rast`. Then, a bedrock elevation map was estimated subtracting the soil depth raster from the DEM using the GRASS module `r.mapcalc`. To simulate the potential of historic agroforestry landscapes in mitigating downslope soil erosion and deposition in the face of climate change, the global rainfall erosivity projections for the year 2050 [91] were employed in the modelling process. Moreover, the C and the P factors were developed using the HLC data about historic land-use changes that occurred during the 20th century CE. The cover management factor was obtained by associating the European Soil Data Centre (ESDAC) C Factor numerical values [87] to the corresponding Regional LULC categorical data [65] for each HLC chronological period. The ESDAC C factor values were chosen to potentially extend the reproducibility of this protocol in other European regions. Lastly, the support practices factors (i.e., P factors) were developed using the equation proposed in Brandolini et al., 2023 [47], which calculates the effectiveness of historic landscape features (i.e., terraces and field boundaries) in reducing soil erosion hazards according to their state of conservation and regional topography (Figure 3).

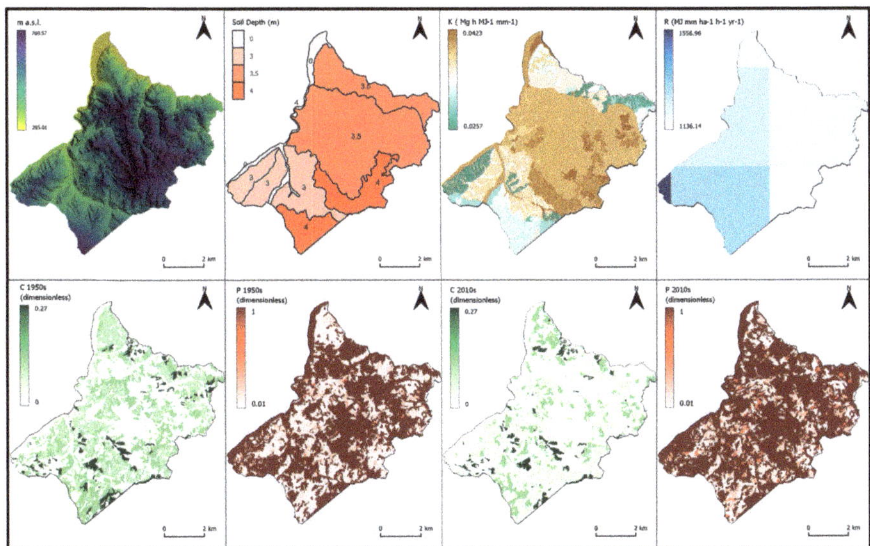

Figure 3. Parameters employed in the GIS geomorphic modelling. From top left to bottom right: DEM, soil depth, K factor, R factor, C factor 1950s, P Factor 1950s, C Factor 1950s, and C Factor 2010s.

4. Results

4.1. Historic Agroforestry Landscape: Genesis, Characteristics, and Decline

In Northern Italy, agroforestry systems have been documented in written sources since Roman times [92]. In his "*Naturalis Historia*", Pliny the Elder mentioned the use of agroforestry to grow grapes and fruit trees together. The same technique was reported by Varro ("*De re rustica*", first century BCE), Columella ("*De re rustica*", first century CE), and finally by Palladius in his "*Opus agriculturae*" (fourth century CE). These authors provide evidence that agroforestry was a common practice in Northern Italy during the Roman Empire [55,92]. The 14th century the Italian agronomist Pietro de' Crescenzi wrote about the practice of planting trees among crops to provide shade and shelter for animals in his book "*Ruralia commoda*" [55]. In 1674, the agronomist Vincenzo Tanara still described this polyculture strategy with the Latin terms *arbustum gallicum* and *arbustum italicum* [56]. In these agroforestry systems, fields were divided into long arable strips separated by rows of vines trained on the trees with intercrops of cereals, vegetables, or forage [37]. The *alberata* (*arbustum italicum*) differs from the *piantata* (*arbustum gallicum*) in terms of the field's extension (15–30 m), the width of the arable strips (4–6 m), and the species (elm—*Ulmus minor*; mulberry—*Morus nigra*; maple—*Acer campestre*) used to sustain the branches of the vines woven from one tree to another along the same row. The *Piantata* was typically adopted in the Po-Venetian Plain, while the *alberata* (aka, *alberata emiliana* and *alberata tosco-umbro-marchigiana*) was largely employed in the Northern Apennines [55,93,94].

The integration of different plants in the same field provides multiple benefits, in addition to food production such as hay and tree fodder to feed animals, domestic fuel (logs), and construction materials (timber) [55]. In the study area, the *alberata* was largely adopted since the 13th century CE, along with a sharecropping system [93,95,96]. In 1800, the agronomist Filippo Re described [57] this rural land management system which appeared to have persisted largely unaltered until the end of the 19th century CE [97–99]. However, in the early 20th century CE, this traditional agroforestry system experienced a substantial decline of around 20% between 1913 and 1957 [55].

The GIS HLC mapping process enables the identification of the relics of *alberata* and quantification of its progressive reduction over the last 70 years (1950s, 1970s, 2000s, and 2010s). Indeed, in the absence of reliable land-use data for the 19th century chronological phase, this period is not considered in this research. Information retrieved in the sources available (Table 1) enabled the compilation of detailed information about the land-use changes that occurred since 1954.

In the mid-20th century, the extent of agroforestry in the study area began to decline. Between the 1950s and the 1970s, the area covered by polyculture fields decreased from 10.2% to 3.7%. This decline continued over the years; today, the historic agroforestry landscape has almost completely disappeared, representing only 0.1% of the study area (Figure 4).

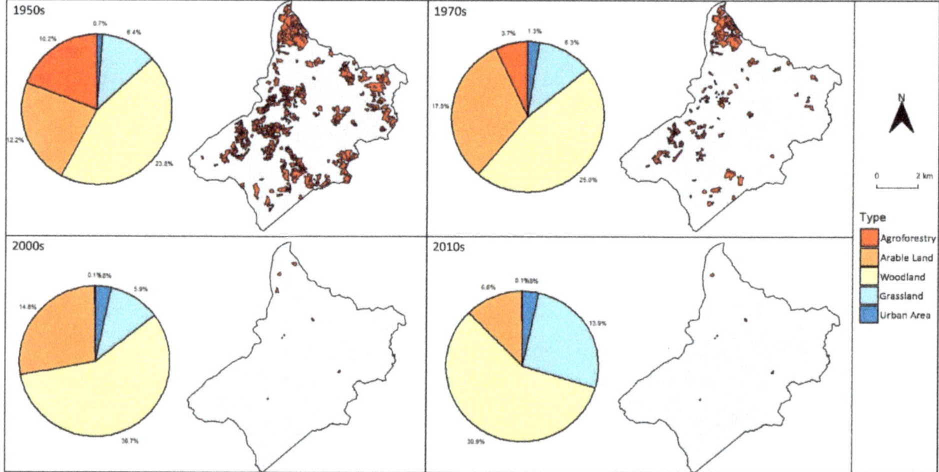

Figure 4. Pie charts summarising the land-use changes in the study area since the 1950s showing the gradual disappearance of *alberata emiliana* (i.e., agroforestry). To highlight the progressive decline in agroforestry, key ESDAC land-use types were merged in the pie charts as follows: woodland (*"mixed forest"* and *"transitional woodland/shrub"*); grassland (*"natural grassland"* and *"pastures"*), and arable land (*"complex cultivation patterns"* and *"non-irrigated arable land"*). The maps show schematically where agroforestry was located in the study area for each time period.

Relics of *alberata* near Vetto d'Enza conserve some of the main characteristics of this historic agroforestry landscape as they were described in historical sources: small fields (max 15–20 m side) divided into strips by regular rows (ca. 5 m distance) of vines trained on maple trees (*Acer campestre*). The tree management and shade-regulation operation employed is pollarding, and all branches are cut at a height of ca. 2 m above ground (Figure 5). Dendrochronological analysis was performed on these remnants of historic agroforestry systems detected remotely through the GIS HLC mapping process. The maples (*Acer campestre*) sampled on the historical agricultural terraces of Vetto d'Enza were likely planted between 1949 and 1980, marking the last phase of the historic agroforestry landscape in the area.

Figure 5. Relics of agroforestry on the historic agricultural terraces of Vetto d'Enza (RE, Italy): (**A**) aerial view of a traditional *alberata emiliana* field; (**B**) detail of a pollarded maple tree; (**C**) ground view of vines trained on pollarded maple trees (© F.B.).

4.2. GIS Modelling of Slope Geomorphic Processes

Three models were developed to simulate changing slope geomorphic processes across the landscape in three different scenarios. Four environmental parameters were constant (DEM, bedrock elevation map, K factor, and R factor) in all the three models, while the P and C factors were changed to reproduce three different scenarios. The first model (model "1950") considers the historic rural landscape as it was in the 1950s before changes due to 20th century socioeconomic dynamics. In this model, agroforestry is still the dominant component of the rural landscape (Figure 4). The second model (model "2020") reflects present-day land management practices in the area, with the C and P factors adjusted accordingly. In this model, the historic agroforestry landscape has almost completely disappeared, and the rural activities are now primarily devoted to forage production for the local dairy industry (Figure 4). The third model (model "AFS") represents a scenario in which the present-day rural landscape is occupied only by historic polyculture activities (i.e., *"agroforestry"*). In this model, the historic agroforestry landscape replaces all other agro-pastoral land-use types (i.e., *"complex cultivation patterns"*, *"non-irrigated arable land"*, *"natural grassland"*, and *"pastures"*) (Figure 6).

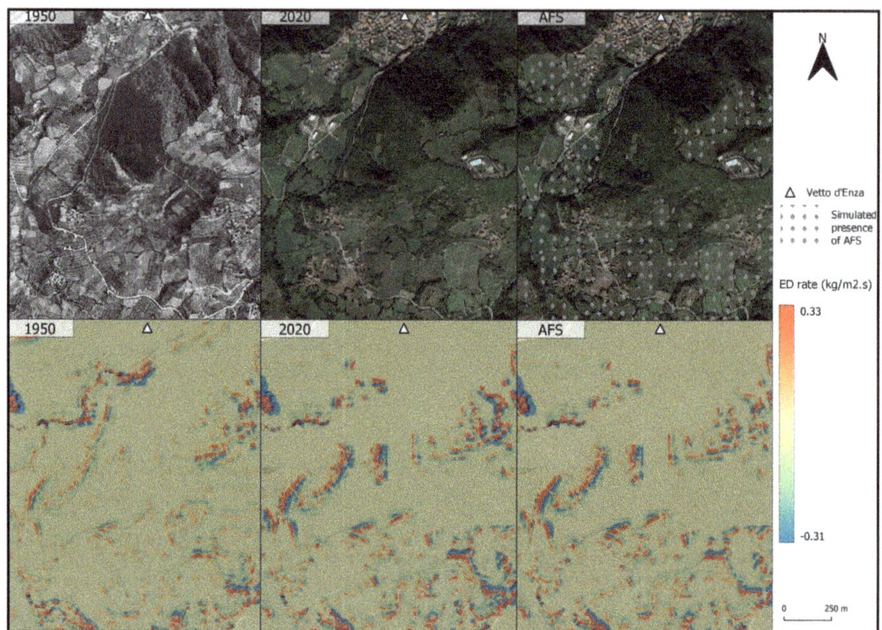

Figure 6. Details of the three GRASS models (1950, 2020, AFS) are shown in correspondence with the area south of Vetto d'Enza (RE, Italy). The estimates of erosion/colluvial deposition rates vary significantly across the three models due to differences in the distribution of agroforestry systems in the study area. Shades of red indicate areas of soil colluvial deposition, while shades of blue represent zones prone to erosion. The full rasters covering the entire study area are provided in the Supplementary Materials.

As displayed in Figure 6, the estimates of erosion/deposition rates varied significantly across the three models due to differences in the distribution of agroforestry systems in the study area. In particular, shades of red indicate areas of soil colluvial deposition, while shades of blue represent zones prone to erosion. The results of the GIS geomorphic modelling conducted in the study area revealed that Model "1950" exhibited the highest soil loss score (Table 2). Although the total soil loss in Model "2020" was 8% lower than in Model "1950", it remained slightly higher than that in Model "AFS" (+1.5%) (Figure 7).

Table 2. Simulated soil loss (i.e., the result of erosion and deposition processes) (t·ha^{-1}·year^{-1}) in the three models ("1950", "2020", and "AFS") for each land-use type and totals. Positive numbers indicate depositions, while negative numbers indicate erosion.

Land Use Type	Model "1950"		Model "2020"		Model "AFS"	
	Area (ha)	Soil Loss (t·ha^{-1}·year^{-1})	Area (ha)	Soil Loss (t·ha^{-1}·year^{-1})	Area (ha)	Soil Loss (t·ha^{-1}·year^{-1})
Agroforestry	1024.83	0.93	5.66	0.02	1730.15	−1.54
Arable land	1220.20	−4.65	657.00	−1.55	nd	nd
Grassland	253.66	0.63	1067.49	−0.89	nd	nd
Rough ground	190.02	−3.29	190.74	−6.50	190.74	−6.55

Table 2. *Cont.*

Land Use Type	Model "1950"		Model "2020"		Model "AFS"	
	Area (ha)	Soil Loss ($t{\cdot}ha^{-1}{\cdot}year^{-1}$)	Area (ha)	Soil Loss ($t{\cdot}ha^{-1}{\cdot}year^{-1}$)	Area (ha)	Soil Loss ($t{\cdot}ha^{-1}{\cdot}year^{-1}$)
Urban area	65.24	−0.28	184.26	−0.45	184.26	−0.30
Woodland	2577.43	2.86	3226.22	5.90	3226.22	4.98
Total	5331.37	−3.80	5331.37	−3.47	5331.37	−3.41

Figure 7. Estimation of the total soil loss ($t{\cdot}ha^{-1}{\cdot}year^{-1}$) in the three models (1950, 2020, and AFS). The simulation indicates that replacing current agro-pastoral land management with agroforestry systems would reduce the soil loss by 40%. Images in .tiff format covering the entire study area are provided in the Supplementary Materials.

In Model "1950" and Model "2020", soil loss generated by rural activities accounted for 57% and 69% of the total, respectively. Thus, focusing only on rural areas can help in highlighting the effects of historic landscape changes for slope geomorphic processes. To ensure consistency in the comparisons among the three models, we only considered the present rural area. Indeed, farmland extension has experienced a decline since the 1950s, primarily replaced by urban areas and woodlands (Figure 4). The present-day agro-pastoral activities represent 32.4% of the study area with an extension of 1731.72 ha (Figure 7). Interestingly, in this context, Model "2020" presented the highest soil loss with a percentage increase of 9.7% compared to Model "1950". Conversely, as displayed in Model "AFS", the simulated restoration of historic agroforestry systems generated a reduction in the total amount of soil loss in rural areas by 40% (Figure 7).

To evaluate the efficacy of restoring historic agroforestry land use in mitigating downslope soil erosion and deposition in the face of climate change, two additional models were developed. Climate change projections indicate that the increasing severity of storms will result in greater water erosion and soil loss in the future compared to the past [40,41]. In light of this, we updated Model "2050" and Model "AFS" with global rainfall erosivity projections for 2050 [91]. Model "2050LU20" estimated erosion and deposition rates in the study area by simulating the interaction between existing environmental factors and land management practices with rainfall erosivity projections for 2050. Similarly, Model "2050AFS" evaluated the response of agroforestry systems in mitigating soil loss with the same forecasts of rainfall-induced erosion.

The results of these two models showed the same trends as observed in Models "2050" and "AFS". When considering the entire region, the benefits of replacing current agro-pastoral systems with agroforestry strategies to reduce erosion and deposition rates appear to be limited. However, when looking only at the current rural area, the total soil loss due to future rainfall-induced erosion was projected to be 36% lower in "Model 2050AFS" than in Model "2050LU20" (Figure 8).

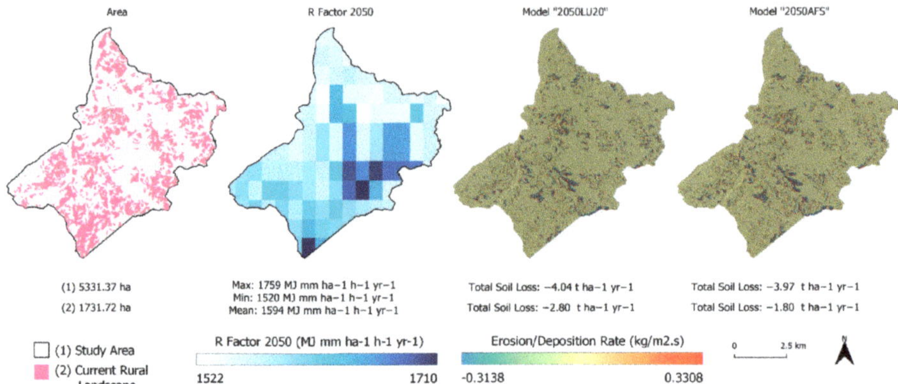

Figure 8. Estimation of the total soil loss ($t{\cdot}ha^{-1}{\cdot}year^{-1}$) in Model "2050LU20" and in Model "2050AFS" considering the rainfall erosivity projections for the year 2050. The simulation indicates that restoring agroforestry systems would reduce the soil loss of rural area by 36%. Images in .tiff format covering the entire study area are provided in the Supplementary Materials.

5. Discussion

The combination of historical sources and landscape archaeological mapping enabled the analysis of agroforestry systems in the Northern Apennines. This ancient historic rural strategy is mentioned in Roman and Mediaeval documents, and it was still widely employed in the 19th century as reported by the agronomist Filippo Re in 1800. The GIS HLC retrogressive analysis enabled a quantitative assessment of the progressive abandonment of agroforestry and the resulting historic landscape changes which occurred during the 20th century CE. The last phase of use of *alberata emiliana* occurred in the 1980s as suggested by preliminary dendrochronological analysis. As registered in other European regions [100], the decline in polyculture strategies such as the *alberata* in the study area appears to be a consequence of post-World War II socioeconomic dynamics such as the rapid modernisation and mechanisation of agriculture and the expansion of urban areas (Figures 3 and 9). Furthermore, in the study area, the progressive decline of agroforestry systems since the mid-20th century CE reflects further local socioeconomic trends: the progressive reduction in rural activities in mountainous regions, the consequent process of rewilding abandoned agricultural areas [101], and the need for forage for the regional dairy industry [102]. These dynamics led the agroforestry fields to be replaced mainly by woodland, pastures, and grassland (Figures 3 and 9).

The disappearance of historic agricultural practices had relevant consequences on land vulnerability. The results of the GIS geomorphic modelling showed how the current overall erosion/deposition rate in the study area (Model "2020") was lower than in the past (Model "1950") (Figure 7), a situation likely due to the afforestation process that occurred in the last 70 years. Indeed, the extension of woodland in the area, higher than in the mid-20th century, has positive benefits in mitigating land degradation (Table 2). Furthermore, the overall soil loss scores of the three models seem to suggest that the reconversion of the present-day agro-pastoral activities to agroforestry appear to be limited when considering the entire region (Figure 7). Nevertheless, the interpretation of the geomorphic simulations cannot ignore the fact that the current rural area accounts for more than 60% of the total soil loss registered, despite being only one-third of the entire region (Table 2). In addition, the current extent of the agricultural area is lower than in the 1950s, but the simulation returned a higher soil loss score than in the past (Table 2). The potential benefits of restoring the historic agroforestry systems (still widely employed in the mid-20th century) in place of the current agro-pastoral strategies were highlighted by Model "AFS". The third simulation returned a significant reduction of potential soil loss in the area, giving valuable indications for future landscape management strategies. Even in the future scenario of increasingly severe storm intensity (Model "2050LU20" and Model "2050AFS"), the replacement of current agro-pastoral strategies with agroforestry seems to be particularly effective in mitigating downslope soil erosion and deposition (Figure 8).

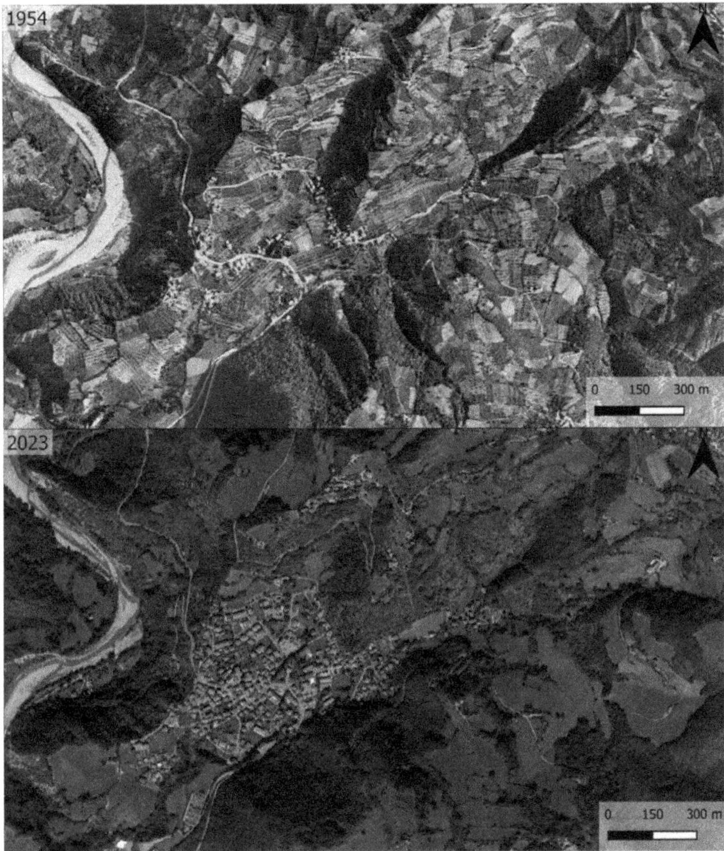

Figure 9. The area of the town of Vetto d'Enza (RE, Italy) as it appeared in 1954 [70] (**top**), compared to the current situation (**bottom**) [64].

The abandonment of agroforestry in the area also had negative consequences for the cultural values of landscape. Indeed, in addition to an environmental deterioration, the progressive loss of farmers' know-how of such rural practices correspond to deep physical modifications of the regional landscape heritage [21]. In the last 10 years, both EU and regional policies have encouraged repopulation in rural mountain areas by providing economic incentives to newcomers who choose to relocate from cities [103,104]. The aim of such a policy is to limit the process of depopulation, thereby avoiding the loss of cultural identity in rural regions while lowering population pressure in urban areas [105]. Furthermore, according to the latest Horizon Europe Strategic Plan [106] and the Agricultural European Innovation Partnership workshop, the major challenge facing current and future agricultural systems is reconciling production with sustainable land management [107]. To address this challenge, EU agencies encourage the development of innovative landscape management approaches that utilise nature-based solutions. Additionally, cultural heritage is identified as a potential driver for improving rural wellbeing and long-term socioeconomic prospects [106].

Furthermore, the restoration of historic AFS has the potential to mitigate the negative impacts of anthropogenic landscape fragmentation and support the conservation of biodiversity in mountain ecosystems. Anthropogenically induced landscape fragmentation, which results from direct physical transformations such as deforestation, agriculture, urbanisation, and road building, is widely acknowledged as a significant threat to biodiversity [108,109]. However, the depopulation of rural mountain areas and the consequent abandonment of traditional land management practices have emerged as major driving forces behind changes in mountain ecosystems in Europe. This has led to decreased landscape connectivity, negatively impacting fauna associated with abandoned traditional agro-pastoral habitats [110]. Several studies have shown that AFS can play a crucial role in improving landscape connectivity. By creating corridors and connecting fragmented habitats, AFS can enhance biodiversity, improve ecosystem services, and promote ecological resilience [111,112].

On the other hand, environmentally sustainable rural LU types very often do not provide an immediate and desirable economic return to farmers and these solutions need to be implemented and adapted to meet site-specific needs at the local scale by inviting stakeholders to contribute to policy development [113]. Nevertheless, EU-funded research focused on AFS strategies have shown a return-on-investment time of about 5 years to recover from workforce training and machinery costs [114], and it has been demonstrated that even the partial integration of historical AFS within arable lands and pastures can contribute strongly to lowering environmental pressures on landscape [20].

Thus, the development of transdisciplinary strategies is crucial to inform an environmentally sustainable conservation of landscape heritage [115–118]. Synergies between disciplines can actively contribute to achieving this goal, deciphering the ecological and historical background of traditional and multifunctional cultural landscapes [119] such as the *alberata emiliana*.

The workflow adopted in this research represents an effective interdisciplinary approach to inform potential holistic landscape strategies. By combining historical sources, dendrochronology, and GIS retrogressive analysis, we were able to reconstruct the long-term sociocultural values of AFS in the area. Furthermore, GIS modelling permitted us to simulate the effects of the abandonment of traditional LU systems on slope instability. However, assessing the accuracy of model predictions is still challenging. Direct measurements on the fields are the most effective way to validate this type of model [46,120]. Even when validation data can be collected, it is usually limited to a small number of sites and a short period, which may not be representative of the entire region or different climatic conditions. As a result, the validation process may not capture the full range of spatial and temporal variability [121]. Nevertheless, GIS modelling tools such as `r.landscape.evol` represent invaluable multiscale sources to simulate long-term annual land degradation across different land management activities and environmental conditions. The scenarios proposed in this study can simulate the effects of historic landscape transformation for the cumulative soil erosion and deposition rates in the area to highlight the potential environmental benefits of restoring historic AFS. Therefore, the development of landscape modelling approaches such as that adopted in this research responds to the need of transdisciplinary tools for landscape management. Indeed, the integration of retrogressive analysis within GIS modelling tools has the potential to yield invaluable insights for embracing natural and cultural values of landscapes as components of the same holistic landscape plans.

6. Conclusions

This study integrated historical documents, landscape archaeology, and GIS tools to trace the evolution of the historic agroforestry (*alberata emiliana*) in the Northern Apennines. It also quantified the decline in *alberata emiliana* during the 20th century CE and assessed its impact on cumulative soil erosion and deposition. Moreover, the GIS HLC mapping process was utilised to create a comprehensive geospatial dataset to monitor the current state of traditional AFS and provide recommendations for their future maintenance and revitalisation while preserving the region's historical landscape characteristics. Considering the potential benefits of agroforestry systems in mitigating downslope soil erosion and deposition, the GIS modelling showed that restoring the historic agroforestry landscape could significantly reduce the land degradation rate in the area. This interdisciplinary approach could inform the development of transdisciplinary management plans that balance mitigating land degradation with preserving the landscape character and cultural identity of a region. The workflow adopted in this research is potentially reproducible in other areas with similar sociocultural and environmental characteristics, and it helps in making historical knowledge useful for future landscape management.

Supplementary Materials: The following supporting information can be downloaded at https://www.mdpi.com/article/10.3390/land12051054/s1: the GIS modelling rasters are provided in a .zip folder.

Author Contributions: Conceptualisation, F.B.; methodology, F.B.; software, F.B.; validation, F.B.; formal analysis, F.B.; investigation, F.B.; resources, F.B. and S.T.; dendrochronological analysis, C.C.; geomorphological review and supervision, M.P.; data curation, F.B. and C.C.; writing—original draft preparation, F.B.; writing—review and editing, M.P. and S.T.; visualisation, F.B.; supervision, F.B.; project administration, F.B. and S.T.; funding acquisition, F.B. and S.T. All authors have read and agreed to the published version of the manuscript.

Funding: This project received funding from the European Union's Horizon 2020 research and innovation programme under the Marie Skłodowska-Curie grant agreement No. 890561, Historic Landscape and Soil Sustainability (HiLSS) and from the TerraSAgE project (Terraces as Sustainable Agricultural Environments) from UKRI-Arts and the Humanities Research Council (AHRC ref: AH/T000104/1).

Data Availability Statement: The HLC dataset used to develop the C and P factors is available on Zenodo: doi: 10.5281/zenodo.6622607 (accessed on: 1 May 2023). The raw raster products derived from the GRASS GIS simulation are available upon request from the corresponding author.

Acknowledgments: The authors would like to acknowledge the help of Paola Tarocco (Area Geologia, Suoli e Sismica, Settore Difesa del Territorio, Sviluppo e Analisi dei Dati sui Suoli a Supporto delle Politiche Agricole, Ambientali e Urbanistiche—Regione Emilia-Romagna) that provided the regional geodata for the USPED equation. Furthermore, the authors wish to thank Stefano Costanzo (Università degli Studi di Napoli "L'Orientale") and Luca Forti (Università degli Studi di Milano) for their suggestions about the geolithological setting of the study area. Lastly, the authors thank all of their colleagues from the Newcastle University Spatial Humanities (NUSH) group for constant enriching discussions during the development of the GIS modelling.

Conflicts of Interest: The authors declare no conflict of interest.

References

1. Vanwalleghem, T.; Gómez, J.A.; Infante Amate, J.; González de Molina, M.; Vanderlinden, K.; Guzmán, G.; Laguna, A.; Giráldez, J.V. Impact of Historical Land Use and Soil Management Change on Soil Erosion and Agricultural Sustainability during the Anthropocene. *Anthropocene* **2017**, *17*, 13–29. [CrossRef]
2. Pica, A.; Vergari, F.; Fredi, P.; Del Monte, M. The Aeterna Urbs Geomorphological Heritage (Rome, Italy). *Geoheritage* **2016**, *8*, 31–42. [CrossRef]
3. Brandolini, F.; Cremaschi, M. The Impact of Late Holocene Flood Management on the Central Po Plain (Northern Italy). *Sustain. Sci. Pract. Policy* **2018**, *10*, 3968. [CrossRef]
4. Srivastava, A.; Kinnaird, T.; Sevara, C.; Holcomb, J.A.; Turner, S. Dating Agricultural Terraces in the Mediterranean Using Luminescence: Recent Progress and Challenges. *Land* **2023**, *12*, 716. [CrossRef]
5. Tarolli, P.; Preti, F.; Romano, N. Terraced Landscapes: From an Old Best Practice to a Potential Hazard for Soil Degradation due to Land Abandonment. *Anthropocene* **2014**, *6*, 10–25. [CrossRef]
6. Montanarella, L.; Panagos, P. The Relevance of Sustainable Soil Management within the European Green Deal. *Land Use Policy* **2021**, *100*, 104950. [CrossRef]
7. Schmidt, J.; Usmar, N.; Westphal, L.; Werner, M.; Roller, S.; Rademacher, R.; Kühn, P.; Werther, L.; Kottmann, A. Erosion Modelling Indicates a Decrease in Erosion Susceptibility of Historic Ridge and Furrow Fields near Albershausen, Southern Germany. *Land* **2023**, *12*, 544. [CrossRef]

8. Plieninger, T.; Muñoz-Rojas, J.; Buck, L.E.; Scherr, S.J. Agroforestry for Sustainable Landscape Management. *Sustain. Sci.* **2020**, *15*, 1255–1266. [CrossRef]
9. Zerbe, S. Traditional Cultural Landscapes—A Theoretical Framework. In *Restoration of Multifunctional Cultural Landscapes: Merging Tradition and Innovation for a Sustainable Future*; Zerbe, S., Ed.; Springer International Publishing: Cham, Switzerland, 2022; pp. 3–17, ISBN 9783030955724.
10. Joffre, R.; Vacher, J.; de los Llanos, C.; Long, G. The Dehesa: An Agrosilvopastoral System of the Mediterranean Region with Special Reference to the Sierra Morena Area of Spain. *Agrofor. Syst.* **1988**, *6*, 71–96. [CrossRef]
11. Correia, T.P. Threatened Landscape in Alentejo, Portugal: The "montado" and Other "agro-Silvo-Pastoral" Systems. *Landsc. Urban Plan.* **1993**, *24*, 43–48. [CrossRef]
12. Zacharová, J.; Riezner, J.; Elznicová, J.; Machová, I.; Kubát, K.; Holcová, D.; Holec, M.; Pacina, J.; Štojdl, J.; Grygar, T.M. Historical Agricultural Landforms—Central European Bio-Cultural Heritage Worthy of Attention. *Land* **2022**, *11*, 963. [CrossRef]
13. Herzog, F. Streuobst: A Traditional Agroforestry System as a Model for Agroforestry Development in Temperate Europe. *Agrofor. Syst.* **1998**, *42*, 61–80. [CrossRef]
14. Lacerda, A.E.B.; Hanisch, A.L.; Nimmo, E.R. Leveraging Traditional Agroforestry Practices to Support Sustainable and Agrobiodiverse Landscapes in Southern Brazil. *Land* **2020**, *9*, 176. [CrossRef]
15. Lentz, D.L.; Hockaday, B. Tikal Timbers and Temples: Ancient Maya Agroforestry and the End of Time. *J. Archaeol. Sci.* **2009**, *36*, 1342–1353. [CrossRef]
16. Stahl, P.W.; Pearsall, D.M. Late Pre-Columbian Agroforestry in the Tropical Lowlands of Western Ecuador. *Quat. Int.* **2012**, *249*, 43–52. [CrossRef]
17. Souza, J.G.d.; de Souza, J.G.; Robinson, M.; Yoshi Maezumi, S.; Capriles, J.; Hoggarth, J.A.; Lombardo, U.; Novello, V.F.; Apaéstegui, J.; Whitney, B.; et al. Climate Change and Cultural Resilience in Late Pre-Columbian Amazonia. *Nat. Ecol. Evol.* **2019**, *3*, 1007–1017. [CrossRef]
18. Augère-Granier, M.-L. *Agroforestry in the European Union*; European Parliamentary Research Service: Brussels, Belgium, 2020.
19. Nair, P.K.R.; Kumar, B.M.; Nair, V.D. Carbon Sequestration and Climate Change Mitigation. In *An Introduction to Agroforestry: Four Decades of Scientific Developments*; Nair, P.K.R., Kumar, B.M., Nair, V.D., Eds.; Springer International Publishing: Cham, Switzerland, 2021; pp. 487–537, ISBN 9783030753580.
20. Kay, S.; Rega, C.; Moreno, G.; den Herder, M.; Palma, J.H.N.; Borek, R.; Crous-Duran, J.; Freese, D.; Giannitsopoulos, M.; Graves, A.; et al. Agroforestry Creates Carbon Sinks Whilst Enhancing the Environment in Agricultural Landscapes in Europe. *Land Use Policy* **2019**, *83*, 581–593. [CrossRef]
21. Brunori, E.; Maesano, M.; Moresi, F.V.; Matteucci, G.; Biasi, R.; Scarascia Mugnozza, G. The Hidden Land Conservation Benefits of Olive-based (*Olea Europaea* L.) Landscapes: An Agroforestry Investigation in the Southern Mediterranean (Calabria Region, Italy). *Land Degrad. Dev.* **2020**, *31*, 801–815. [CrossRef]
22. Purwaningsih, R.; Sartohadi, J.; Setiawan, M.A. Trees and Crops Arrangement in the Agroforestry System Based on Slope Units to Control Landslide Reactivation on Volcanic Foot Slopes in Java, Indonesia. *Land* **2020**, *9*, 327. [CrossRef]
23. Hairiah, K.; Widianto, W.; Suprayogo, D.; Van Noordwijk, M. Tree Roots Anchoring and Binding Soil: Reducing Landslide Risk in Indonesian Agroforestry. *Land* **2020**, *9*, 256. [CrossRef]
24. Nair, P.K.R.; Kumar, B.M.; Nair, V.D. Soil Conservation and Control of Land-Degradation. In *An Introduction to Agroforestry: Four Decades of Scientific Developments*; Nair, P.K.R., Kumar, B.M., Nair, V.D., Eds.; Springer International Publishing: Cham, Switzerland, 2021; pp. 445–474, ISBN 9783030753580.
25. Foster, G.R.; Young, R.A.; Römkens, M.J.M.; Onstad, C.A. Processes of Soil Erosion by Water. In *Soil Erosion and Crop Productivity*; American Society of Agronomy, Crop Science Society of America, Soil Science Society of America: Madison, WI, USA, 2015; pp. 137–162, ISBN 9780891182511.
26. Bollati, I.; Reynard, E.; Palmieri, E.L.; Pelfini, M. Runoff Impact on Active Geomorphosites in Unconsolidated Substrate. A Comparison Between Landforms in Glacial and Marine Clay Sediments: Two Case Studies from the Swiss Alps and the Italian Apennines. *Geoheritage* **2016**, *8*, 61–75. [CrossRef]
27. Bollati, I.M.; Masseroli, A.; Mortara, G.; Pelfini, M.; Trombino, L. Alpine Gullies System Evolution: Erosion Drivers and Control Factors. Two Examples from the Western Italian Alps. *Geomorphology* **2019**, *327*, 248–263. [CrossRef]
28. Crosta, G.B.; Frattini, P. Rainfall-Induced Landslides and Debris Flows. *Hydrol. Process.* **2008**, *22*, 473–477. [CrossRef]
29. Geertsema, M.; Highland, L.; Vaugeouis, L. Environmental Impact of Landslides. In *Landslides—Disaster Risk Reduction*; Sassa, K., Canuti, P., Eds.; Springer: Berlin/Heidelberg, Germany, 2009; pp. 589–607, ISBN 9783540699705.
30. Kjekstad, O.; Highland, L. Economic and Social Impacts of Landslides. In *Landslides—Disaster Risk Reduction*; Sassa, K., Canuti, P., Eds.; Springer: Berlin/Heidelberg, Germany, 2009; pp. 573–587, ISBN 9783540699705.
31. Canuti, P.; Casagli, N.; Catani, F.; Fanti, R. Hydrogeological Hazard and Risk in Archaeological Sites: Some Case Studies in Italy. *J. Cult. Herit.* **2000**, *1*, 117–125. [CrossRef]
32. Masseroli, A.; Bollati, I.M.; Trombino, L.; Pelfini, M. The Soil Trail of Buscagna Valley, an Example of the Role of Soil Science in Geodiversity and Geoheritage Analyses. In *Visages of Geodiversity and Geoheritage*; Kubalíková, L., Coratza, P., Pál, M., Zwoliński, Z., Irapta, P.N., Wyk de Vries, B., Eds.; Special Publications; Geological Society: London, UK, 2023; Volume 530.

33. Borrelli, P.; Robinson, D.A.; Panagos, P.; Lugato, E.; Yang, J.E.; Alewell, C.; Wuepper, D.; Montanarella, L.; Ballabio, C. Land Use and Climate Change Impacts on Global Soil Erosion by Water (2015–2070). *Proc. Natl. Acad. Sci. USA* **2020**, *117*, 21994–22001. [CrossRef]
34. Jaedicke, C.; Van Den Eeckhaut, M.; Nadim, F.; Hervás, J.; Kalsnes, B.; Vangelsten, B.V.; Smith, J.T.; Tofani, V.; Ciurean, R.; Winter, M.G.; et al. Identification of Landslide Hazard and Risk "hotspots" in Europe. *Bull. Eng. Geol. Environ.* **2014**, *73*, 325–339. [CrossRef]
35. Papa, G.L.; Dazzi, C.; Némethy, S.; Corti, G.; Cocco, S. Land Set-up Systems in Italy: A Long Tradition of Soil and Water Conservation Sewed up to a Variety of Pedo-Climatic Environments. *Ital. J. Agron.* **2020**, *15*, 281–292. [CrossRef]
36. Salvatico, A. *L'economia Dell'alteno: Viticoltura e Cerealicoltura nel Roero e Nelle Langhe tra il Basso Medioevo e la Prima età Moderna*; Marco Valerio Editore: Turin, Italy, 2004; ISBN 9788888132266.
37. Paris, P.; Camilli, F.; Rosati, A.; Mantino, A.; Mezzalira, G.; Dalla Valle, C.; Franca, A.; Seddaiu, G.; Pisanelli, A.; Lauteri, M.; et al. What Is the Future for Agroforestry in Italy? *Agrofor. Syst.* **2019**, *93*, 2243–2256. [CrossRef]
38. Ferrario, V. Learning from Agricultural Heritage? Lessons of Sustainability from Italian "Coltura Promiscua". *Sustain. Sci. Pract. Policy* **2021**, *13*, 8879. [CrossRef]
39. Montrasio, L.; Valentino, R.; Losi, G.L. Shallow Landslides Triggered by Rainfalls: Modeling of Some Case Histories in the Reggiano Apennine (Emilia Romagna Region, Northern Italy). *Nat. Hazards* **2012**, *60*, 1231–1254. [CrossRef]
40. Olsson, L.; Barbosa, H.; Bhadwal, S.; Cowie, A.; Delusca, K.; Flores-Renteria, D.; Hermans, K.; Jobbagy, E.; Kurz, W.; Li, D.; et al. Land Degradation. In *Climate Change and Land: An IPCC Special Report on Climate Change, Desertification, Land Degradation, Sustainable Land Management, Food Security, and Greenhouse Gas Fluxes in Terrestrial Ecosystems*; Shukla, P.R., Skea, J., Calvo Buendia, E., Masson-Delmotte, V., Pörtner, H., Roberts, D.C., Zhai, P., Malley, J., Eds.; Intergovernmental Panel on Climate Change (IPCC): Geneva, Switzerland, 2019.
41. Panagos, P.; Ballabio, C.; Himics, M.; Scarpa, S.; Matthews, F.; Bogonos, M.; Poesen, J.; Borrelli, P. Projections of Soil Loss by Water Erosion in Europe by 2050. *Environ. Sci. Policy* **2021**, *124*, 380–392. [CrossRef]
42. Pieri, L.; Bittelli, M.; Wu, J.Q.; Dun, S.; Flanagan, D.C.; Pisa, P.R.; Ventura, F.; Salvatorelli, F. Using the Water Erosion Prediction Project (WEPP) Model to Simulate Field-Observed Runoff and Erosion in the Apennines Mountain Range, Italy. *J. Hydrol.* **2007**, *336*, 84–97. [CrossRef]
43. Guerra, V.; Lazzari, M. Geomorphic Approaches to Estimate Short-Term Erosion Rates: An Example from Valmarecchia River System (Northern Apennines, Italy). *Water* **2020**, *12*, 2535. [CrossRef]
44. Ghosal, K.; Das Bhattacharya, S. A Review of RUSLE Model. *J. Indian Soc. Remote Sens.* **2020**, *48*, 689–707. [CrossRef]
45. Borrelli, P.; Märker, M.; Panagos, P.; Schütt, B. Modeling Soil Erosion and River Sediment Yield for an Intermountain Drainage Basin of the Central Apennines, Italy. *Catena* **2014**, *114*, 45–58. [CrossRef]
46. Borrelli, P.; Märker, M.; Schütt, B. Modelling Post-Tree-Harvesting Soil Erosion and Sediment Deposition Potential in the Turano River Basin (Italian Central Apennine). *Land Degrad. Dev.* **2015**, *26*, 356–366. [CrossRef]
47. Brandolini, F.; Kinnaird, T.C.; Srivastava, A.; Turner, S. Modelling the Impact of Historic Landscape Change on Soil Erosion and Degradation. *Sci. Rep.* **2023**, *13*, 4949. [CrossRef]
48. Brandolini, F.; Turner, S. Revealing Patterns and Connections in the Historic Landscape of the Northern Apennines (Vetto, Italy). *J. Maps* **2022**, *18*, 663–673. [CrossRef]
49. Mariani, G.S.; Brandolini, F.; Pelfini, M.; Zerboni, A. Matilda's Castles, Northern Apennines: Geological and Geomorphological Constrains. *J. Maps* **2019**, *15*, 521–529. [CrossRef]
50. Kottek, M.; Grieser, J.; Beck, C.; Rudolf, B.; Rubel, F. World Map of the Köppen-Geiger Climate Classification Updated. *Meteorol. Z.* **2006**, *15*, 259–263. [CrossRef]
51. Haller, A.; Bender, O. Among Rewilding Mountains: Grassland Conservation and Abandoned Settlements in the Northern Apennines. *Landsc. Res.* **2018**, *43*, 1068–1084. [CrossRef]
52. Bini, C. Geology and Geomorphology. In *The Soils of Italy*; Costantini, E.A.C., Dazzi, C., Eds.; Springer: Dordrecht, The Netherlands, 2013; pp. 39–56, ISBN 9789400756427.
53. Piacentini, D.; Troiani, F.; Daniele, G.; Pizziolo, M. Historical Geospatial Database for Landslide Analysis: The Catalogue of Landslide OCcurrences in the Emilia-Romagna Region (CLOCkER). *Landslides* **2018**, *15*, 811–822. [CrossRef]
54. Geoportale ER. Available online: https://geoportale.regione.emilia-romagna.it/ (accessed on 12 February 2023).
55. Sereni, E. *Storia del Paesaggio Agrario Italiano*; Editori Laterza: Bari, Italy, 1961.
56. Tanara, V. *L'economia del Cittadino in Villa, del Signor Vincenzo Tanara Libri 7. Riueduta, ed Accresciuta in Molto Luoghi dal Medesimo Auttore, con L'aggiunta delle Qualita del Cacciatore*; Appresso Bortolo Zerletti: Venezia, Italy, 1713.
57. Re, F. *Viaggio Agronomico per la Montagna Reggiana E Dei Mezzi Di Migliorare L'agricoltura Delle Montagne Reggiane*; Casali, C., Ed.; Officine Grafiche Reggiane: Reggio Emilia, Italy, 1927.
58. Wheatley, D.; Gillings, M. *Spatial Technology and Archaeology: The Archaeological Applications of GIS*; CRC Press: Boca Raton, FL, USA, 2013; ISBN 9781466576612.
59. Brandolini, F.; Domingo-Ribas, G.; Zerboni, A.; Turner, S. A Google Earth Engine-Enabled Python Approach to Improve Identification of Anthropogenic Palaeo-Landscape Features. *Open Res. Eur.* **2021**, *1*, 22. [CrossRef]
60. Otto, J.-C.; Prasicek, G.; Blöthe, J.; Schrott, L. 2.05—GIS Applications in Geomorphology. In *Comprehensive Geographic Information Systems*; Huang, B., Ed.; Elsevier: Oxford, UK, 2018; pp. 81–111, ISBN 9780128047934.

61. Steiniger, S.; Hay, G.J. Free and Open Source Geographic Information Tools for Landscape Ecology. *Ecol. Inform.* **2009**, *4*, 183–195. [CrossRef]
62. Turner, S.; Shillito, L.-M.; Carrer, F. Landscape Archaeology. In *The Routledge Companion to Landscape Studies*; Howard, P., Thompson, I., Waterton, E., Atha, M., Eds.; Routledge: London, UK, 2018; pp. 155–165.
63. Dabaut, N.; Carrer, F. Historic LAndscape CHaracterisation: Technical Approaches beyond Theory. *Landscapes* **2020**, *21*, 152–167. [CrossRef]
64. Turner, S. Historic Landscape Characterisation: An Archaeological Approach to Landscape Heritage. In *Routledge Handbook of Landscape Character Assessment*; Routledge: London, UK, 2018; pp. 37–50.
65. Geoportale ER. Database Uso Del Suolo. Available online: https://geoportale.regione.emilia-romagna.it/approfondimenti/database-uso-del-suolo (accessed on 4 April 2023).
66. Yutzler, J.; Daise, P. OGC®—GeoPackage Encoding Standard. 2021. Available online: https://portal.ogc.org/files/?artifact_id=74225 (accessed on 8 May 2023).
67. NextGIS. QuickMapServices. Available online: https://nextgis.com/blog/quickmapservices/ (accessed on 3 March 2019).
68. QGIS Development Team. QGIS Geographic Information System; Open Source Geospatial Foundation Project. 2023. Available online: https://qgis.org/en/site/ (accessed on 8 May 2023).
69. CTR. Available online: https://geoportale.regione.emilia-romagna.it/catalogo/dati-cartografici/cartografia-di-base/cartografia-tecnica/layer-1 (accessed on 23 June 2021).
70. CGR. Available online: https://geoportale.regione.emilia-romagna.it/catalogo/dati-cartografici/cartografia-di-base/immagini/layer-4 (accessed on 23 June 2021).
71. AGEA. Available online: https://geoportale.regione.emilia-romagna.it/catalogo/dati-cartografici/cartografia-di-base/immagini/layer-2 (accessed on 23 June 2021).
72. AGEA. Available online: https://geoportale.regione.emilia-romagna.it/catalogo/dati-cartografici/cartografia-di-base/immagini/layer-3 (accessed on 23 June 2021).
73. CGRA. Available online: https://www.provincia.re.it/aree-tematiche/pianificazione-territoriale/sistema-informativo-territoriale/archivio-cartografico/foto-aeree/volo-cgra-1976-78-scala-113-500/ (accessed on 23 June 2021).
74. USGS. EROS Archive—Declassified Data—Declassified Satellite Imagery-3. Available online: https://www.usgs.gov/centers/eros/science/usgs-eros-archive-declassified-data-declassified-satellite-imagery-3 (accessed on 13 December 2021).
75. IGM. Available online: https://www.igmi.org/ (accessed on 11 July 2021).
76. Nuovo Catasto Terreni. Available online: https://www.provincia.re.it/aree-tematiche/pianificazione-territoriale/sistema-informativo-territoriale/archivio-cartografico/cartografia-storica/carta-catastale-di-impianto-della-provincia-di-reggio-emilia-scala-12-000-11-000-edizione-fine-xix-secolo/ (accessed on 4 January 2021).
77. Carta Storica Regionale. Available online: https://geoportale.regione.emilia-romagna.it/catalogo/dati-cartografici/cartografia-di-base/cartografia-storica/layer-1 (accessed on 4 January 2021).
78. Timár, G.; Molnár, G.; Székely, B.; Biszak, S.; Varga, J.; Jankó, A. *Digitized Maps of the Habsburg Empire—The Map Sheets of the Second Military Survey and Their Georeferenced Version*; Arcanum: Budapest, Hungary, 2006; ISBN 9789637374333.
79. Ostafin, K.; Pietrzak, M.; Kaim, D. Impact of the Cartographer's Position and Topographic Accessibility on the Accuracy of Historical Land Use Information: Case of the Second Military Survey Maps of the Habsburg Empire. *ISPRS Int. J. Geo-Inf.* **2021**, *10*, 820. [CrossRef]
80. Guida, D.; Pelfini, M.; Santilli, M. Geomorphological and Dendrochronological Analyses of a Complex Landslide in the Southern Apennines. *Geogr. Ann. Ser. A Phys. Geogr.* **2008**, *90*, 211–226. [CrossRef]
81. Bollati, I.; Vergari, F.; Del Monte, M.; Pelfini, M. Multitemporal Dendrogeomorphological Analysis of Slope Instability in Upper Orcia Valley (Southern Tuscany, Italy). *Geogr. Fis. Din. Quat.* **2016**, *39*, 105–120.
82. Gebrekirstos, A.; Bräuning, A.; Sass-Klassen, U.; Mbow, C. Opportunities and Applications of Dendrochronology in Africa. *Curr. Opin. Environ. Sustain.* **2014**, *6*, 48–53. [CrossRef]
83. Tronti, G.; Vergari, F.; Bollati, I.M.; Belisario, F.; Del Monte, M.; Pelfini, M.; Fredi, P. From Landslide Characterization to Nature Reserve Management: The "Scialimata Grande Di Torre Alfina" Landslide Geosite (Central Apennines, Italy). *J. Mt. Sci.* **2023**, *20*, 585–606. [CrossRef]
84. Grissino-Mayer, H.D. A Manual and Tutorial for the Proper Use of an Increment Borer. *Tree-Ring Res.* **2003**, *59*, 63–79.
85. GRASS Development Team. *Geographic Resources Analysis Support System (GRASS)*; Open Source Geospatial Foundation: Beaverton, OR, USA, 2023.
86. Mitasova, H.; Michael Barton, C.; Ullah, I.; Hofierka, J.; Harmon, R.S. 3.9 GIS-Based Soil Erosion Modeling. In *Treatise in Geomorphology: Vol. 3 Remote Sensing and GI Science in Geomorpholog*; Shroder, J.F., Ed.; Academic Press: San Diego, CA, USA, 2013; Volume 3, pp. 228–258, ISBN 9780080885223.
87. Panagos, P.; Borrelli, P.; Meusburger, K.; Alewell, C.; Lugato, E.; Montanarella, L. Estimating the Soil Erosion Cover-Management Factor at the European Scale. *Land Use Policy* **2015**, *48*, 38–50. [CrossRef]
88. Ullah, I.; Barton, M.; Mitasova, H. *r.landscape.evol*. 2023. Available online: https://grass.osgeo.org/grass82/manuals/addons/r.landscape.evol.html (accessed on 8 May 2023).
89. Encyclopedia Britannica. The Editors of Encyclopedia Britannica Momentum. 2022. Available online: https://www.britannica.com/science/momentum (accessed on 8 May 2023).

90. Soil Erosion (1:50k) Map—Emilia Romagna Region. Available online: https://mappegis.regione.emilia-romagna.it/gstatico/documenti/dati_pedol/download/erosione_idrica_2019_raster.zip (accessed on 1 November 2022).
91. Panagos, P.; Borrelli, P.; Matthews, F.; Liakos, L.; Bezak, N.; Diodato, N.; Ballabio, C. Global Rainfall Erosivity Projections for 2050 and 2070. *J. Hydrol.* **2022**, *610*, 127865. [CrossRef]
92. Lelle, M.A.; Gold, M.A. Agroforestry Systems for Temperate Climates: Lessons from Roman Italy. *For. Conserv. Hist.* **1994**, *38*, 118–126. [CrossRef]
93. Rombaldi, O. Dalla Mezzadria Nel Reggiano, a Proposito Del Saggio Sopra La Storia dell'Agricoltura Di Filippo Re. *Riv. Stor. Dell'agricoltura* **1965**, *5*, 22–48.
94. Cingolani, G. L'Italia Delle Colline. In *Storia e Problemi Contemporanei*; CLUEB: Bologna, Italy, 2004; pp. 1–7.
95. Montanari, M. Colture, lavori, tecniche, rendimenti. In *Storia Dell'agricoltura Italiana—IL Medioevo E L'età Moderna*; Pinto, G., Poni, C., Tucci, U., Eds.; Polistampa: San Fermo Della Battaglia, Italy, 2010; Volume 2, pp. 59–82, ISBN 9788859607656.
96. Piccinni, G. La proprietà della terra, i percettori dei prodotti e della rendita. In *Storia Dell'agricoltura Italiana—IL Medioevo E L'età Moderna*; Pinto, G., Poni, C., Tucci, U., Eds.; Polistampa: San Fermo Della Battaglia, Italy, 2010; Volume 2, pp. 145–170, ISBN 9788859607656.
97. Cafasi, F. L'agricoltura Negli Stati Estensi Nel Periodo Pre-Unitario. *Riv. Stor. Dell'agricoltura* **1980**, *XX*, 79–94.
98. Cazzola, F. Verso la centralità del frumento. In *Storia Dell'agricoltura Italiana—IL Medioevo E L'età Moderna*; Pinto, G., Poni, C., Tucci, U., Eds.; Polistampa: San Fermo Della Battaglia, Italy, 2010; Volume 2, pp. 223–254, ISBN 9788859607656.
99. Rombai, L.; Bomcompagni, A. Popolazione, popolamento, sistemi colturali, spazi coltivati, aree boschive ed incolte. In *Storia Dell'agricoltura Italiana—IL MEDIOEVO E L'ETÀ MODERNA*; Pinto, G., Poni, C., Tucci, U., Eds.; Polistampa: San Fermo Della Battaglia, Italy, 2010; Volume 2, pp. 171–222, ISBN 9788859607656.
100. Nair, P.K.R.; Kumar, B.M.; Nair, V.D. Agroforestry Systems in The Temperate Zone. In *An Introduction to Agroforestry: Four Decades of Scientific Developments*; Nair, P.K.R., Kumar, B.M., Nair, V.D., Eds.; Springer International Publishing: Cham, Switzerland, 2021; pp. 195–232, ISBN 9783030753580.
101. Assini, S.; Filipponi, F.; Zucca, F. Land Cover Changes in an Abandoned Agricultural Land in the Northern Apennine (Italy) between 1954 and 2008: Spatio-Temporal Dynamics. *Plant Biosyst.—Int. J. Deal. All Asp. Plant Biol.* **2015**, *149*, 807–817. [CrossRef]
102. Argenti, G.; Parrini, S.; Staglianò, N.; Bozzi, R. Evolution of Production and Forage Quality in Sown Meadows of a Mountain Area inside Parmesan Cheese Consortium. *Agron. Res.* **2021**, *19*, 344–356.
103. EAFRD. Rural Development Programmes 2014–2020: Key Facts & Figures; European Agricultural Fund for Rural Development. 2014. Available online: https://ec.europa.eu/enrd/sites/default/files/priority-4-summary.pdf (accessed on 8 May 2023).
104. Bando Montagna. 2022. Available online: https://tinyurl.com/yr5m6cxz (accessed on 1 November 2022).
105. Karcagi Kováts, A.; Katona Kovács, J. Factors of population decline in rural areas and answers given in EU member states' strategies. *Food Res. Inst. Stud. Agric. Econ. Trade Dev.* **2012**, *114*, 49–56.
106. European Commission. *Horizon Europe Strategic Plan (2021–2024)*; European Commission Directorate-General for Research and Innovation: Brussels, Belgium, 2021.
107. EIP-AGRI. *Workshop Towards Carbon Neutral Agriculture Workshop Report 24–25 March 2021*; The Agricultural European Innovation Partnership: Brussels, Belgium, 2021.
108. Berger-Tal, O.; Saltz, D. Invisible Barriers: Anthropogenic Impacts on Inter- and Intra-Specific Interactions as Drivers of Landscape-Independent Fragmentation. *Philos. Trans. R. Soc. Lond. B Biol. Sci.* **2019**, *374*, 20180049. [CrossRef]
109. Sutor, S.; McIntyre, N.E.; Griffis-Kyle, K. Characterizing Range-Wide Impacts of Anthropogenic Barriers on Structural Landscape Connectivity for the Sonoran Desert Tortoise (Gopherus Morafkai). *Landsc. Ecol.* **2023**, *4*, 118–126. [CrossRef]
110. Rippa, D.; Maselli, V.; Soppelsa, O.; Fulgione, D. The Impact of Agro-Pastoral Abandonment on the Rock Partridge Alectoris Graeca in the Apennines. *IBIS* **2011**, *153*, 721–734. [CrossRef]
111. Haggar, J.; Pons, D.; Saenz, L.; Vides, M. Contribution of Agroforestry Systems to Sustaining Biodiversity in Fragmented Forest Landscapes. *Agric. Ecosyst. Environ.* **2019**, *283*, 106567. [CrossRef]
112. Bentrup, G.; Hopwood, J.; Adamson, N.L.; Vaughan, M. Temperate Agroforestry Systems and Insect Pollinators: A Review. *Forests* **2019**, *10*, 981. [CrossRef]
113. Santiago-Freijanes, J.J.; Mosquera-Losada, M.R.; Rois-Díaz, M.; Ferreiro-Domínguez, N.; Pantera, A.; Aldrey, J.A.; Rigueiro-Rodríguez, A. Global and European Policies to Foster Agricultural Sustainability: Agroforestry. *Agrofor. Syst.* **2021**, *95*, 775–790. [CrossRef]
114. Messean; Viguier; Paresys Enabling Crop Diversification to Support Transitions towards More Sustainable European Agrifood Systems. *Front. Agric. Sci. Eng.* **2021**, *8*, 474–480.
115. Tress, B.; Tress, G.; van der Valk, A.; Fry, G. *Interdisciplinary and Transdisciplinary Landscape Studies: Potential and Limitations*; Delta Program: Wageningen, The Netherlands, 2003.
116. Piacente, S.; Badiali, F. La Ricerca Transdisciplinare per La Conoscenza E La Valorizzazione Del Territorio. *Ric. Transdiscipl. Conosc. Valoriz. Territ.* **2019**, *1*, 105–116.
117. Turner, S.; Kinnaird, T.; Koparal, E.; Lekakis, S.; Sevara, C. Landscape Archaeology, Sustainability and the Necessity of Change. *World Archaeol.* **2020**, *52*, 589–606. [CrossRef]
118. Brandolini, F. *Tecniche Digitali e Geoarcheologia per lo Studio del Paesaggio Medievale: Uno Studio Interdisciplinare in Pianura Padana Centrale*; British Archaeological Reports Limited: Oxford, UK, 2021; ISBN 9781407358543.

119. Zerbe, S. Merging Traditions and Innovation for Sustainability and Multifunctionality of Cultural Landscapes. In *Restoration of Multifunctional Cultural Landscapes: Merging Tradition and Innovation for a Sustainable Future*; Zerbe, S., Ed.; Springer International Publishing: Cham, Switzerland, 2022; pp. 497–513, ISBN 9783030955724.
120. Ganasri, B.P.; Ramesh, H. Assessment of Soil Erosion by RUSLE Model Using Remote Sensing and GIS—A Case Study of Nethravathi Basin. *Geosci. Front.* **2016**, *7*, 953–961. [CrossRef]
121. Alewell, C.; Borrelli, P.; Meusburger, K.; Panagos, P. Using the USLE: Chances, Challenges and Limitations of Soil Erosion Modelling. *Int. Soil Water Conserv. Res.* **2019**, *7*, 203–225. [CrossRef]

Article

Research on Changsha Gardens in Ming Dynasty, China

Weiwen Li and Chi Gao *

Department of Landscape Architecture, Huazhong Agricultural University, No. 1 Shizishan Street, Hongshan District, Wuhan 430070, China
* Correspondence: gaochi@mail.hzau.edu.cn

Abstract: Despite the growing interest in Chinese gardens, there is a lack of research on Changsha Gardens. Through document retrieval, review, and map analysis, we reconstructed the Changsha Gardens during the Ming Dynasty. Our findings reveal that gardening flourished in Changsha during this period. The royal gardens, dominated by literati aesthetics, set the trend for development, while the landscaping techniques were influenced by Jiangnan Gardens. Private gardens placed more emphasis on artistic mood and cultural implications. Landscape architects incorporated towers, terraces, and suburban mountain gardening to borrow scenery. The use of spring water to create landscapes and the rectangular shapes of the water bodies reflected Neo-Confucianism and practical functions. Planting design focused on meaning over form, with landscapes used to commemorate sages and promote the farming and reading culture and other Confucian values. However, it relied too much on borrowing natural scenery and the pragmatism of the Hunan culture, which might have affected the progress of landscaping techniques and the development of Changsha Gardens. Changsha Gardens were not as skilled as Jiangnan Gardens, and the style was not as prominent as other regional schools of gardens. This might be the reason why Changsha Gardens have not become one of the regional schools of gardens in China.

Keywords: landscape architecture; Changsha Garden; Chinese garden in the Ming Dynasty; Hunan Garden; landscaping technique; landscaping ideas

Citation: Li, W.; Gao, C. Research on Changsha Gardens in Ming Dynasty, China. *Land* **2023**, *12*, 707. https://doi.org/10.3390/land12030707

Academic Editors: Bellotti Piero and Alessia Pica

Received: 17 February 2023
Revised: 10 March 2023
Accepted: 15 March 2023
Published: 18 March 2023

1. Introduction

China is widely recognized as the "Mother of Gardens" [1], and since the 20th century, Chinese gardens, with their ideas and their principles, have gained increasing attention, recognition, and practice worldwide [2]. After the establishment of the Suzhou Garden "Astor Court (明轩)" in the Metropolitan Museum of Art in New York in 1980, China began building numerous Chinese gardens overseas, and UNESCO has listed Chengde Mountain Resort and its surrounding temples, Suzhou classical gardens, and the Summer Palace as World Heritage Sites [3,4]. The academic field of landscape architecture traditionally divides Chinese gardens into three major regional schools: Northern Gardens (北方园林), Jiangnan Gardens (江南园林), and Lingnan Gardens (岭南园林) [5]. In the 21st century, Ba-Shu Gardens (巴蜀园林) have been recognized as the fourth regional school of gardens [6], while some scholars have proposed concepts for gardens in smaller regions, such as Central Plains Gardens (中原园林) (a subtype of Northern Gardens), Fujian and Taiwan Gardens (闽台园林) (a subtype of Lingnan Gardens), Huizhou Gardens (徽州园林) (a subtype of Jiangnan Gardens), and Xu Style Gardens (徐派园林) [7]. However, due to the lack of documentation, drawings, pictures, and other materials, most of the gardens in Changsha were destroyed and are little known, making them absent from the history of gardens. Nonetheless, local chronicles and literati works of Changsha reveal that, during the Ming Dynasty (1368–1644), which was the mature stage of Chinese gardens [5], various types of gardens existed in Changsha that deserve attention and research.

Internationally, research on regional Chinese schools of gardens mainly focuses on Jiangnan Gardens and Lingnan Gardens, while the Northern Garden, Ba-Shu Gardens, and

Xu Style Gardens have received little attention. Regarding Jiangnan Gardens, researchers have taken Suzhou Gardens as an example to compare Chinese garden ideas with Western ideas [8], to explore the spatial distribution of Suzhou Gardens during the Qianlong period of the Qing Dynasty [9], and to analyze Tang Dynasty gardens. They have also examined the distribution and scale evolution of Suzhou Gardens under the urbanization process of the Qing Dynasty [10]. For Lingnan Gardens, researchers have attempted to reconstruct the gardens of the merchant Howqua [11] and explored the characteristics and styles of Lingnan Gardens [12]. As for Northern Gardens, researchers have analyzed the imperial gardens during the Qianlong period of the Qing Dynasty [13]. Regarding Changsha Gardens, researchers have taken Yuelu Academy (a famous Changsha Garden) as an example of Chinese Academic Garden and analyzed the space and design of Chinese, Japanese, and Korean Academy Gardens [14]. However, there is no literature dedicated to the research on Changsha Gardens.

Chinese scholars have conducted limited research on Changsha Gardens. Some have classified Changsha Gardens into historical stages [15–18] and analyzed their characteristics, ideas, and some existing garden layouts [16,18]. Others have classified religious gardens in Changsha and researched the history of existing ones [19]. Most of these studies focus on the analysis of the history, classification, and characteristics of Changsha Gardens, especially existing gardens. Existing research seldom discusses the social and cultural background of the prosperity of Changsha Gardens in the Ming Dynasty, landscaping techniques, and landscaping ideas. The literature interpretation tends to be a simple overview, and there is no conjecture about the restoration of destroyed gardens. At the same time, the researchers have not explored the influence of other regional schools of gardens on Changsha Gardens and have not considered why Changsha Gardens have not become one of the regional schools of gardens in China.

2. Materials and Methods

The authors take the modern Changsha urban area as the research scope and take Changsha Gardens in the Ming Dynasty as an example to research Changsha Gardens in the Ming Dynasty by consulting relative articles, literati works, and local chronicles (Table 1), drawing garden restoration maps, hoping to fill the research gaps in Changsha Gardens, Hunan gardens, and Ming Dynasty gardens of China. We collected literati works and local chronicles from the Chinese Classics and Ancient Books Library (中華經典古籍庫), a large database of ancient books launched by Zhonghua Book Company in 2014, as well as some books from the Hunan University Library. This paper mainly uses document retrieval, document review, and map analysis as research methods.

Table 1. Literati works and local chronicles.

Literati Works	Local Chronicles
Poetry of Southern Chu, 南楚诗纪, 1827	Chronicles and Atlas of Rebuilding Yuelu Academy, 重修岳麓书院图志, 1594
Analects of Tianyue Mountain Mansion, 天岳山馆文钞, 1880 Records of Exploring Ancient Relics in Changsha, 湘城访古录, 1893 Celebrities and Changsha Landscape, 名人与长沙风景, 2012 Memories of a City—Changsha in Old Maps, 一个城市的记忆— 老地图中的长沙, 2018	Chronicles of Changsha Prefecture, 长沙府志, 1639 Chronicles of Newly Revised Yuelu Academy, 新修岳麓书院志, 1687 Brief Chronicles of Yuelu, 岳麓小志, 1932

3. Results

3.1. Overview of Changsha Gardens in the Ming Dynasty

3.1.1. The Social Background of the Rise of Changsha Gardens in the Ming Dynasty

Changsha, the capital of China's Hunan Province, is located in the valley basin where the Xiang and Liuyang rivers meet and is surrounded by mountains (refer to Figures 1 and 2). In the Ming Dynasty, Chinese emperors distributed princes in various places to protect their power, and only prosperous places could guarantee their stable rule. In 1372, Qiu Guang, the Garrison Commander, undertook large-scale repairs of war-torn Changsha City at the end of the Yuan Dynasty. Changsha officials improved water transportation by dredging river channels and renovating wharves, making Changsha the largest cargo distribution center and commercial center in Hunan. The huge expenditure of the princes also supported social stability [20], which brought economic development and cultural prosperity—the political, economic, and cultural conditions necessary for Changsha to build gardens.

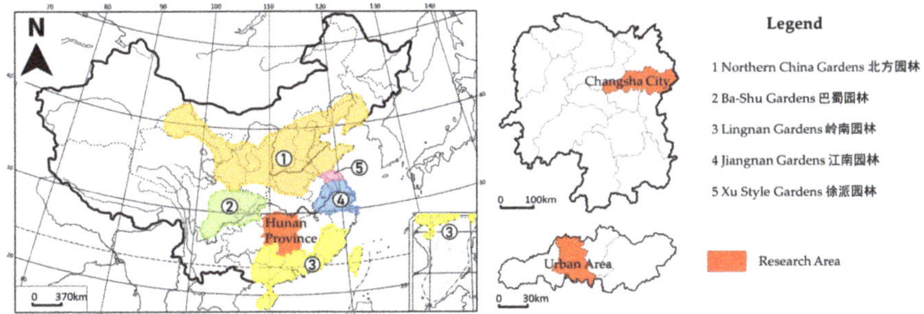

Figure 1. Research Area.

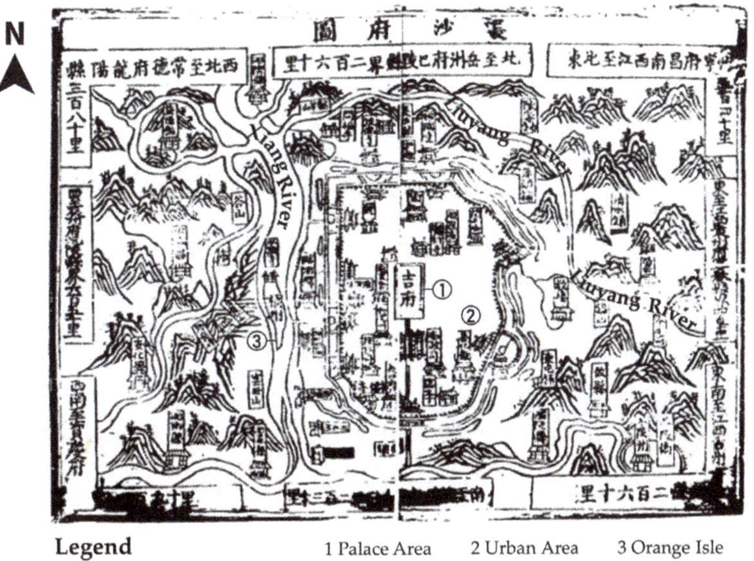

Figure 2. Map of Changsha in the Ming Dynasty (source: Chronicles of Changsha Prefecture, 1639).

3.1.2. The Overall Distribution and Number of Types of Gardens in Changsha in the Ming Dynasty

According to related articles [15–19,21–23], there were 20 gardens in Changsha during the Ming Dynasty, including four royal gardens, four private gardens, three academic gardens, seven religious gardens, one garden of ancestral hall, and one garden of government office.

The gardens belonging to the palace were all within the city for Prince Ji (吉王) to enjoy. The large Palace Gardens (大内园林), such as Zijin Garden and East Garden, were all close to the palace. Phoenix Terrace and Pine and Osmanthus Garden were the Detached Palace Gardens (离宫园林) located near the suburbs. Private gardens, academic gardens, and religious gardens were widely distributed, making it convenient for Changsha citizens to stay at home, read, worship, and engage in recreational activities. Religious gardens were mostly located in suburban mountains, which were conducive to practice in peace. Based on local chronicles, literati works (refer to Table 1), and related articles [15–19], we have drawn a distribution map of garden sites and garden types (refer to Figures 3 and 4).

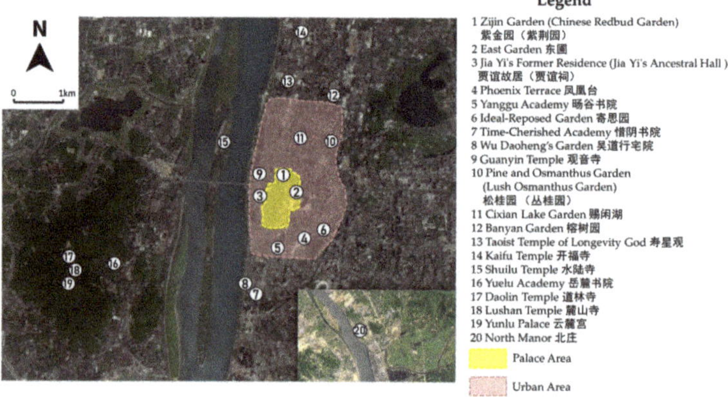

Figure 3. Names and distribution of Changsha Gardens in the Ming Dynasty (source: the author repainted from Google Maps).

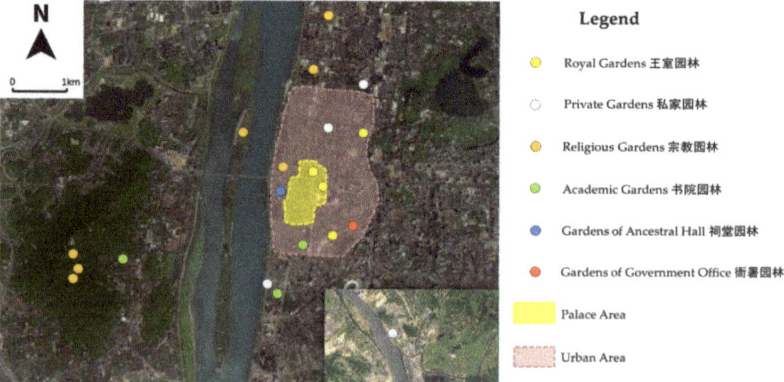

Figure 4. Distribution of Changsha Garden types in the Ming Dynasty (source: the author repainted this from Google Maps).

3.2. Description

3.2.1. Royal Gardens

There were three royal gardens: Royal Gardens (Zijin Garden and East Garden), Phoenix Terrace, and Pine and Osmanthus Garden. The architectural layout of the palace is the same as the Forbidden City in Beijing, the capital of the Ming Dynasty.

(1) Prince's Palace Garden

It was built by Zhu Jianjun (朱见浚), Prince Ji. To the northwest of the palace was the back garden of the palace, named Zijin Garden, also known as the Chinese Redbud Garden because of the many Chinese Redbuds. It was the only large-scale palace garden in Changsha. Zijin Garden had two major landscapes, Zijin Mountain in the east, also known as Chinese Redbud Mountain, and Wanchun Pool in the west. Zijin Mountain was made of Taihu stone, and it is the first large rockery group in the history of Changsha Gardens. There were many slender and winding stone paths on the mountain, as well as many caves for people to rest and view were formed inside [18,21]. It is speculated that the landscape architects used the crevices and caves in the rockery to fill in the soil to plant Chinese Redbud.

Zijin Mountain is one of the commanding heights of the whole city landscape. Wanchun Pond, which had a wide water surface, was a place for gathering and boating. There was a dressing-changing pavilion on the south side and a dressing building on the north side [18]. It could be seen that the garden landscape architecture had various functions, which were set up for the entertainment services of dignitaries (Figures 5 and 6).

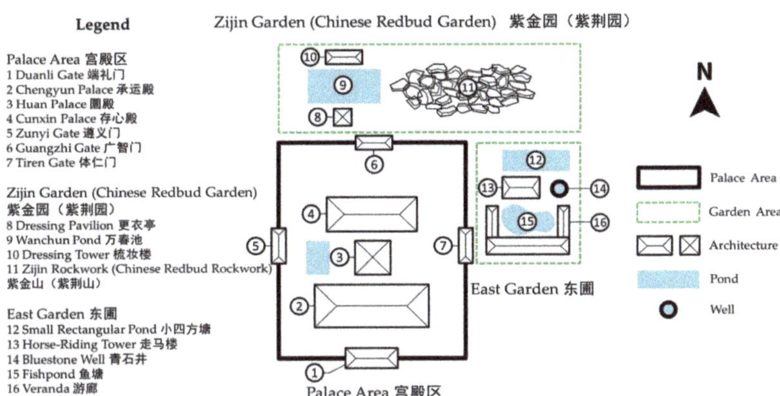

Figure 5. Plan of Prince's palace and gardens.

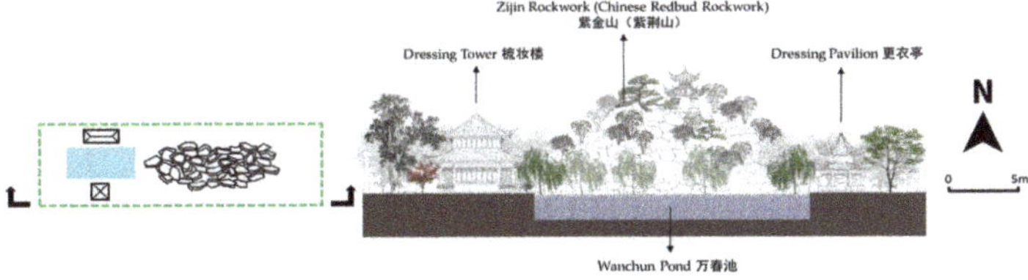

Figure 6. Section of Zijin Garden (Chinese Redbud Garden).

The northeast of the palace is the harem, called the "East Garden". The Horse-Riding Tower is one of the veranda-shaped buildings, which was the place where Prince Ji rode horses and read books. There is a rectangular pool in the north of the building, which is surrounded by palms, later known as the "Small Rectangular Pond". There is a bluestone well on the east side of the building for drawing water. The south of the building and the well is a fishpond, which is enclosed by verandas and terraces [15,17,21]. The fishpond landscape has both production and ornamental value (Figure 7).

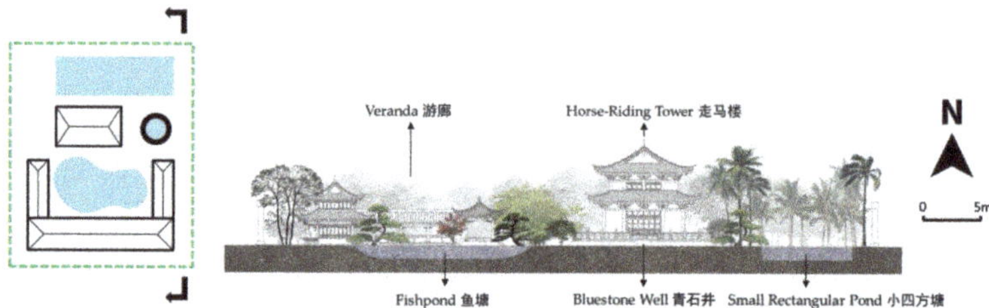

Figure 7. Section of East Garden.

(2) Phoenix Terrace

Adjacent to Tianxin Pavilion, there was a raised terrace in the southeast of the city. It is a dressing tower with a garden built by Prince Ji for his daughter, Princess Phoenix. On the stage, people could overlook the whole city of Changsha, Orange Isle, and opposite Yuelu Mountain and Tianma Mountain, becoming another commanding height in the city [17,21].

(3) Pine and Osmanthus Garden

Also known as Lush Osmanthus Garden, according to the name, it could be speculated that the garden was dominated by pine and osmanthus trees. It was a detached palace garden named after the plant. Prince Ji often entertained celebrities here [15,17,21].

3.2.2. Private Gardens

(1) Banyan Tree Garden

The owner of the garden was Zhuang Tianhe, the Minister of Rites. The garden was named after a plant, and it was dominated by banyan trees, as well as climbing figs, Japanese kerria, willows, bamboos, and other plants. There were many old trees and paths in the bamboo forest for walking. Inside the garden, there was a rectangular pond with pavilions and a reading tower beside it. Lotus flowers were planted in the pond, and ornamental fish were cultivated. A small bridge was set up on the water, and whale stone carvings were placed by the water [15,21,22,24].

(2) North Manor

North Manor was a retreat garden built by Wang Wei, the Vice Minister of the Ministry of War, in the area of Eyang Mountain in the northern suburb of Changsha City. It was the only private garden that belongs to the natural landscape garden of Changsha in the Ming Dynasty. There was a book house in the garden and a fishing terrace by the stream. The landscape architecture made fewer changes to the original natural environment. There was a large Chinese parasol tree planted on the mountain, and there was spring water in the garden. In the park, people could borrow the scenery outside the park [15,18]. The near view next to the mountain is the Lake of the Chu Family, and the distant view was the waterlogged pond surrounded by trees, as well as Yuelu Mountain, Xiang River, and Orange Isle. Wang Wei named North Manor and its surrounding scenery "Eight Views of North Manor" [17,18,22].

(3) Cixian Lake

It was originally a detached palace garden and was later given to Yan Jing, the Huguang inspector who was idle at home, by Prince Ji. There were pavilions, terraces, bridges, waterside pavilions, and other types of buildings in the garden. The pond is rectangular in shape, and there were natural springs in the park, so a pavilion was built for protection [15,21,25].

(4) Wu Daoheng's Garden

The owner of the garden was Wu Daoheng, a famous scholar who has not been an official. There were two ponds in the garden, one for fish and one for lotus. There were various garden buildings, such as hall, veranda, pavilion, tower, and windowed veranda beside the pool. There were many old trees and green trees in the garden, and the plant space is tightly enclosed, making the whole garden seem to be surrounded by valleys. The landscape architecture took the Xiang River as a close-up view and Yuelu Mountain as a distant view by borrowing scenery [15,17].

3.2.3. Academy Gardens

(1) Time-Cherished Academy

Created by Lv Tingjue, the magistrate of Changsha, in 1525, the Time-Cherished Academy features Mingdao Hall at the front and Tao Gong Shrine at the back, which honors Confucian sage Tao Kan (陶侃). The Heart-Washing Pavilion was in the middle, with Frog-Prohibited Ponds on both sides. The central point of the garden was the Talent-Gathering Tower, and Broadcast Benevolence Hall was situated behind it. The surrounding area of hall was a concentrated area of academy gardens, featuring verandas, pools, and pavilions. The Mountain-Overlooking Tower was at the back of the hall, and in front of it was the Mountain-Overlooking Pavilion with school fields, ponds, and large open spaces on the left and right.

The academy's overall layout was symmetrical on the left and right, with the main buildings strictly arranged on the axis from Mingdao Hall to Mountain-Overlooking Tower. The central hill's Talent-Gathering Tower was the commanding height of the whole garden, acting as the climax of the architectural sequence. Broadcast Benevolence Hall was located behind the Talent-Gathering Tower, and its surrounding area was a concentrated area of gardens. The Mountain-Overlooking Tower, located at the end of the axis, was also the commanding height, echoing with the Talent-Gathering Tower and serving as the end of the academy building sequence (Figures 8 and 9) [15–19,26,27].

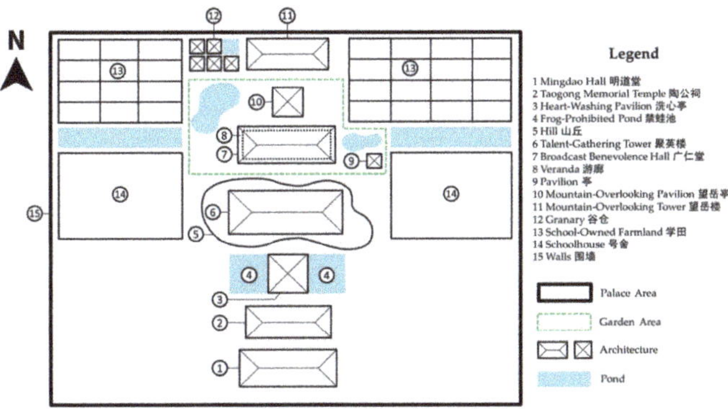

Figure 8. Plan of Time-Cherished Academy.

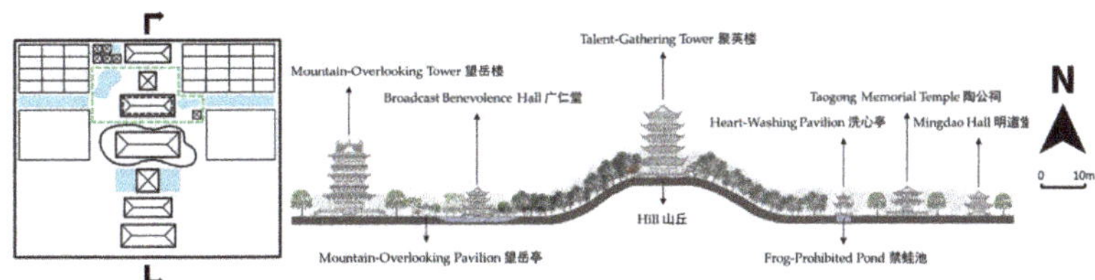

Figure 9. Section of Time-Cherished Academy.

(2) Yanggu Academy

It was a palace school for the children of the prince. The garden was named and designed with the theme of the four seasons, such as the Eight Hexagrams Building (for viewing the spring scenery), Summer Clouds Pavilion, Autumn Osmanthus Windowed Veranda, Severe Winter Pavilion, and other buildings [15,18,21]. It was established earlier than the Ge Garden (个园) in Yangzhou (a Jiangnan Garden in the Qing Dynasty), which is famous for its landscape with the theme of the four seasons.

In addition, in 1509, Wu Shizhong, the garrison commander of Changsha, demolished the Daolin Temple and rebuilt the Yuelu Academy, which was originally built in the Northern Song Dynasty (976).

3.2.4. Religious Gardens

Yunlu Palace was built by Zhu Jianjun, Prince Ji, and consists of three halls: front, middle, and back. The courtyard was filled with pines, cypresses, tung trees, catalpas, bamboos, and other flowers and trees [15–19,27]. The surrounding area was enveloped by natural forests, and the atmosphere of the Taoist temple was solemn and peaceful. Kaifu Temple is a religious garden constructed using borrowed scenery, with the Xiang River and Yuelu Mountain serving as its borrowed scenery and landscape background [15,17,21].

Some gardens were established before the Ming Dynasty. Shuilu Temple was situated north of Orange Isle, and the Circumpolar Tower (拱极楼) was the tallest building among them, serving as a summer resort [15,26–28]. Lushan Temple was constructed in the Western Jin Dynasty (268). Some religious gardens were poorly documented in the literature and difficult to predict, such as the Guanyin Temple, the Taoist Temple of Longevity God, and the Daolin Temple used by Prince Ji [15–18].

3.2.5. Gardens of Ancestral Hall

Currently, Jia Yi's Former Residence is the only garden of ancestral hall that could be examined. It was established during the Western Han Dynasty (202 B.C.—8 A.D.) and has been rebuilt in all the subsequent dynasties. In 1465, Qian Shu, the prefectural magistrate of Changsha, discovered the ancient well of Jia Yi and raised funds to rebuild Jia Yi's former residence. This pattern continued into the Qing Dynasty.

Jia Yi's former residence is divided into the Grand Preceptor Temple of Jia, the house, and Qingxiang Villa (清湘别墅). The garden land is regular, surrounded by verandas and buildings, with a regular pond in the middle and a triangular pond in the northwest, which are separated by a veranda (Figure 10) [15–18,21,29].

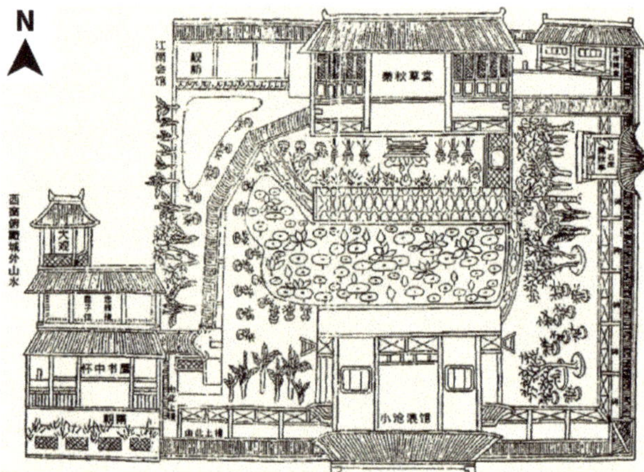

Figure 10. Plan of Jia Yi's former residence in the Qing Dynasty (source: The contemporary significance of Jia Yi's former residence cultural heritage by Wu Song-geng).

3.2.6. Gardens of Government Office

The only known garden of government office is Ideal-Reposed Garden, where the owner expressed his ideas through landscape design. The garden features a rectangular pond excavated by the owner, with a terrace built from the excavated soil and raised to a height of approximately 1.6 m. A bamboo pavilion was set up on the terrace, surrounded by various plants, such as flowers, bamboos, lotus, peach and plum trees, and plum blossoms, with cranes and other animals also present in the park. Each plant and animal in the garden had its meaning and symbolism [15–18,21,22].

3.3. *Analysis of Landscaping Techniques*

3.3.1. The Method of Borrowing Scenery—Borrowing the Scenery of "Mountain, Water, isle and City" and the Hills Surrounding the City

The urban pattern of Changsha is characterized by "mountain, water, isle, and city (山水洲城)", which is unique to the regional environment. The isle refers to the Orange Isle, located in the center of the Xiang River. The mountain refers to Yuelu Mountain, a famous historical and cultural mountain in China, while water refers to the Xiang River, the mother river of Changsha [30]. Changsha is surrounded by hills, the landscape architect made use of the landscape of Mountain, Water, Isle, and City (the urban pattern of Changsha) and the hills surrounding the city and used the borrowed scenery to borrow the scenery into the garden by borrowing the scenery to become a part and background of the garden. For gardens located in urban areas, the landscape background could be divided into four parts: Xiang River and Orange Isle, Yuelu Mountain and other mountains, Miaofeng Mountain and other mountains, and Eyang Mountain and other mountains. Changsha City is a concentrated distribution area of gardens, and the surrounding mountains, Xiang River, and Orange Isle served as the landscape backgrounds of the gardens in the city. Many religious gardens, as well as a few private and academic gardens, were built in this area, and the buildings faintly exposed from the woods complemented each other. For viewers located within the landscape background, other landscape backgrounds and Changsha City also became part of the garden's landscape background. Eyang Mountain and the mountains were far away from Changsha City and became secondary backgrounds. North Manor, a private garden located in it, only offered views of the Xiang River and the remnants of Yuelu Mountain (Figure 11).

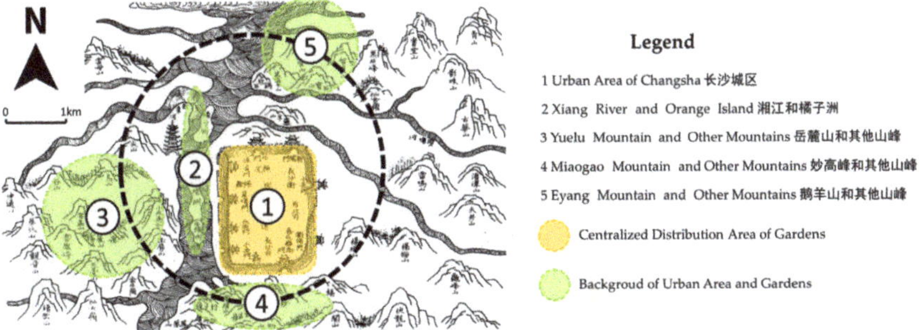

Figure 11. Analysis of borrowed scenery (source: the authors repainted from the Mountains and Rivers Map of Changsha in the Ming Dynasty, from Memories of a City—Changsha in Old Maps by Shen Xiaoding).

To obtain a distant view, it is necessary to climb to a high place. The gardener borrows the distant view in two ways:

(1) Building Terraces

Most of the gardens in Changsha city were built with terraces for viewing. People looking from a high-rise building could not only overlook the entire garden, but also the terraces of other gardens in the distance. The terraces offer close-up views of palaces and Changsha City, medium views of Xiang River, Orange Isle, and Circumpolar Tower, and distant views of Wenchang Tower and Towers on Yuelu Mountain.

(2) Mountaineering

Gardens in the mountains have the best conditions for scenery borrowing. Viewers located in Yuelu Mountain and the temple complex could see Yuelu Academy and farmland from a close view, Orange Isle and Circumpolar Tower from a middle view, and Changsha City and the Palace of Prince Ji from a distant view.

The field of vision was wide, and the landscape was rich in layers. The mountains, waters, isles, and cities were all visible. At North Manor, which was located on Eyang Mountain, viewers could see the general outline of Changsha City, Yuelu Mountain, Xiang River, Orange Isle, Circumpolar Tower, and five of the eight scenic spots were borrowed. Visitors felt as though the garden was indistinguishable from the inside and outside.

3.3.2. Landscaping Techniques

(1) Springs and Wells

Changsha is a city with many springs, and the spring water enjoys the reputation of "not overflowing and inexhaustible" [29]. The landscape architect of Banyan Tree Garden used natural spring water to create waterfalls on rockeries, which became important landscapes. The landscape architect of North Manor and Cixian Lake built wells to tap into spring water, serving both production and landscape functions. The wells of Cixian Lake were protected by pavilions and became a landscape combined with garden architecture. The garden of Jia Yi's former residence was even built around a well, which has become a symbol of Jia Yi's spirit.

(2) Rectangular Ponds

Most of the water bodies in the Ming Dynasty Changsha Gardens were rectangular, such as the Wanchun Pond, commonly known as the Big Rectangular Pond, the Cixian Lake, known as the Small Rectangular Pond, and the rectangular pond in the Ideal-Reposed Garden.

(3) Rockeries

Taihu stone is produced in the Taihu Basin of Jiangsu and Zhejiang provinces, which are part of Jiangnan in China [31]. Hunan is not the origin of the Taihu stone. Based on the characteristics of the rockery described in the literature, it is likely that the Taihu stone used in the rockery was shipped from Jiangsu and Zhejiang. During the Ming Dynasty, a large number of people, including those from Jiangnan, came to settle in Hunan [32], and commodity activities such as food trafficking increased population movement between Jiangnan and Hunan [33]. Therefore, the landscape architect and craftsmen of Zijin Garden were likely from Jiangnan and used the techniques of Jiangnan gardens to create large rockeries. The landscape architect of Banyan Tree Garden replaced the common islands in the water body of Chinese gardens with stones shaped as whales, which innovatively broke the traditional Chinese gardening method of "Three Mountains in a Pool" that symbolized the ocean and islands [5].

4. Discussion

4.1. Eight Views Culture

Chinese scenic spots often summarize the scenic characteristics by numbers, such as "eight views", "ten views" and so on. This expression forms the "Eight Views Culture" (八景文化) [34]. The Eight Views Culture originated in Hunan and covered the essence of traditional landscaping from all over China [35]. It depicts and showcases people's perception of nature, as well as patterns and customs of leisure [36]. The owner of the North Manor divided the visible scenery from the garden into eight scenes. The landscape architect borrowed the scenery to incorporate the landscape of Orange Isle and Xiang River into the garden so that the garden owner could enjoy the "Evening Snow Blending River and Sky" (江天暮雪), which is one of the "Eight Views of Xiaoxiang (潇湘八景)" in winter.

4.2. Literati Tastes

Private garden landscapes were entrusted with spiritual sustenance. From the literati's inscriptions on Yanggu Academy, it is evident that the Eight Diagrams Building was a reading building that one could climb to enjoy a distant view. The Summer Clouds Pavilion had pillow seats and curtains for drinking and resting, and rockeries and goldfish ponds adorned its surroundings. The Autumn Osmanthus Pavilion had osmanthus trees planted around it, making it a perfect place to enjoy the autumn scenery. The Severe Winter Pavilion was a place for scholars to drink and write poems after school. It had rockeries and noble and clean plants, such as plums and chrysanthemums in its vicinity. The owner of Banyan Tree Garden regarded the garden as a place of seclusion, where he could enjoy the moon, drink, play Guqin (古琴), sing and read with friends, and write inscriptions on the garden's scenery.

Landscape architects mostly use plants with noble meanings in Chinese cultures, such as pine, bamboo, peach, plum, banyan, sunflower, kerria japonica, and sycamore. However, none of the literature describes the plant's shape, leading us to speculate that it was not taken seriously. Some gardens were named after plants, such as Banyan Garden. The owner of the Garden of Contemplation entrusted bamboo and lotus to his parents, sunflowers to symbolize respect for light, and peaches and plums to cultivate talent. He also realized the beauty of the artistic mood (意境) created by the changing seasons of the landscape. The owner of the North Manor resigned and went into seclusion, building a garden on Eyang Mountain, which was far away from Changsha City. There were many Chinese parasol trees on the mountain, symbolizing modesty and loneliness.

4.3. Confucianism

(1) Carry forward the Farming and Reading Culture

Confucian literati advocated the lifestyle and value orientation of "farming on one side and studying on the other side, the heirloom of cultivating and studying", forming a farming and reading culture [37]. Among the academy gardens in Hunan, Time-Cherished Academy still had farmland inside, even though there was a school field outside. We speculate that the landscape architect deliberately used the farmland as part of the landscape, not only to meet needs, but also to make the academy a place to promote the farming and reading culture.

(2) Commemorating Famous Sages and Educated People

The government regards ancestral gardens and academic gardens as centers for the dissemination of Confucian culture. The ancestral hall commemorating Tao Kan, the Duke of Changsha County in the Eastern Jin Dynasty, was expanded into Time-Cherished Academy in the Ming Dynasty, and Tao Kan Temple became the center of the academy. Jia Yi's former residence was officially regarded as a place to promote the Confucian culture of "loyalty and filial piety" [38], and the restoration of Jia Yi's former residence as a cultural holy place had become a way to educate the people.

(3) Allegory of Confucianism

The Hunan school of thought was an important school of Neo-Confucianism during the Song and Ming Dynasties, and it has a profound impact on Hunan culture [39]. The rectangular pond has been a symbol of Neo-Confucianism since the Song Dynasty. Influenced by traditional ideas, such as Feng Shui and the Five Elements, the construction of rectangular ponds was very popular from the Song Dynasty to the early and mid-Ming Dynasty [40]. Therefore, most of the water bodies in Changsha Gardens were rectangular ponds, which may have been both the trend at that time and the influence of Neo-Confucianism.

Legend has it that when Zhang Shi (张栻), a famous Confucianist in the Southern Song Dynasty, was giving lectures at Chengnan Academy (城南书院), he stayed overnight in the academy. Due to the noise of frogs in the pond, Zhang Shi picked up his writing desk and threw an inkstone at it, and the sound of frogs died down. Therefore, the Forbidden-Frog Pond in Time-Cherished Academy is likely named not only to commemorate the ancient sages, but also to express the hope that the academy's environment will be peaceful by naming the landscape.

4.4. The Needs and Political Purposes of the Royal and the Literati

Prince Ji built several royal gardens for leisure and enjoyment. In addition to their basic viewing and recreational functions, these gardens had other purposes, such as housing books, hosting horse racing events, and serving as fish farms. The garden area of the prince's palace was vast, including Zijin Mountain, which was made entirely of Taihu Lake and Wanchun Pond, which were large enough for boating. These gardens showcased the royal family's style and taste. Prince Ji won over the gentry group by rewarding private gardens and inviting celebrities to feasts and events in the royal gardens. In the Ming Dynasty, Taoism allowed the prince to enjoy a degree of freedom and autonomy beyond his social role restricted by the vassal prohibition system. Local Taoist temples took advantage of their close relationship with the prince to expand their scale and influence [41]. Through the construction and reconstruction of temples, Prince Ji established himself as a local ruler who cared for the people while satisfying his religious beliefs, and he promoted the revival of religious gardens. Prince Ji was an avid book collector and built a library in the Royal Academy. He also placed great importance on education and supported the construction of the academy, which promoted the development of the academic garden.

The academy served as the material carrier of Confucian education, which was the official ideology [42]. After the mid-Ming Dynasty, academic education began to revive [43], a phenomenon known as "the turning of culture and education" [42]. According to available information, the Ideal-Reposed Garden was the only garden of government office in Changsha, while there were three gardens in the academy, reflecting Changsha's "the turning of culture and education". Besides Yanggu Academy, dedicated to the royal family, Yuelu Academy and Time-Cherished Academy were built during this period. It is evident that the academic gardens in Changsha have developed under the revival of the academy and the attention of the royal family and literati.

5. Conclusions

Changsha's landscape architects in the Ming Dynasty were adept at utilizing the landscape and the landscape of hills surrounding the city as the background of the garden. The gardens above it borrowed scenery from each other. Hunan, the birthplace of the Eight Views Culture, was deeply influenced by it. Landscape architects built structures, such as pavilions and terraces, to enjoy the "Evening Snow Blending River and Sky", one of the eight views of Xiaoxiang, by borrowing the scenery of Orange Isle and Xiang River into the garden. Some garden owners even divided the visible scenery in the garden into eight views.

The abundant spring water resources of Changsha were utilized to create waterfalls on rockeries, wells as productive landscapes, or gardens with wells at the center. The design of ponds took into account the practicality influenced by Neo-Confucianism, Feng Shui, Five Elements, and other traditions. The ponds were mostly rectangular in shape, and the landscaping predominantly used native plants. The design emphasized meaning over the shape. Landscape architects utilized the scenery to promote Confucianism, demonstrate the farming and reading culture, commemorating the sages, and educating the people.

In the Ming Dynasty, Changsha had a prosperous gardening atmosphere, and the development trend of Changsha Gardens was led by the royal gardens. The prince who ruled Changsha introduced the gardening techniques of Jiangnan Gardens, which influenced Changsha Gardens. Gardening was a manifestation of identity, and the landscape architects of royal gardens showed the style of the royal family through large-scale landscapes with diverse functions. The owners of private gardens were mostly civil servants and literati, and they paid more attention to creating artistic mood and cultural implications in the landscape. This may be because they lacked the power and financial resources to obtain a large area of the site and create a large-scale landscape or because they were more concerned with spiritual connotations. We have not found any private gardens whose owners were businessmen, which is very different from Yangzhou gardens, a subtype of Jiangnan Gardens that had many businessmen. We speculate that this is because Changsha's economy was not yet developed, and the status of the merchant class was not strong.

The development of Changsha Gardens in the Ming Dynasty was greatly influenced by politics. The prince created a good image of local rulers by inviting literati officials to garden gatherings, rewarding gardens, and building new religious gardens. The academic gardens and the gardens of ancestral hall were mainly promoted by civil servants and literati, but they were also inseparable from the support of the royal family. The academic gardens flourished under the social background of "the turning of culture and education" and were driven by the royal family, civil servants, and literati.

As with Jiangnan Gardens, Changsha Gardens were also dominated by literati aesthetics, but they were simpler and heavier, emphasizing practicality and meaning over modeling. During the Ming Dynasty, Changsha was blessed with abundant natural scenery resources, and all the gardens relied on borrowing natural scenery for their aesthetic appeal, which might have hindered the development of landscaping techniques in the gardens. Hunan culture, which was deeply influenced by Neo-Confucianism in the Song and Ming Dynasties, emphasized "pragmatism (经世致用)" [39]. Landscape architects might have focused too much on practical functions and artistic mood, neglecting the innovation of landscaping techniques. These factors might have contributed to the underdevelopment of gardening in Changsha during the Ming Dynasty. As a result, Changsha Gardens were not as sophisticated as the Jiangnan Gardens of the time, and their style was not as distinctive as other regional schools of gardens. This might explain why Changsha gardens, and even Hunan gardens in general, have not been considered as one of the regional schools of gardens in China.

"Landscape" and "history" are determined by specific social, cultural, and economic backgrounds [44]. The cultural expression of landscape, especially the expression of the visual type, is an excellent tool to understand the historical evolution of the region [45]. Unfortunately, all of Changsha Gardens were destroyed during the war in the late Ming Dynasty, and only a few were rebuilt and have survived to this day [46]. The dense high-rise buildings in Changsha now obstruct the view of the city. "Evening Snow Blending River and Sky" could only be seen on the banks of the Xiang River and Orange Isle. The hills surrounding the city have been flattened due to urban expansion, destroying the landscape background, and the newly constructed garden landscapes do not pay attention to the use of borrowed scenery. Most of the springs and ancient wells have been buried, and landscape architects have not considered the use of spring water or the digging of wells. The plant configurations seldom consider cultural significance, and the shrubs are mostly trimmed neatly, resulting in a monotonous form and high maintenance costs. Changsha's gardens lack local characteristics. People have an affective tie with the physical environment, and the ties will be more complex in their hometown [47]. In order to strengthen the tie between people and the environment, increase people's sense of pride and belonging to their hometown, and inherit regional culture, such as the local garden culture, we suggest that landscape architects use Changsha's abundant groundwater to create landscapes and use water wells as a unique feature of Changsha's landscape. Some classical gardens could be reconstructed, new parks could be built in the mountains around Changsha, and buildings, such as terraces, could be constructed to provide views and borrow scenery. The mountain parks could also become a part of the urban landscape background. Plant design should pursue a more simple and natural style, focusing on cultural meaning rather than regular appearance. For other places with few existing gardens such as Changsha, we suggest that landscape architects excavate the characteristics of local classical gardens through relevant literature, historical maps, and other sources, restore classical gardens, and construct new landscapes as public leisure spaces and displays of garden culture, while also protecting urban modernization and historic sites.

The research has certain limitations. Firstly, there are no extant remains of Changsha gardens in the Ming Dynasty. Only five existing gardens are known, which comprise only a quarter of the total number. These gardens are Jia Yi's former residence (the garden of ancestral hall), Kaifu Temple, Yuelu Academy, Lushan Temple, and Yunlu Palace (religious gardens), which belong to two types. The five existing gardens have all been rebuilt and repaired on a large scale during the Qing Dynasty (1644—1911) and modern times. As with the literature, paintings and maps are visual representations of landscape in cultural form, and maps are a complex and culturally constructed means of representing knowledge [48,49]. However, due to the lack of visual data, such as historical maps, garden plans, and garden paintings of the Ming Dynasty, it is difficult to verify the style and appearance of the Ming Dynasty gardens. The other four types of gardens, namely, royal gardens, private gardens, academic gardens, and gardens of government office, could not be researched through relics. Secondly, the historical documents of Changsha gardens in the Ming Dynasty might be incomplete. The introduction of gardens in the documents is often brief, and some documents tend to be lyrical, which might limit the understanding of the history of Changsha gardens in the Ming Dynasty.

Author Contributions: Conceptualization, writing—original draft preparation and investigation, methodology, W.L.; supervision, C.G. All authors have read and agreed to the published version of the manuscript.

Funding: This research received no external funding.

Data Availability Statement: Not applicable.

Conflicts of Interest: The authors declare no conflict of interest.

References

1. Wilson, E. *China: Mother of Gardens*; The Stratford Company: Rockingham, VA, USA, 1929.
2. Zhao, J.; Zhao, W.; Shen, Z. Research on the Construction and Operation Mechanism of Overseas Chinese Gardens in the Past 40 Years. *Chin. Landsc. Archit.* **2021**, *37*, 46–51.
3. Chen, X.; Wu, J. Sustainable Landscape Architecture: Enlightenment and Transcendence of Chinese Philosophy of 'Heaven and Man'. *Landsc. Ecol.* **2009**, *24*, 1015–1026. [CrossRef]
4. World Heritage List: Chinese Garden. UNESCO World Heritage Centre. Available online: https://whc.unesco.org/en/list/?search=chinese+garden&order=country (accessed on 5 October 2022).
5. Zhou, W. *History of Chinese Classical Gardens*; Tsinghua University Press: Beijing, China, 1999; pp. 351–764.
6. Zhu, C. Comparison of Ba-Shu Gardens and Jiangnan Gardens. Master's Thesis, Nanjing Forestry University, Nanjing, China, 2006.
7. Liu, T. Enriching the system of Chinese garden culture and art with in-depth research on local gardens. *Garden* **2021**, *38*, 1.
8. Bullington, J. East-West relational imaginaries: Classical Chinese gardens & self cultivation. *Educ. Philos. Theory* **2021**, *53*, 1–6.
9. Yun, J.; Yu, W.; Wang, H. Exploring the Distribution of Gardens in Suzhou City in the Qianlong Period through a Space Syntax Approach. *Land* **2021**, *10*, 659. [CrossRef]
10. Zhang, T.; Lian, Z. Research on the Distribution and Scale Evolution of Suzhou Gardens under the Urbanization Process from the Tang to the Qing Dynasty. *Land* **2021**, *10*, 281. [CrossRef]
11. Richard, J. Uncovering the garden of the richest man on earth in nineteenth-century Canton: Howqua's garden in Honam, China. *Gard. Hist.* **2015**, *43*, 168–181.
12. Zheng, W. Analysis of Gardening Characteristics and Style Formation of Lingnan Gardens. *J. Landsc. Res.* **2019**, *11*, 130–134.
13. Siu, V. *Gardens of a Chinese Emperor: Imperial Creations of the Qianlong Era, 1736–1796*; Lehigh University Press: Bethlehem, PA, USA, 2013.

14. Zhu, Y.; Lee, S.-H.; Choi, K.-R. Analysis of the Space and Design of Chinese, Japanese, and Korean Academy Gardens. *Land* **2022**, *11*, 1474. [CrossRef]
15. He, L. Research on the History of Changsha Garden. Master's Thesis, Central South University of Forestry and Technology, Changsha, China, 2007.
16. Jiang, X. Research on the Inheritance and Innovation of Changsha Classical Gardens. Master's Thesis, Central South University of Forestry and Technology, Changsha, China, 2011.
17. Liu, F. Research on the Development of Hunan Gardens. Master's Thesis, Central South University of Forestry and Technology, Changsha, China, 2014.
18. Sun, Y. Research on Changsha Garden. Master's Thesis, Central South Forestry College, Changsha, China, 2005.
19. Wang, D. Research on the Development History and Environmental Characteristics of Traditional Religious Gardens in Changsha. Master's Thesis, Central South University of Forestry and Technology, Changsha, China, 2013.
20. Wang, W. Discussion about Why Changsha Became a Princedom Four Times during the Ming Dynasty. *Hunan Prov. Mus.* **2017**, *1*, 571–578.
21. Chen, Y. *Records of Exploring Ancient Relics in Changsha*; Yuelu Publishing House: Changsha, China, 2009.
22. Zhang, X. *Celebrities and Changsha Landscape*; Hunan People's Publishing House: Changsha, China, 2012.
23. Lei, Q. *Chronicles of Changsha Prefecture*; Yuelu Publishing House: Changsha, China, 2008.
24. Peng, K. *Poetry of Southern Chu*; Chinese Classics and Ancient Books Library; Zhonghua Book Company: Beijing, China, 2014.
25. Li, Y. *Analects of Tianyue Mountain Mansion*; Chinese Classics and Ancient Books Library; Zhonghua Book Company: Beijing, China, 2014.
26. Chen, L. *Chronicles and Atlas of Rebuilding Yuelu Academy*; Chinese Classics and Ancient Books Library; Zhonghua Book Company: Beijing, China, 2014.
27. Zhao, N. *Chronicles of Newly Revised Yuelu Academy*; Chinese Classics and Ancient Books Library; Zhonghua Book Company: Beijing, China, 2014.
28. Cheng, J. *Brief Chronicles of Yuelu*; Hunan News Agency: Changsha, China, 1932.
29. Li, B.; Li, Y.; Yi, B. From Tradition to Modern: The Change to the Urban Ancient Wells' Folklore. *J. Hunan Univ. Soc. Sci.* **2007**, *1*, 110–115.
30. Zhang, T. A preliminary research on the landscape form analysis of Changsha's historical features. In Transformation and Reconstruction: In Proceedings of the 2011 China Urban Planning Annual Conference, Chongqing, China, 20 September 2011; pp. 7575–7584.
31. Stones Commonly Used in Rockery Engineering (Stones Commonly Used in Rockery Engineering). Available online: http://chla.com.cn/html/c116/2010-03/52396.html (accessed on 3 November 2022).
32. Zhang, J. On the Population Change of Huguang in Ming Dynasty. *Econ. Rev.* **1994**, *2*, 82–89.
33. Zhang, S.; Feng, H. The Migration of Commercial Population and the Development of Local Culture in the Ming Dynasty. *Acad. J. Zhongzhou* **2019**, *11*, 132–137.
34. Wu, Q. Landscape collective culture in China. *Huazhong Archit.* **1994**, *2*, 23–25.
35. Geng, X.; Li, X.; Zhang, J. On the Cultural Consciousness of Chinese Garden in the View of Chinese "Eight Sceneries". *Chin. Landsc. Archit.* **2009**, *25*, 34–39.
36. Li, K.; Woudstra, J.; Feng, W. 'Eight Views' versus 'Eight Scenes': The History of the Beijing Tradition in China. *Landsc. Res.* **2010**, *35*, 83–110. [CrossRef]
37. Zou, D. "Farming and Reading Culture" in China. *Agric. Hist. China* **1996**, *4*, 61–63.
38. Wu, S. The contemporary significance of Jia Yi's former residence cultural heritage. *Hunan Prov. Mus. Society. Collect. Work. Museol.* **2014**, *10*, 113–123.
39. The Formation and Development of Huxiang School. Available online: https://www.xtol.cn/culture/2019/0722/5271732.shtml (accessed on 12 December 2022).
40. Bao, Q. Study on Square Pools in Song Dynasty Gardens. *Chin. Landsc. Archit.* **2012**, *28*, 73–76.
41. Wang, G. *The Ming Prince and Daoism: Institutional Patronage of an Elite*; Shanghai Classics Publishing House: Shanghai, China, 2019.
42. Yuan, Y. The Manifestations and Causes of "Cultural and Educational Turn" in the Urban Landscape of Changsha Prefecture in Ming and Qing Dynasties. Master's Thesis, Shenzhen University, Shenzhen, China, 2017.
43. Zhou, J.; Zhu, X. The concept of Chinese Academy Education and Its Modern Enlightenment. *Mod. Educ. Sci.* **2009**, *3*, 39–44.
44. Maderuelo, J. *Paisaje e Historia*; Abada Editores: Madrid, Spain, 2009.
45. Corredera, S.R. Methodological proposal for the study of cultural representations of landscape: The case of the western Costa del Sol (Málaga, Spain)". *Ería Rev. Cuatrimest. Geogr.* **2022**, *42*, 213–241.
46. Chu, W. Prequel to Changsha West Chang Street: Built in imitation of Beijing West Chang'an Street. Available online: https://moment.rednet.cn/pc/content/2019/11/02/6179783.html (accessed on 23 December 2022).
47. Tuan, Y.; Blake, B.F. *Topophilia: A Study of Environmental Perception, Attitudes and Values*; Prentice Hall: Hoboken, NJ, USA, 1974; p. 323.

48. Schwartz, J.; Ryan, J. *Picturing Place: Photography and the Geographical Imagination*; I. B. Tauris: London, UK, 2003; pp. 3–4.
49. Hawkins, H. Geography's creative (re) turn: Toward a critical framework. *Prog. Hum. Geogr.* **2019**, *43*, 963–984. [CrossRef]

 land

Article

The Multidisciplinary Approach in the Study of Landscape Evolution: The Fluvial Capture of the San Donato Creek (Gubbio, Central Italy), an Example of Hydrological Regime and Hydrogeological Risk Changes

Corrado Cencetti [1,*], Filippo Paciotti [2], Ettore A. Sannipoli [3] and Ubaldo E. Scavizzi [4]

1 Department of Physics and Geology, University of Perugia, 06123 Perugia, Italy
2 Independent Researcher (Archaeologist), 06024 Gubbio, Italy
3 Independent Researcher (Art Historian), 06024 Gubbio, Italy
4 Independent Researcher (Geologist), 06024 Gubbio, Italy
* Correspondence: corrado.cencetti@unipg.it

Abstract: Historical maps, especially those at a small scale and rich in detail (e.g., the old "Cadastres"), represent an exceptionally important tool for understanding the recent historical evolution of landscapes. The note describes the example of the territory of Gubbio, in Umbria (Central Italy), where a map from the end of the 16th century shows a drawing of the hydrographic network partially different from the current one. A multidisciplinary study based on field surveys, observations of satellite images, archaeological discoveries, and archival research proved useful to confirm what was reported by the cartographer at the time. The possible causes that led to this variation of the surface hydrography up to its current configuration are then discussed in the light of other documentary finds from the archives, taken from the chronicles of the time, which have made it possible to identify, with sufficient approximation, the period where this change occurred. All this leads to a highlighting of a profound evolution of fluvial and slope morphogenetic processes that have affected the study area in recent centuries, in which the regulation of surface waters and afforestation, conducted during the 20th century, have played a decisive role.

Keywords: historic cartographic documents; landscape evolution; fluvial dynamics; Umbria (central Italy)

Citation: Cencetti, C.; Paciotti, F.; Sannipoli, E.A.; Scavizzi, U.E. The Multidisciplinary Approach in the Study of Landscape Evolution: The Fluvial Capture of the San Donato Creek (Gubbio, Central Italy), an Example of Hydrological Regime and Hydrogeological Risk Changes. *Land* **2023**, *12*, 694. https://doi.org/10.3390/land12030694

Academic Editors: Bellotti Piero and Alessia Pica

Received: 23 February 2023
Revised: 13 March 2023
Accepted: 14 March 2023
Published: 16 March 2023

1. Introduction

The study of landscape evolution, especially in the case of areas characterized by anthropic presence since ancient times, requires a multidisciplinary approach due to the great variety of sources to be used (historical maps, archival documents, field investigation data both geological-geomorphological and archaeological, the latter being connected with anthropological aspects). The morphogenetic processes often underwent changes in historical times, almost always induced by human activity, i.e., the upsets of the hydrographic network due to the building of dams and the hydraulic works in the riverbeds and hydrographic basins, conducted mainly during the 19th and 20th centuries [1–4]; the land use change from forest to agriculture, or from agriculture to urbanization [5–7]. These interventions have frequently led to a mitigation of the hydrogeological risk but, at times, also to its increase [8–12].

Historical cartographic documents have been an important source of data for identifying the changes, natural and/or induced, which have characterized the evolution of the landscape over the last few centuries [13–19]. Among these, large-scale and handwritten administrative maps are particularly useful, as they constitute the most reliable and rich in content original sources compared to printed maps [20]. In Italy, for example, between the end of the 18th and the beginning of the 19th century, various "Cadastres" were produced which, although they only report information of a planimetric type, provide excellent information, especially about the configuration of streams, before the important hydraulic works that significantly modified the planimetric track of the streams themselves. The Napoleonic Cadastre was published in 1812 for northern Italy. The Gregorian Cadastre (1816–1835) and the Lorraine Cadastre (1776–1832) were published for central Italy; the Bourbon Cadastre, also created in the first half of the 19th century were published for southern Italy. Furthermore, modern territorial analysis techniques that use geographic information systems (GIS) allow comparisons between historical maps and the current conformation of natural and man-made environments with a high degree of accuracy [21–24].

This study describes an emblematic case that highlights the importance of historical cartographic analysis in understanding the recent evolution of the landscape. One of the main and oldest historical maps of the territory of Gubbio (Umbria, Central Italy) was examined; according to scholars, this is the first example of the use of mapping for demographic survey purposes [25]. From the map, dated to the 16th century, albeit stylized, a hydrographic network emerges within the plain of Gubbio, which is partially different from the current one. This induced research includes the use of a field survey of geomorphological clues and historical finds. It also prompted the analysis of documents and other subsequent cartographic sources of causes that may have produced the transformation of the hydrographic network itself up to its current configuration. The identification of these causes implies that the recognition of morphogenetic processes is significantly different from those in progress today, with the consequent modifications of the hydrogeological risk conditions.

2. Materials and Methods

The Gubbio basin (Umbria, Central Italy) is examined, of which the geological-geomorphological peculiarities are described below, as well as the "Georgi's Map" (1570), one of the most important historical cartographic documents, which is specifically the object of the note.

2.1. Geological-Geomorphological Framework of the Study Area

To understand the evolution of the landscape in the study area, it is essential to refer to its geological-structural characteristics, which have strongly controlled the "footprint" of the landscape itself, as normally occurs in areas affected by recent tectonics. Indeed, the Gubbio Plain represents one of the classic intermontane basins located on the western side of the Apennine Chain in Central Italy (Figure 1) [26].

This is a fold mountain belt, verging towards the Adriatic Sea, which was formed due to compressive tectonic induced by the counterclockwise rotation of the Italian peninsula. The compressive tectonic phase, still underway on the eastern (Adriatic) side of the chain, was superimposed in a spatial-temporal sense by an extensional phase of crustal thinning which produced the formation of tectonic trenches, i.e., graben and semi-graben [27–30]. The Apennine intermontane basins—of which that of Gubbio is the most recent and easternmost, together with Gualdo Tadino—represent the geomorphological expression of this extensional tectonic phase which begun in the Pliocene and still affects the western slope of the Apennine Chain. It is thus responsible for the seismicity present in this area [31–33].

The two tectonic phases have left evident traces in the territory of Gubbio. The compressive phase produced the formation of a brachyanticline (Figure 2) [34], where the marine environment formations, mainly calcareous and calcareous-marly (dating from the Jurassic to the Miocene) of the Umbria-Marche stratigraphic series, outcrop and today constitute the structure of the relief that overlooks the city.

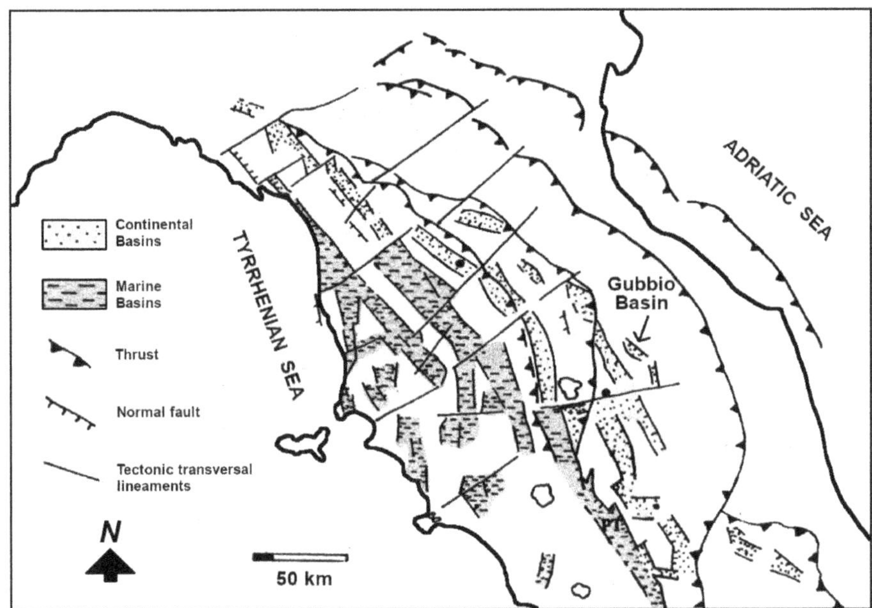

Figure 1. The marine and continental basins (graben and semi-graben) that characterize the Tyrrhenian side of the Apennine Chain in central Italy. The Gubbio basin is the easternmost; therefore, it is the most recent together with the Gualdo Tadino basin (immediately to the side). From [14], modified.

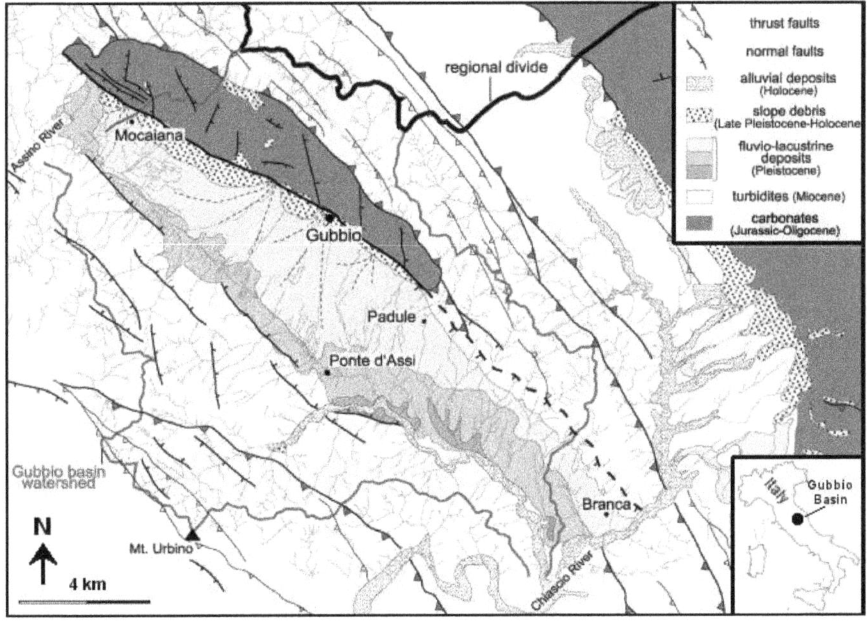

Figure 2. Geological map of Gubbio basin (from [34], modified).

The extensional phase was superimposed from the previous one, cutting the aforesaid anticline precisely in correspondence with its nucleus via an important normal fault [35]. It produced a large tectonic trench, i.e., the Gubbio Plain, which spans about 24 km in a

NW-SE direction. The subsidence of this basin, with endorheic characteristics, initially allowed the formation of a lake, in the Plio-Pleistocene age, subsequently drained and filled by sediments in lacustrine and fluvial-lacustrine facies transported by streams flowing from both sides (north-eastern and south-western) of the basin itself [36,37]. The establishment of the graben generated a local base level of erosion for all these rivers which, from the area in subsidence, began their process of regressive erosion [38]. The result, from the geomorphological point of view, is the alignment of a series of reliefs (Gubbio Mountains), structurally constituted by what remains of the anticline fold and divided from each other by narrow valleys, produced from the streams that flow into the Gubbio Plain forming large alluvial fans (Figure 3) [39].

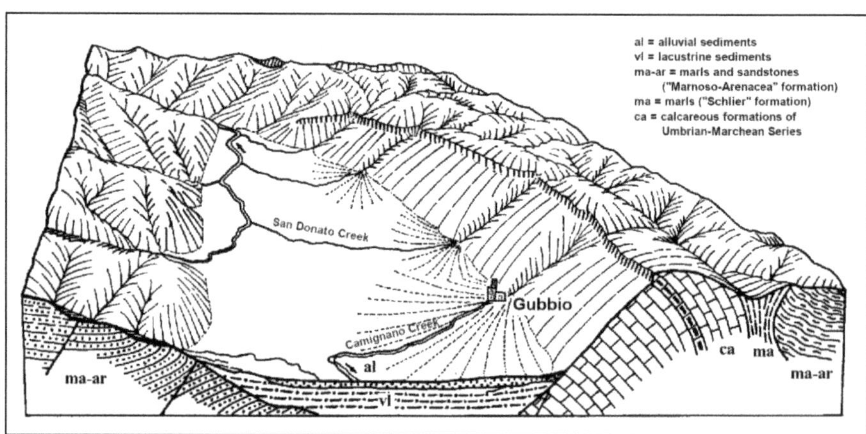

Figure 3. Block diagram of Gubbio basin (from [39], modified).

There are concerns regarding the hydrographic network in the Gubbio Plain. It is bordered to the NW and SW, and at its ends, by two rivers: the Assino River, which drains at the northern portion of the plain, and the Chiascio River, which drains at its central-southern portion. The divide between the two basins is located properly inside the Gubbio Plain, in correspondence with the large, low-sloping alluvial fan formed by the San Donato Creek at its outlet in the plain area. This aspect is important and will be discussed later.

2.2. The Georgi's Map

The "Georgi's Map"—also titled "Diocese della città di Ugubbio, descritta dal R.mo don Ubaldo Georgii clerico Eugubino. Ad istanza di monsignore Ill.mo Et Re.mo Mariano Savello vescovo dignissimo" (Figure 4)—was made in the second half of the 16th century and represents one of the most important historical cartographic documents of the Gubbio area. One specimen is kept in the Municipal Museum of Gubbio (inv. No. 2414 (6505)). It consists of two copper plates, each measuring 475 × 680 mm, and owes its origin to the decisions taken within the Council of Trent (1563) during which a document was approved asking parish priests to draw up a status animarum of their own parish through a real demographic census [25,40–43].

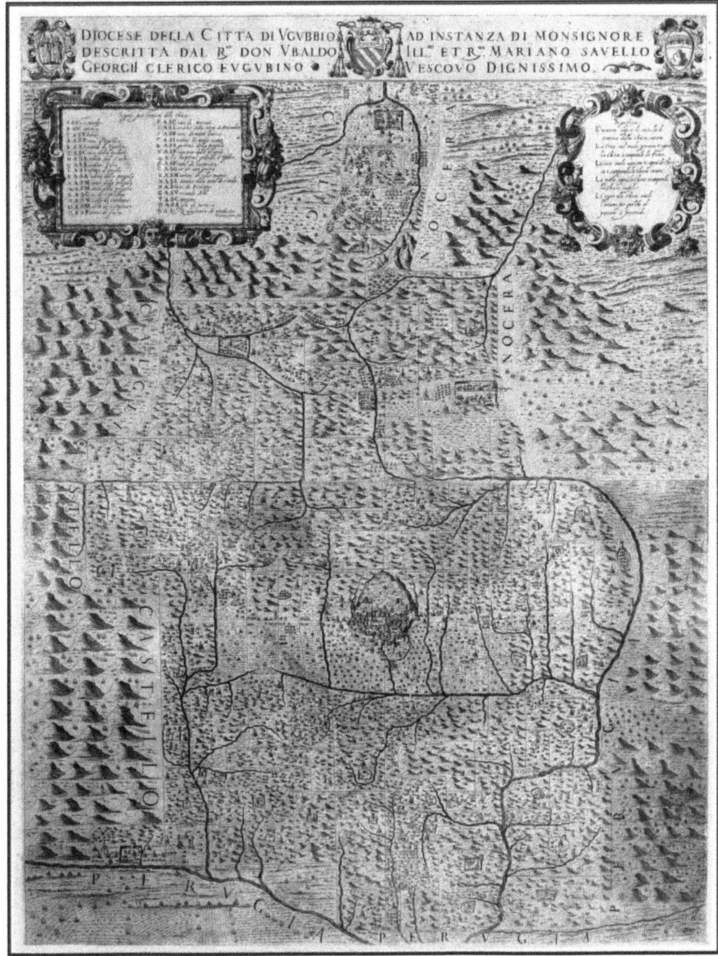

Figure 4. The Georgi's Map—Municipal Museum of Gubbio, inv. No. 2414 (6505).

In 1567, the bishop of Gubbio (mons. Mariano Savelli) ordered all parish priests of the Diocese to indicate the number of families belonging to their own parish and the distances that separated them from their respective churches. In 1570, the bishop commissioned the cleric don Ubaldo Georgi, who was considered to be an excellent geographer [44], to graphically transpose all the information received. From the data found by Allegrucci [25], it can be deduced that in 1571 Georgi went to Venice with the design of the map and stayed there for a year to have the map engraved; the "coppers" were then brought to Gubbio.

The document, a real topographic map, depicts a relatively large territory, at a scale of extreme detail for the period in which it was drawn up (normally, the maps of this period are at a significantly smaller scale), and it is also a thematic map, considering that it represents the population of the Diocese. In fact, the territory is divided into squares and rectangles, each of which delimits, albeit approximately, the various parish districts. The particularity of the map consists in the fact that it represents a unicum, as in no other Diocese the surveys promoted by the Council of Trent were graphically converted into a map. Furthermore, it represents the first cartography created specifically in the Gubbio area.

2.3. The Anomaly of the Hydrographic Network Represented in the Georgi's Map

By carefully examining Georgi's Map, a substantial difference in the hydrographic network represented by the cartographer compared to the current situation can be observed. The map, in its lower portion, represents the plain of Gubbio, bordered on the right (SE) by the Chiascio River and on the left (NW) by the Assino River. The divide between the two hydrographic catchments, which drain towards the Assino and Chiascio Rivers, respectively, is moved much more to the left (NW) than the current one, so much so that, to the NW of the town of Gubbio, the San Donato Creek, although represented in a stylized way, unequivocally flows into the Saonda River, a tributary of the Chiascio River. Currently, it is an integral part of the hydrographic network which belongs to the Assino River. The same applies to a minor tributary, the creek located between Nerbisci and S. Martino in Colle, on the opposite side of the Gubbio basin: it too, today flowing into the San Donato stream, is represented in Georgi's Map as a tributary of the Saonda River (Figure 5).

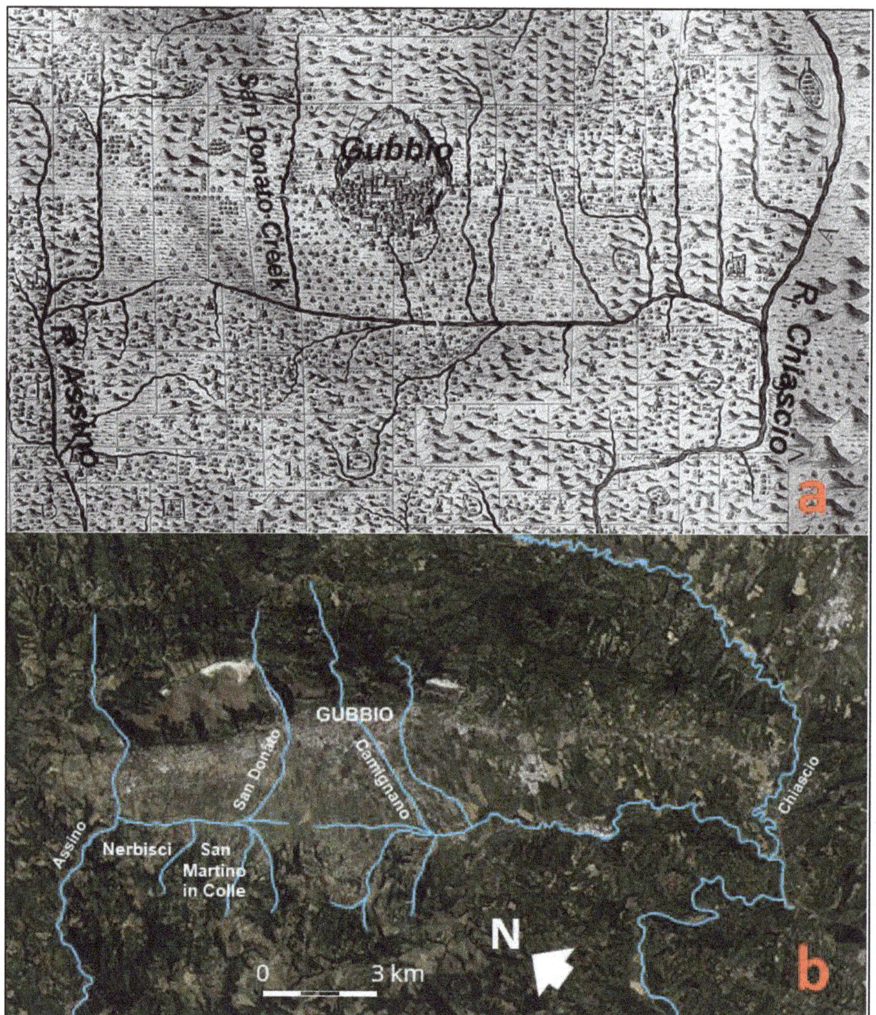

Figure 5. Comparison between (**a**) Georgi's Map and (**b**) the present situation (from Google Earth—https://earth.google.com/, highlighting the main hydrographic network).

An error by the cartographer can reasonably be excluded, considering the extreme detail with which the natural and anthropic elements are represented and above all the proximity of the place to the city of Gubbio (therefore in an area that is not impervious and easily accessible). The problem then arises of the search for clues, from land and documents, which testify to the presence of a fluvial track of the San Donato Creek different from the current one, as well as the causes and the period in which the diversion, which led to the present situation, may have occurred.

2.4. The Data Available

The study was carried out following two lines of research: (1) The field investigation, which attempted to search for geomorphological clues and archaeological finds that could highlight the presence of a fluvial track of the San Donato Creek different from the current one; (2) the search for archival material (cartographic and possibly documentary) on the subject.

2.4.1. The Field Investigation

The direct investigation in the field proved difficult, considering that the area in question has always been developed and is used mainly for agriculture. Therefore, there are no outcrops of autochthonous and unaltered material along the presumed ancient fluvial track of the San Donato Creek, which can be attributed to infill of sediments of fluvial origin.

On the other hand, the analysis of aerial photos proved to be more important and useful, from which it is possible to identify, through a wider aerial view, any indications of a paleo-riverbed. In fact, the identification of ancient river tracks, especially in lowland areas, is facilitated by the fact that the sub-surface channel deposits are clearly evident, above all following precipitation events, for the different colors they assume due to their degree of humidity and their permeability, different from the neighboring areas. Therefore, the search for aerial photos via Google Earth was very useful, exploiting its potential and above all the fact that the photos present refer to different periods (which can be inspected via the menu item "show historical images"). Among the photos available in this area, there is one in particular whose acquisition date is 29 May 2017. It shows with extraordinary evidence the indications of a drainage line placed along the San Donato Creek at its outlet in the plain of Gubbio, which most likely follows the ancient fluvial track of this stream, as reported in Georgi's Map (Figure 6), i.e., directed towards the Saonda River.

Figure 6. Evidence (red arrows) of San Donato Creek's paleo-riverbed (from Google Earth—https://earth.google.com/).

Furthermore, there is another important element in favor of this interpretation. During the works carried out in 2014 for the construction of the new church of the Madonna del Ponte (church of the Mother of the Savior), today located precisely according to the direction of the presumed paleo-riverbed of the San Donato Creek, very interesting archaeological and geological data emerged. The preparatory excavation works for the construction of the new church have brought to light an ancient artifact consisting of a small chapel which had probably incorporated a pre-existing rural aedicule dating back to the 15th century (Figure 7). This is located close to the place where the old church of the Madonna del Ponte was built during the 16th century.

Figure 7. The ruins of the artifact, probably a small chapel (**a**) with an aedicule inside, detailed in (**b**).

In particular, the excavations of 2014 highlighted the presence of a sedimentary body, which transversely cut the entire construction site at a depth of 0.5–1 m from ground level. This lithosome was clearly distinguishable from the neighboring alluvial sediments and consisted of well-processed pebbles with a good degree of rounding, with an average size of 6–8 cm (also used for the construction of the perimeter walls of the aedicule) in an abundant silty-sandy matrix (Figure 8).

The position of this sedimentary body, located right along the line of the hypothetical paleo-riverbed of the San Donato Creek, as represented by Georgi, and its lithological-sedimentary characteristics make the presence of a river channel probable in this area.

Currently, on the main altar of the church of the Madonna del Ponte, near the bridge over the San Donato Creek, there is a detached block fresco coming from the chapel described above. It is assumed that the aedicule, contained in it, was placed near the bank of the San Donato Creek to invoke protection from any damage that the stream could cause with its floods, as was the custom in that period and in the following centuries [45]. See, for example, the still present aedicule, located on the right bank of the Camignano River, near the bridge that crosses it in the San Martino district, within the historic center of Gubbio (Figure 9). The fresco in question (Figure 10) depicts the Madonna and Child enthroned between St. Bernardino and a holy bishop (perhaps St. Ubaldo or St. Donato) and is attributed to the Gubbio painter Jacopo Bedi who realized it shortly after the mid-15th century.

Figure 8. (**a**) The trench made during the excavation of the building in Figure 7; (**b**) pebbles of clear fluvial origin used to build the artifact.

Figure 9. Aedicule located on the right bank of the Camignano Creek, close to the bridge that crosses it in the San Martino quarter, in the historic center of Gubbio.

Figure 10. Fresco detached in block currently located in the Church of Madonna del Ponte.

Some archival documents relating to Jacopo Bedi, dated between 1460 and 1478 [46], ensure that the painter owned land in the Villa of Spognola, located on the right bank of the San Donato Creek, near the bridge already called at that time "of San Donato".

It cannot be excluded, though it is possible, that the majesty or the small chapel where the fresco was originally placed, was located precisely in one of the lands owned by the 15th-century painter from Gubbio near the San Donato bridge. Even the dimensions of the small building found in the area now occupied by the new church of the Mother of the Savior, especially those of its back wall, are compatible with those of the detached fresco (150 × 150 cm).

2.4.2. The Historical-Cartographic Investigation

Next, we analyzed historical maps subsequent to Georgi. We referred to the same area of study to search for the one closest in time to Georgi's Map, in which the San Donato Creek was depicted in the position in which it is currently located. This was done to identify the time interval in which the diversion phenomenon of the San Donato Creek may have occurred. The maps examined are listed in Table 1, in increasing order of age.

We found the examined cartography, which was produced before Georgi's Map (1570), to be unhelpful, as it is not detailed enough to confirm that how the San Donato Creek flowed into the Saonda River (Chiascio River basin). These maps are drawn on too small a scale, which does not allow for a degree of detail that can correctly identify the hydrograph network of the Gubbio Plain (Figure 11).

Table 1. The historical maps examined.

Title	Author	Year	Location	Notes
Tavola Nuova della Marca d'Ancona	Girolamo Ruscelli	1561	In: "*Geographia*" of Claudio Tolomeo	printed
Novo et vero dissegno della Marca di Ancona con li sui confini	Ferrando Bertelli	1565	British Museum, London	printed
Marcha Anconae olim Picenum	Abramo Ortelio	1572	Municipal Library of Fermo	printed
Urbini Ducatus	Egnatio Danti	1580–1582	Vatican Museums	wall painting
Marchia Anconitana cum Spoletano Ducatu	Gerhard Mercatore	1589	private collection	printed
Ducatus Urbini nova et exacta descriptio	Abramo Ortelio	1606	"Federiciana" Library of Fano	printed
Marca d'Ancona olim Picenum	Giovanni Antonio Magini	1620	"Oliveriana" Library of Pesaro	printed
Urbini Ducatus	Giovanni Antonio Magini	1620	"Federiciana" Library of Fano	printed
Gli Stati dei Serenissimi Duchi della Rovere	Francesco Mingucci	1626	Vatican manuscript code 4434	hand drawn
Urbini Ducatus	Jodoco Hondio	1627	"Federiciana" Library of Fano	printed
Stato d'Urbino	Marco Ferrante Gerlassa	circa 1630	National Gallery of Marche, Urbino	hand drawn
Ducato di Urbino	Henricus Hondius	1635	"Federiciana" Library of Fano	printed
Marchia Anconitana olim Picenum	Henricus Hondius	1640	Municipal Library of Civitanova Marche	printed
Legatione del Ducato d'Urbino	Filippo Titi	1697	"Federiciana" Library of Fano	printed
Marca d'Ancona	Vincenzo Coronelli	1708	Franciscan and Picene Historical Library of Falconara	printed
Provincia Piceni	Giovanni di Montecavallero	1711	Franciscan and Picene Historical Library of Falconara	printed
Marca Anconae, olim Picenum	Lasor a Varea	1713	"Federiciana" Library of Fano	printed
Ducato d'Urbino		1723	In: "*Memorie istoriche concernenti la Devoluzione dello Stato di Urbino alla Santa Sede Apostolica*" (Riveriana)	printed
Nuova Carta Geografica dello Stato Ecclesiastico	Ruđer Josip Bošković e Christoper Maire	1755	National Library of Florence	printed
Nuova Delineazione della Legazione di Urbino	Christoper Maire	1757	"Oliveriana" Library of Pesaro	printed
Catasto geometrico-particellare del territorio di Gubbio	Giuseppe Maria Ghelli	1768	State Archive Section of Gubbio	27 hand drawn maps and 48 "brogliardi" (descriptive registers)

Table 1. *Cont.*

Title	Author	Year	Location	Notes
Lo Stato della Chiesa	Antonio Zatta	1782	Franciscan and Picene Historical Library of Falconara	printed
Legazione d'Urbino e Governo di Città di Castello	Antonio Zatta	1783	Franciscan and Picene Historical Library of Falconara	printed
Catasto gregoriano	AA.VV.	1816–1835	State Archive of Perugia	unit 287, hand drawn

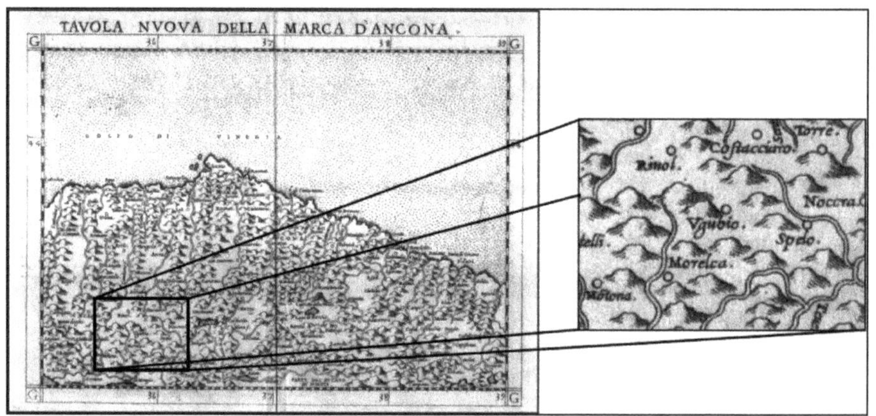

Figure 11. *"Tavola Nuova della Marca d'Ancona"* (Ruscelli, 1561). In the box: the enlargement of the Gubbio area.

Some maps produced after Georgi's present the same problem because the Gubbio area is located on the edges of the cartographic representation and is therefore not well sketched (Figure 12). Others, which are in any case the majority, all show the San Donato Creek with a track like the current one, i.e., flowing into the Assino River (Figure 13). Among the latter, the closest in time to Georgi's Map is *"Urbini Ducatus"* by Egnatio Danti, produced between 1580 and 1582 and preserved as a mural painting in Rome, in the Vatican Museums (Figures 14 and 15).

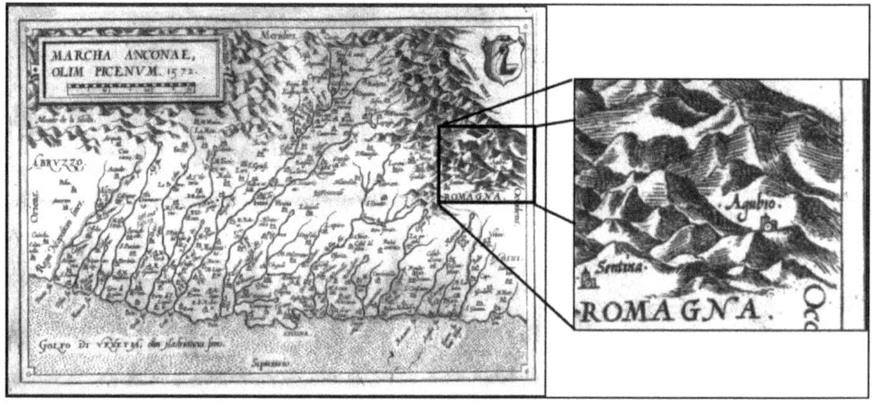

Figure 12. *"Marcha Anconae, olim Picenum"* (Ortelio, 1572). In the box: the enlargement of the Gubbio area.

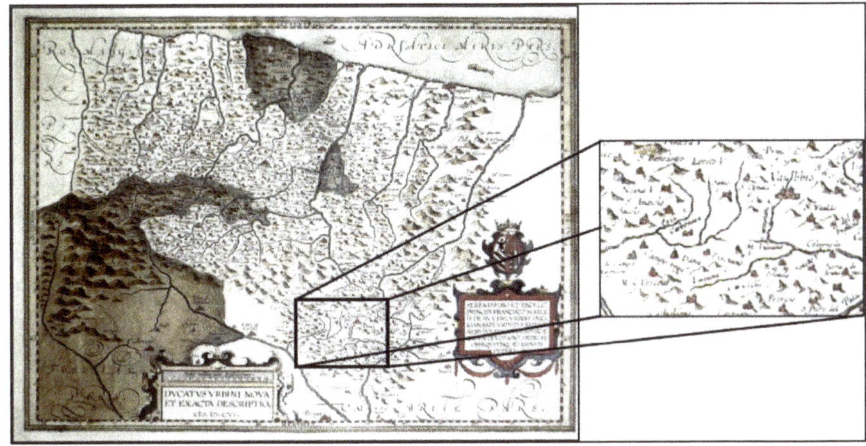

Figure 13. *"Ducatus Urbini nova et exacta descriptio"* (Ortelio, 1606). In the box: the enlargement of the Gubbio area.

Figure 14. *"Urbini Ducatus"* (Egnatio Danti, 1580–1582). In the square: the Gubbio area (highlighted in Figure 15.

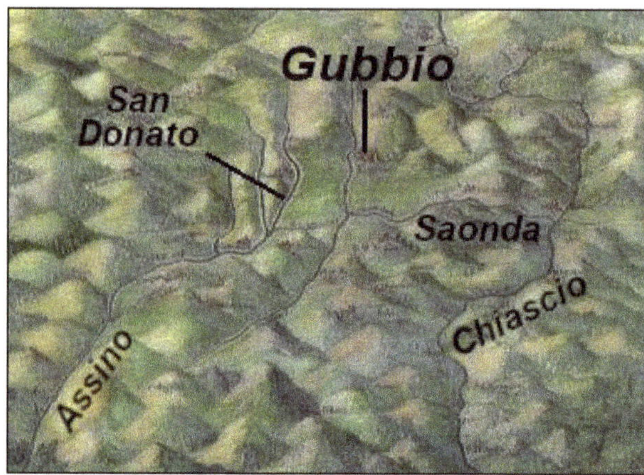

Figure 15. A detail of *"Urbini Ducatus"* by Egnatio Danti (1580–1582)—see Figure 14.

3. Results

From the available field and historical-cartographic data, two important elements emerge. The first is that there are numerous clues of a geological and geomorphological nature linked to the discovery of architectural and pictorial finds, which seem to confirm what Georgi represented regarding a track of the San Donato Creek different from the current one. The second concerns the period in which this diversion of the track may have occurred. It can reasonably be assumed that this occurred between 1571 (the year Georgi went to Venice to have the "coppers" of his paper engraved, as reported by Allegrucci [25]) and 1582 (the year Danti finished his painting).

Some questions remain unresolved, such as what the causes are that could produce the modification of the river track—i.e., natural or anthropogenic causes?

The Causes of the Modification of the Fluvial Track of the San Donato Creek

The possibility of an error by Georgi in drawing the fluvial track of the San Donato Creek on his map can be excluded. This hypothesis, in fact, appears unlikely, both taking note of the recognized competence and professionalism of Georgi himself, and because the area was easily accessible even for those times, and just as directly "verifiable." Furthermore, the purpose of the map must be considered: the abundance of details with which it was drawn up does not make credible that it could be an error, which would have been too obvious to remain unnoticed.

Two possible hypotheses remain:

- a "forced" deviation, artificially induced by hydraulic works in the riverbed, which would therefore have constrained it to modify its track;
- a natural cause, whereby the stream would have changed direction "spontaneously" caused by a flood event, such as often occur on alluvial fans (such as the one that the San Donato Creek forms), at their mouth on a plain.

The first of the two hypotheses mentioned above seems scarcely credible. In fact, no archival document has been found in the *"Riformanze"* contained in the Historical Archive of Gubbio (dating from 1326 to 1815, which faithfully report any works of public utility carried out in the area) that testify to the implementation of hydraulic works in that period. This was a period in which Gubbio, under the regency of the Della Rovere, was in a non-florid state of the municipality and its economy [47,48]. Furthermore, the reasons that would have led to carrying out works of this kind are absent (i.e., modifying a riverbed and making it flow towards another hydrographic basin). The drainage and rectification works of the riverbed of the San Donato Creek, as seen today, must have taken place no earlier than the 19th century. Graphic indications relating to a possible rectification project of some reaches, for hydraulic protection purposes, are already present in the Ghelli's Cadastre (1768; Figure 16). The San Donato Creek still has a sinuous pattern; plans of adjustment of the San Donato Creek riverbed are present in the Gregorian Cadastre, affixed after the creation of the map, probably after 1880, when its maps were "redrawn" [49] (Figure 17).

Figure 16. The San Donato Creek in the "Ghelli's Cadastre" (1768). The dotted lines, highlighted by red arrows in the figure, are plans for canalizing the stream, where it states "land in danger" due to frequent flooding affecting this area.

Figure 17. Extract from the Gregorian Cadastre of the Madonna del Ponte area, redrawn at the end of the 19th century.

Instead, the second hypothesis, which attributes the diversion of the San Donato Creek to natural causes, seems realistically not only possible but probable. The alluvial fan formed by the San Donato Creek has a morphology characterized by a gentle slope, the lowest found in the Gubbio Plain at the mouth of the streams which, crossing the Gubbio hills, form these systems. It is precisely this feature that makes the riverbed extremely mobile: in such cases, even a single flood is sufficient to cause a diversion of the track. It should also be remembered that the alluvial fan of the San Donato Creek, despite its gentle morphology, constitutes, within the Gubbio plain, a divide between the hydrographic basins of the Assino River and the Chiascio River: a diversion of a few degrees could easily have diverted the stream into one or the other catchment area. On the other hand, it is sufficient to observe any "active" fan (i.e., growing by sedimentation; Figure 18) to verify how diversions, even sensitive ones, of the feeder channel can occur very quickly, especially as alluvial fans have low slopes.

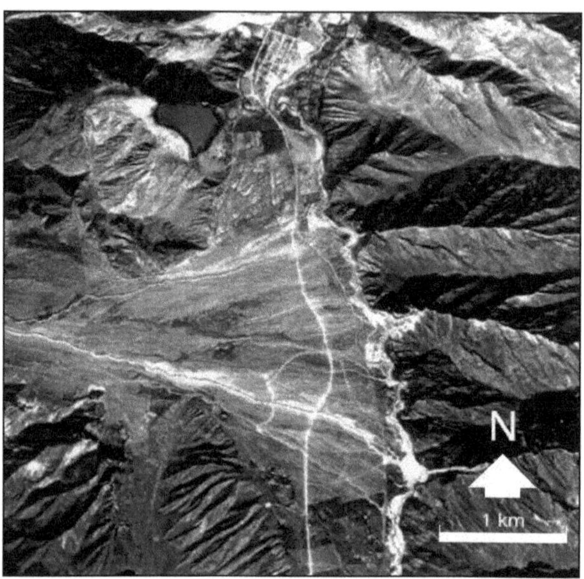

Figure 18. A typical alluvial fan (Arroyo del Medio, province of Jujuy, northwestern Argentina). The various channels that grow the cone are evident.

In essence, it would be that phenomenon which in geomorphology is defined as "river capture by spill (or overflow)" [50] (Figure 19): a stream, changing direction due to a sudden sedimentary input, can cause its waters to flow into a neighboring hydrographic basin that is different from the one in which it was located. Figure 20 shows an essential geomorphological map of the study area, where the San Donato fluvial perfectly follows what is reported in the sketch of Figure 19.

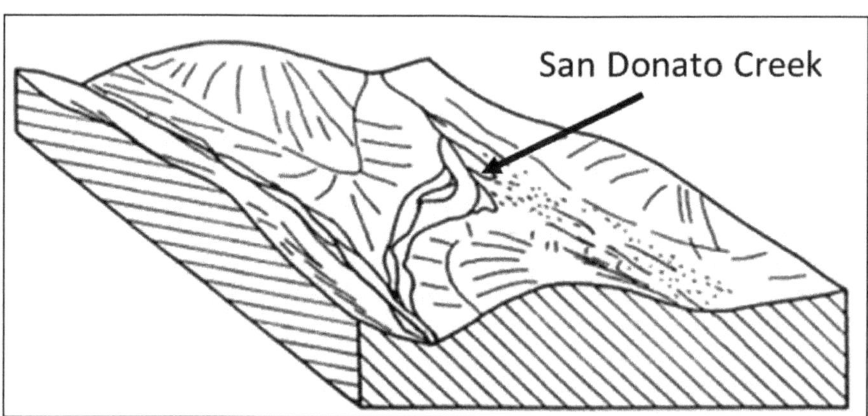

Figure 19. Fluvial capture by spill (or overflow). From [50], modified.

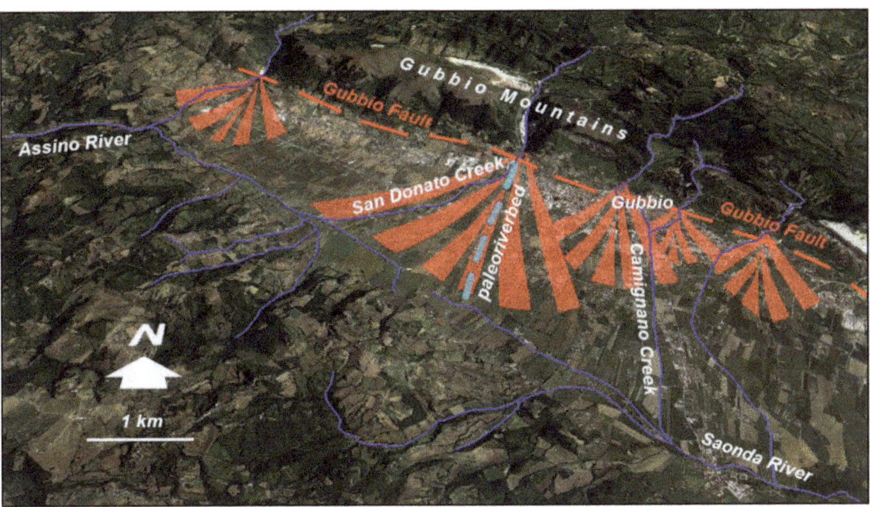

Figure 20. Geomorphological sketch of the Gubbio Plain.

This remains the most probable hypothesis and seems to be supported by an archival document.

In "Memoirs" (written from 1557 to 1604) by don Paris Montanari, the Gubbio parish priest at the time, consulted in the State Historical Archive, Gubbio Section, it states "1571. I remember that on July 18th the River known as Camignano made a huge flood, and did huge damage by flooding houses and churches, and passed over the bridges of the city: the closest houses along the river, those above St. Biagio, those of the Market, the Macelli, and the Houses of Borgo: and the churches it flooded were St. Biagio, the Mastadella, St. Bernardino, St. Rocco, and the Monastery of Good Jesus: and it carried very large beams, a frightening and terrible thing: and I don Paris was in St. Belardino with some friends of mine"—(Sezione Archivio Storico di Gubbio i.e., State Historical Archive, Gubbio Section—SASG, Fondo Armanni, II.B.13).

The "river known as Camignano" is the one that crosses the city of Gubbio (Figure 3), parallel to, and a short distance from the San Donato Creek which, evidently, suffered the same effects.

Could this be the flood event that caused the diversion of the route? The date should also be noted: 18 July 1571, indicated by don Paris as the day on which the event occurred. Georgi must have already drawn up his map if, as Allegrucci says [25], precisely on that year the cartographer went to Venice to have the "coppers" of his work engraved.

Beyond the fact that the diversion of the San Donato Creek can be attributed to this specific event described by don Paris Montanari, documentary evidence highlights an important element, namely flood events in the streams that cross the Gubbio Mountains, were certainly present in that historical period (see also Table 2). This is no longer the case today, thanks to the works of hydraulic regulation of the running waters and the arrangement of the slopes carried out in later times.

As evidence that such flood events were frequent in the Gubbio Plain, the text of the epigraph is reported, affixed after two important alluvial events in Camignano River occurred. One occurred in the aedicule (already mentioned, located on the right bank of the Camignano River and the other in the district of San Martino, which is located within the historic center of Gubbio) (Figure 21). The citizens of Gubbio thank the effigy of the Madonna with the words, translated from the Italian: "To you, Great Virgin Help of Christians who, motionless in this poor location among the ruins of repeated floods in 1859 and 1866, wanted to save the lives of so many citizens in danger, the elite of the districts, as to a dear, most vigilant, compassionate mother, as a sign of filial gratitude, this memory they put."

Table 2. List of the most important flood events in the Gubbio area which are traced in the historical documentation. The most cited refer to the Camignano River, as it crosses the city of Gubbio, but the references often extend to the surrounding rivers.

Data	Fonte	Note
1388 ca.	Luongo [51], 1.11	Flood of Camignano River
18 July 1571	Montanari, Memorie [52]	Flood of Camignano River
20 October 1613	SASG, Fondo Pesci, n. 12, p. 696	Flood of Camignano River
25 August 1616	SASG, Fondo Pesci, b. 12, Effemeridi Orsaioli, p. 446	Great flood of San Donato Creek
September 1619	SASG, Fondo Pesci, n. 12, p. 505	Flood of Camignano River
3 June 1772	SASG, Fondo Comunale, Registri Parrocchiali (san Giovanni e sant'Agostino)	"*per impetuosa pioggia si rompì* [. . .] *il Bottaccione*"; flood and damages in Gubbio, 9 dead
End of July, 1792	Cece, personal communication, 2020	Flood of Camignano River
8 September 1835	Lucarelli, Memorie [53], pp. 196–197	"*Tutti i fiumi hanno straripato in modo tale che non vi era memoria di casi simili*"; flood of Camignano River and other streams
18 September 1859	Lucarelli, Memorie [53], p. 344	"*il debordamento de fiumi ha recato immensi danni non solo nelle campagne, ma benanche in città*"; generalized floods
26 May 1866	ADG, Fondo del Capitolo dei Canonici, I.C. 21, p. 136	Flood of Camignano River

Figure 21. Detail of the aedicule shown in Figure 9, which highlights the epigraph located at the base.

Such events could easily have ben produced between 1570 and 1582. The phenomenon of "capture by overflow" was identified based on the comparison between Georgi's Map and the current situation.

4. Discussion

From the data and results described, different conditions and morphogenetic factors emerge between the period under consideration (16th century) and the current situation. Different from past periods, the streams flowing from the Gubbio Mountains towards the Gubbio basin present a regime strongly controlled by the riverbed and slope arrangements. This, presumably, only began in the last century, up to the paving of the Camignano River, which took place near the end of the 1970s (Figure 22).

Figure 22. The paving works of the riverbed of the Camignano Creek (around the end of 1970s) in a photo from the time.

Proof of this is the photographic evidence that, already from the 1800s and up to the 1950s, showed slopes bare of vegetation and affected by debris flows (Figures 23 and 24), such as that of 1341, described by Luongo [38]. The detrital material easily poured into the streams, which presented a greater solid transport than the current one (Figure 25). Moreover, in the absence of regulation of rainwater and runoff on the slopes, the rivers were subject to sudden floods, causing frequent and severe damage to the structures and the resident population. Historical testimonies document these conditions of hydrogeological risk (Table 2).

Figure 23. The slopes of the Gubbio Mountains, substantially bare of vegetation, as they appeared in a postcard from the 1950s.

Figure 24. The detrital material was once easily mobilizable, considering the absence of vegetation and the steep slopes, so as to allow the triggering of debris flows that poured into the streams.

Figure 25. The streams (in the photo the Camignano Creek in the urban reach of Gubbio) were still subject to strong solid transport of detrital material coming from the bare and steep valley sides in the 1800s. The point of view is from upstream towards downstream.

From the point of view of hydraulic risk, the present situation is profoundly different. The afforestation carried out on the hills facing the city (Figure 26) has had the effect of greatly reducing the possibility of triggering debris flows and, above all, has increased the run-off time and the concentration time within the hydrographic basins of the streams flowing into the lowland area, effectively preventing the onset of catastrophic alluvial events such as the one described by don Paris Montanari in his memoirs. The channeling and construction of embankments in the river reaches of the plain have to be added (as well as the paving of the stream which crosses the city of Gubbio, the Camignano River), which have allowed for greater hydraulic efficiency of the outflow sections, thus avoiding inundation and flooding in the urban reaches.

Profound changes of the landscape, the evidence of which was made possible precisely by the historical-cartographic and documentary analysis.

Figure 26. Gubbio at present. The slopes of the Gubbio Mountains appear today densely revegetated.

5. Conclusions

Georgi's Map, which depicts the territory of Gubbio (Umbria, Central Italy) at the end of the 16th century, shows a design of the hydrographic network that is partly different from the current one. In particular, the San Donato Creek, today a left tributary of the Assino River in the northern portion of the Gubbio basin, is represented in this map as a tributary of the Saonda River, which belongs to another hydrographic basin, that of the Chiascio River, which occupies the central-southern portion of the basin itself.

The investigation carried out through direct surveys in the field, also with the support of the observation of satellite images, seems to confirm what Georgi has represented. This hypothesis is also supported by other elements, thanks to the excavations carried out for the construction of a new cult building, precisely in the hypothetical direction of flow of the paleo-riverbed of the San Donato Creek. These, in fact, have brought to light a sedimentary body of evident fluvial origin and an archaeological find, in particular a votive aedicule, dating back to the 15th century, which are similar to those that are still present today along the banks of the streams in the territory.

Furthermore, the study has allowed us to hypothesize the period in which this variation of the hydrographic network may have occurred, and the causes that may have produced it.

About this period, the comparison between Georgi's Map and the ones created subsequently allowed us to identify a precise time slice between 1570 (presumed date of creation of the Georgi's Map) and 1582 (date in which it was completed the mural painting by Egnatio Danti, preserved in the Vatican Museums, which represents a drawing of the hydrographic network substantially like the current one).

With regard to the causes of the modification of the track of the San Donato Creek, this suggests natural processes, in particular a diversion of the stream (river capture by overflow) induced by an alluvial event. This seems to be confirmed by the chronicles of time that highlight more severe hydrogeological risk conditions, characterized by a higher frequency of flood events than the current situation.

Therefore, a substantial evolution of the morphogenetic processes (fluvial and slope) in the study area should be highlighted, due both to the afforestation carried out along the slopes of the Gubbio Mountains and to the numerous surface water regulation interventions, both carried out during the 20th century. These have greatly reduced erosion processes, the triggering of debris flows, and the consequent solid transport of the streams that drain the slopes of the Gubbio hills, contributing to a decisive mitigation of the hydrogeological risk connected to the flooding phenomena of the lowland area.

Therefore, the present study is an excellent example of the validity of a multidisciplinary study method in the reconstruction of landscape evolution in areas of ancient and persistent human presence.

Author Contributions: Conceptualization, C.C., E.A.S., F.P. and U.E.S.; methodology, C.C., E.A.S. and F.P.; software, C.C.; validation, C.C., E.A.S., F.P. and U.E.S.; field investigation, C.C., E.A.S., F.P. and U.E.S.; archival researches, F.P. and E.A.S.; resources, C.C.; data curation, C.C., E.A.S., F.P. and U.E.S.; writing—original draft preparation, C.C., E.A.S. and F.P.; writing—review and editing, C.C., E.A.S. and F.P.; visualization, C.C., E.A.S. and F.P.; supervision, C.C., E.A.S. and F.P.; project administration, C.C.; funding acquisition, C.C. All authors have read and agreed to the published version of the manuscript.

Funding: This research received no external funding; the APC was funded by Corrado Cencetti with resources at his disposal furnished by Department of Physics and Geology, University of Perugia.

Data Availability Statement: All data are freely available in the State Archives and National Libraries mentioned in the text.

Acknowledgments: The authors thank Fabrizio Cece for providing some documentary data.

Conflicts of Interest: The authors declare no conflict of interest.

References

1. Williams, G.P.; Wolman, M.G. *Downstream Effects of Dams on Alluvial Rivers*; U.S. Geological Survey Professional Paper; US Government Printing Office: Washington, DC, USA, 1984.
2. Wyżga, B. River response to channel regulation: Case study of the Raba River, Carpathians, Poland. *Earth Surf. Process. Landf.* **1993**, *18*, 541–556. [CrossRef]
3. Surian, N.; Rinaldi, M. Morphological response to river engineering and management in alluvial channels in Italy. *Geomorphology* **2003**, *50*, 307–326. [CrossRef]
4. Surian, N.; Rinaldi, M. Channel adjustments in response to human alteration of sediment fluxes: Examples from Italian rivers. In *Sediment Transfer through the Fluvial System*; Golosov, V., Belyaev, V., Walling, D.E., Eds.; IAHS Publ.: Moscow, Russia, 2004; Volume 288, pp. 276–282.
5. Romano, S.; Cozzi, M. Cambiamenti nell'uso del suolo: Analisi e comparazione di mappe storiche e recenti. Il caso della Valle dell'Agri, Basilicata, Italia. *Aestimum* **2007**, *51*, 63–89. [CrossRef]
6. Gardi, C.; Dall'Olio, N.; Salata, S. *L'insostenibile Consumo di Suolo*; EdicomEdizioni: Monfalcone, Italy, 2013.
7. *Consumo di Suolo, Dinamiche Territoriali e Servizi Ecosistemici*; Report SNPA 32/22; Munafò, M. (Ed.) ISPRA: Roma, Italy, 2022.
8. Bravard, J.P.; Kondolf, G.M.; Piegay, H. Environmental and societal effects of channel incision and remedial strategies. In *Incised River Channels: Processes, Forms, Engineering and Management*; Darby, S.E., Simon, A., Eds.; Wiley: London, UK, 1999; pp. 303–341.
9. Liebault, F.; Piegay, H. Assessment of channel changes due to long-term bedload supply decrease, Roubion River, France. *Geomorphology* **2001**, *36*, 167–186. [CrossRef]
10. Surian, N. Effects of human impact on braided river morphology: Examples from Northern Italy. In *Braided Rivers*; Smith, G.H., Best, J.L., Bristow, C., Petts, G.E., Eds.; IAS Special Publication 36; Wiley-Blackwell: Hoboken, NJ, USA, 2006; pp. 327–338.
11. Da Canal, M.; Comiti, F.; Surian, N.; Mao, L.; Lenzi, M.A. Studio delle variazioni morfologiche del F. Piave nel Vallone Bellunese durante gli ultimi duecento anni. *Quad. Di Idronomia Mont.* **2007**, *27*, 259–271.
12. Surian, N.; Ziliani, L.; Cibien, L.; Cisotto, A.; Baruffi, F. Variazioni morfologiche degli alvei dei principali corsi d'acqua veneto-friulani negli ultimi 200 anni. *Alp. Mediterr. Quat.* **2008**, *21*, 279–290. Available online: https://amq.aiqua.it/index.php/amq/article/view/361 (accessed on 2 March 2023).
13. Koeman, C. The study of early maps: Methodology—Levels of Historical Evidence in Early Maps (With Examples). *Int. J. Hist. Cartogr.* **1968**, *22*, 75–80. [CrossRef]
14. Kent, A.J. The Value of Historical Maps: Art, Accuracy and Society. Doctoral Thesis, Oxford Brookes University, Oxford, UK, 1998. [CrossRef]
15. Cencetti, C. La cartografia storica come strumento per lo studio dell'evoluzione degli alvei fluviali. In Proceedings of the 6a Conferenza Nazionale ASITA, Perugia, Italia, 5–8 November 2002; Volume 1, pp. 757–762.

16. Zámolyi, A.; Székely, B.; Draganits, E.; Timár, G. Historic maps and landscape evolution: A case study in the Little Hungarian Plain. In Proceedings of the European Geosciences Union General Assembly, Vienna, Austria, 19–24 April 2009.
17. Lambrick, G.; Hind, J.; Wain, I. *Historic Landscape Characterisation in Ireland: Best Practice Guidance*; The Heritage Council of Ireland Series: Dublin, Ireland, 2013.
18. Coomans, T. *Mapping Landscapes in Transformation: Multidisciplinary Methods for Historical Analysis*; Leuven University Press: Leuven, Belgium, 2019.
19. Sobczynski, D.; Karsznia, I. Landscape evolution in the area of Kazimierski Landscape Park. *Pol. Cartogr. Rev.* **2019**, *51*, 81–94. [CrossRef]
20. Rombai, L. Problems related to the use of the history cartography. *Boll. Ass. It. Cart.* **2010**, *138*, 69–89.
21. Statuto, D.; Cillis, G.; Picuno, P. Using Historical Maps within a GIS to Analyze Two Centuries of Rural Landscape Changes in Southern Italy. *Land* **2017**, *6*, 65. [CrossRef]
22. Branduini, P.; Laviscio, R.; L'Erario, A.; Toso, F.C. Mapping Evolving Historical Landscape System. In Proceedings of the 2nd International Conference of Geomatics and Restoration, Milan, Italy, 8–10 May 2019; pp. 277–284.
23. Picuno, P.; Cillis, G.; Statuto, D. Investigating the time evolution of a rural landscape: How historical maps may provide environmental information when processed using a GIS. *Ecol. Eng.* **2019**, *139*, 105580. [CrossRef]
24. Piovan, S. Historical Maps in GIS. In *The Geographic Information Science & Technology Body of Knowledge, 1st Quarter 2019 Ed.*; Wilson, J.P., Ed.; Digital version; UCGIS: Ithaca, NY, USA, 2019. [CrossRef]
25. Allegrucci, F. Lettura cartografica dell'Eugubino dalla Carta del Georgi a quella del Coronelli. In *Gubbio e la sua Storia – Sez. III. Gubbio nel Rinascimento*; Biblioteca Sperelliana: Gubbio, Italia, 1999; 7p.
26. Ambrosetti, P.; Basilici, G. Il Plio-Pleistocene Umbro. In *Guide Geologiche Regionali. Appennino Umbro-Marchigiano (Società Geologica italiana)*, 1st ed.; BE.MA.: Milano, Italy, 1994; pp. 46–47.
27. Elter, P.; Giglia, G.; Tongiorgi, M.; Trevisan, L. Tensional and compressional areas in the recent (Tortonian to Present) evolution of the Northern Apennines. *Boll. Geofis. Teor. Appl.* **1975**, *65*, 3–18.
28. Carmignani, L.; Kligfield, R. Crustal extension in the Northern Apennines: The transition from compression to extension in the Alpi Apuane core complex. *Tectonics* **1990**, *9*, 1275–1303. [CrossRef]
29. Boccaletti, M.; Cerrina Feroni, A.; Martinelli, P.; Moratti, G.; Plesi, G.; Sani, F. L'alternanza distensione-compressione nel quadro evolutivo dei bacini neogenici dell'Appennino Settentrionale. In *Studi Preliminari All'acquisizione Dati del Profilo CROP 03 Punta Ala—Gabicce*; Pialli, G., Barchi, M., Menichetti, M., Eds.; Studi Geologici Camerti; Università di Camerino: Camerino, Italy, 1991; pp. 187–192.
30. Cinque, A.; Patacca, E.; Scandone, P.; Tozzi, M. Quaternary kinematic evolution of the Southern Apennines. Relationships between surface geological features and deep lithospheric structures. *Ann. Geofis.* **1993**, *36*, 249–260. [CrossRef]
31. Amato, A.; Selvaggi, G. Terremoti crostali e sub-crostali nell'Appennino settentrionale. In *Studi Preliminari All'acquisizione Dati del Profilo CROP 03 Punta Ala—Gabicce*; Pialli, G., Barchi, M., Menichetti, M., Eds.; Studi Geologici Camerti; Università di Camerino: Camerino, Italy, 1991.
32. Boncio, P.; Brozzetti, F.; Ponziani, F.; Barchi, M.; Lavecchia, G.; Pialli, G. Seismicity and extensional tectonics in the northern Umbria-Marche Apennines. *Mem. Soc. Geol. It* **1998**, *52*, 55.
33. Barchi, M.; Beltrando, M. The Neogene-Quaternary evolution of the Northern Apennines: Crustal structure, style of deformation and seismicity. *J. Virtual Explor.* **2010**, *36*, 1–24. [CrossRef]
34. Collettini, C.; Barchi, M.; Chiaraluce, L.; Mirabella, F.; Pucci, S. The Gubbio fault: Can different methods give pictures of the same object? *J. Geodyn.* **2003**, *36*, 51–66. [CrossRef]
35. Menichetti, M.; Minelli, G. Extensional tectonics and seismogenesis in Umbria (Central Italy) the Gubbio area. *Boll. Soc. Geol. It.* **1991**, *110*, 857–880.
36. Cattuto, C.; Cencetti, C.; Gregori, L. Elementi geomorfologici dell'area compresa tra l'Umbria nord-orientale e la regione marchigiana. *Mem. Descr. Della Carta Geol. D'italia* **1989**, *39*, 37–44.
37. Cattuto, C.; Cencetti, C.; Gregori, L. Il Plio-Pleistocene nell'area medio-alta del bacino del F. Tevere: Possibile modello morfotettonico. *Studi Geol. Camerti* **1992**, 103–108.
38. Cencetti, C. Evoluzione del reticolo idrografico in un tratto umbro-marchigiano dello spartiacque principale dell'Appennino. *Geogr. Fis. Dinam. Quat.* **1988**, *11*, 11–24.
39. Sestini, A. Sull'origine della rete idrografica e dei bacini intermontani nell'Appennino centro-settentrionale. *Riv. Geogr. It.* **1950**, *57*, 249–256.
40. Desplanques, H. Carte du peuplement dans la région de Gubbio au XVIe siècle. *Mediterranèe* **1963**, *4*, 5–13. [CrossRef]
41. Allegrucci, F. La popolazione della diocesi di Gubbio in una carta topografica del XVI secolo. *Propos. E Ric.* **1985**, *15*, 51–60.
42. Armeni, C.; Falcucci, C. *Museo Comunale di Gubbio. Incisioni*; Electa: Firenze, Italy, 1993; 156p.
43. Graziani, M.O. *La Carta del Georgi*; L'Arte Grafica: Gubbio, Italy, 2005; 28p.
44. Menichetti, P.L. *Storia di Gubbio Dalle Origini all'Unità d'Italia*; Petruzzi Editore: Città di Castello, Italy, 1987; Volume 2, pp. 94–95.
45. Charalambous, K.; Bruggeman, A.; Bakirtzis, N.; Lange, M.A. Historical Flooding of the Pedieos River in Nicosia, Cyprus. *Water Hist.* **2016**, *8*, 191–207. [CrossRef]
46. Sannipoli, E.A. Documenti inediti su Jacopo Bedi pittore eugubino (1a parte). *Gubbio Arte* **1994**, *12*, 21–23.
47. Paci, R. La crisi del Comune popolare di Gubbio nel Cinquecento. *Quad. Stor. Delle Marche* **1967**, *2*, 457–507.

48. Biscarini, P.E.P. Splendore e decadenza di una città: Gubbio tra la fine del Quattrocento e la prima metà del Cinquecento. Alcune note di storia. In Proceedings of the La maiolica Italiana del Cinquecento. Il Lustro Eugubino e L'istoriato del Ducato di Urbino, Gubbio, Italy, 21–23 September 1998; Centro Di: Firenze, Italy, 2002; pp. 31–48.
49. Cece, F.; (Independent Researcher, Gubbio, Italy). Personal communications. 2020.
50. Panizza, M. *Elementi di Geomorfologia*; Pitagora: Bologna, Italy, 1978; 188p.
51. Luongo, A. *Gubbio nel Trecento. Il comune Polare e la Mutazione Signorile (1300–1404)*; Viella: Roma, Italy, 2016; 726p.
52. Cece, F. Memorie di don Paris Montanari. In *Il Diario di Giacomo Armanni (1622–1631). Gubbio Negli Ultimi Dieci Anni del Ducato di Urbino*; Self-Published: Gubbio, Italy, 2006; pp. 53–62.
53. Cece, F. *Luigi Lucarelli. Memorie*; Media Video: Gubbio, Italy, 2011; 400p.

 land

Article

Where the Second World War in Europe Broke Out: The Landscape History of Westerplatte, Gdańsk/Danzig

Wojciech Samól [1,2], Szymon Kowalski [3], Arkadiusz Woźniakowski [3] and Piotr Samól [3,*]

1 Museum of the Second World War, Bartoszewski Sq 1, 80-862 Gdansk, Poland
2 Insitute of History, University of Gdańsk, Wit Stwosz St. 55, 80-308 Gdansk, Poland
3 Faculty of Architecture, Gdansk University of Technology, Narutowicz St. 11/12, 80-233 Gdansk, Poland
* Correspondence: piosamol@pg.edu.pl

Abstract: The article describes the landscape history of the Westerplatte Peninsula in Gdańsk (Poland) from the 17th to the 20th century presented as a complex process of the landscape's environmental, urban and military transformations. Westerplatte is known as the symbolic place where the Second World War in Europe broke out, and for this reason the current discourse is mainly concentrated on that period. Nonetheless, the history of Westerplatte includes many other important events involving Polish, German, Russian and even French politics over the last three centuries. Thanks to its location at the entrance of one of the main harbours on the Baltic Sea, it is cartographically the best-documented part of the Vistula river estuary. A comprehensive archival survey conducted in the Polish and German archives and cartographical analysis of over 200 selected historical maps allowed the authors to reconstruct its spatial history over three centuries. This case study of Westerplatte can be regarded as an example of the research modus operandi of a historical landscape which has been transformed multiple times. It might form the basis for establishing a new policy for its preservation, allowing a balance to be kept between fluctuations of the current historical politics and more universal requirements for the protection of tangible and intangible heritage. The article also stresses the importance of a holistic and interdisciplinary approach in the analysis of a historical landscape and the necessity of proper selection and critical verification of sources.

Keywords: HUL; cultural landscape; anthropopression; landscape transformation; heritage conservation; spatial history; the Second World War; coast; Baltic Sea; Gdansk

Citation: Samól, W.; Kowalski, S.; Woźniakowski, A.; Samól, P. Where the Second World War in Europe Broke Out: The Landscape History of Westerplatte, Gdańsk/Danzig. *Land* **2023**, *12*, 596. https://doi.org/10.3390/land12030596

Academic Editors: Bellotti Piero and Alessia Pica

Received: 10 February 2023
Revised: 27 February 2023
Accepted: 28 February 2023
Published: 2 March 2023

1. Introduction

The history of Westerplatte—a peninsula between the old Vistula mouth (current port canal) and the Gulf of Gdansk—began over three hundred years ago. The event that gave this place a unique status in Polish historiography as a symbolic place where World War II broke out was the dramatic defence of the Polish outpost at Westerplatte in the first days of the war. On this basis, the Westerplatte peninsula was awarded the status of the Monument of History—the highest form of monument protection in Poland, covering slightly more than 100 objects of outstanding importance to the cultural heritage of Poland. Thus, it was the subject of several historical elaborations on the margin of the synthesis of World War II [1] as well as studies on Gdansk [2] and a particular monograph on the Polish military base [3–7], which emphasized its importance in this context [8]. However, the history and the spatial transformations of Westerplatte are more complicated, and they encompass geographical, political and economic issues tangled together in a small area which often seems to be overlooked or treated superficially in contemporary literature [Figure 1].

The aim of this article is to reconstruct the spatial history of Westerplatte over the last 400 years. Analysing data onvarious characteristics from such a long period enables the combination of different methods (including tedious, traditional analysis of sources) in research on landscape transformations, depending on the chronological and territorial scope of the study area.

Figure 1. Location of the research topic: Gdańsk as a historical port city at the mouth of the Vistula river.

Literature Review

In the increasingly popular study of landscape evolution, possible perspectives refer to various scientific disciplines: geography, spatial management and urban planning, but also archaeology and architecture. Most recent studies on the geomorphology of anthropocenic landscape transformation focus on the scale 100:000–10:000 [9–11]. Meanwhile, spaces such as Westerplatte need more detailed analyses, taking into account the limited area of their multitemporal transformations. For this reason, the literature review is divided according to the subjects of interest: battlefields, urban (suburban) areas, coastline formation and estuaries. It is worth emphasizing that all these issues, usually related to analyses at various scales, appear in the analysed example of Westerplatte.

Battlefield studies are currently a developing field of research on historical landscape evolution. Studies of this type have recently been published concerning Austerlitz [12], Verdun [13], Cabezo de Alcalá [14] and Flanders [15]. They were most often focused on archaeological research and artefacts discovered during them [16], as well as the use of new research technologies such as photogrammetry and LiDAR [13,17,18]. A broader view, including changes in land cover and landform, also for understandable reasons, oscillates around one particular event in the history of the studied area [12]. However, historical events such as battles are violent and brief actions; therefore, their analysis allows us to reconstruct the landscape only at this specific point in time. This has been pointed out, for example, by W. Altizer in the context of research on the Santiago Campaign of 1898, based on the theory of "time perspectivism" by G. Bailey [19].

A particular problem is built environment analysis (urban studies), where changes have continuous character and the landscape, formed as a result of anthropopression and environmental changes, is still undergoing transformation. Historical Urban Landscapes (HULs) are the subject of research on conservation and heritage preservation issues. The idea of conservation—keeping the monument in the best possible condition—cannot be applied as rigorously in the case of HULs as it can in the case of individual objects [20–22]. For this reason, awareness of the change and documenting it as a process is of value in itself.

Although attempts to map the process of historical spatial transformations of urban areas have been carried out in heritage studies of Venice [23,24], Naples [25], New York [26], London [27], San Francisco [28] and Tokyo [29], the analysis was still usually limited to the particular time due to the representativeness of the sources. The extensiveproject aimed at analysing historical transformation—"European Atlas of Historic Cities"—has been conducted for over five decades and is still far from complete [30].

Rivers and waterways in the pre-industrial era constituted the transportation network. Not surprisingly, studies of these, especially of the alluvial zones (estuaries), are also the subject of research by landscape historians [31,32]. Studies on formation processes and geological changes have also been conducted for the Vistula estuary [33,34]; however, the complex analysis of historical written sources and archaeological surveys was not implemented. This article presents the history of the formation of the Vistula estuary in the modern era since the end of the 17th century.

When the landscape and spatial transformations are analysed in architectural, urban or geographical terms, depending on the size of the researched object, it is quite rare, and it is difficult to combine the methods used for each of the three scales as each of them applies different solutions [35]. However, the role of historical cartography in such research might be illustrated by the history of Bombay [36]. It shows that combining interdisciplinary sources allows one to overcome the limitations of a particular scientific discipline and might help analyse other cases [37]. As it appears, one of the main problems for historians, archaeologists or geographers [31,33,34,38] dealing with the study of historical space, which can greatly limit the scope of research, is access to sources, as well as environmental changes, both natural and man-made, irreversibly destroying the object of the study itself.

The conditions that the authors encountered in the case of Westerplatte fall between the study of the urban landscape and the natural one, marked by human activity. These two trends in research on historical space may meet during the analysis of the urban environment exposed to several natural, economic and military changes. Westerplatte is an excellent example of such a place.

In recent years, discourse on the history of Westerplatte became strongly linked with its conservation and heritage protection issues [7]. The future development of the peninsula is widely disputed, and two main strategies might be observed. On the one hand, some heritage conservators propose reconstructing the former military base space as it was in the summer of 1939 (just before World War II broke out). On the other hand, local politicians and architectural societies would like to create a modern museum on the periphery of the peninsula with full respect to the monument's unhistorical conception from the 1960s. A half-century later, that vision reflected the political struggle between the local municipality and the new directorship of the Museum of the Second World War [39,40].

However, both visions do not consider the complex history of Westerplatte with its numerous transformations and links to three centuries of rough relationships between Poland, Germany, Russia and even France. Therefore, the analysis presented below contributes to the discourse on the future of this significant area and allows us to question how the cultural landscape should be protected today. Therefore, the authors proposed the analysis based on defining nine periods of the area's history, from the end of the 17th century to the turn of the 21st century. Their purpose is to present the way that in-depth interdisciplinary studies enable a better understanding of the past and how the complex database of cartographical sources, compared with archaeological surveys, might be used in reconstructing the whole process of spatial transformation in a strictly defined area. The theoretical base for such studies was set by the so-called spatial turn in history, which has been evolving since the late 1980s [41–43].

2. Methods

The reconstruction of the space of Westerplatte required the application of historical methods (analysis and criticism of historical sources, deductive and retrogressive methods) and field studies (archaeological, inventory of landscape on an urban scale), supplemented by the results of geomorphological (geological history of the Vistula estuary) and geographical studies (confronting the results of historical studies with knowledge of the formation of the Vistula estuary). Only non-invasive studies, i.e., geo-radar or electrical resistivity technique, were not included, due to the experience required for the analysis and the much greater difficulty of interpreting the results of the studies.

First of all, the authors collected the evidence base of over 200 maps and aerial photos from the rich collections (over 3500 items in Gdańsk), including Architekturmuseum der Technischen Universität Berlin (Architectural Museum of the Technical University of Berlin, Berlin, Germany), Archiwum Państwowe w Gdańsku (State Archive in Gdańsk, Gdańsk, Poland), Bundesarchiv-Militärarchiv (Federal Archive-Military Archive, Freiburg im Breisgau, Germany), Biblioteka Gdańska Polskiej Akademii Nauk (Polish Academy of Sciences, Gdańsk Library, Gdańsk, Poland), Centralne Archiwum Wojskowe (Central Military Archive, Warsaw, Poland), Geheimes Staatsarchiv Preußischer Kulturbesitz (Secret State Archives Prussian Cultural Heritage Foundation, Berlin, Germany), National Archives and Records Administration, College Park, Maryland, United States of America, and Urząd Miasta Gdańska, Wydział Geodezji (Gdańsk Municipal Administration, Department of Geodesy, Gdańsk, Poland) [Chart 1]. During the archival research, a representative group of such sources were selected. Although the first records concerning the Vistula estuary (the depth of the fairway) date back to 1583, the authors focused on maps showing the formation of Westerplatte isle, its transformation into a peninsula and subsequent development and transformations. Therefore, Chart 1 shows that the structural division of maps, plans and photos is accurate and illustrative for reconstructing the space of Westerplatte for the entire period after 1698. The particular references to those archival records are linked with relevant items in the mentioned list.

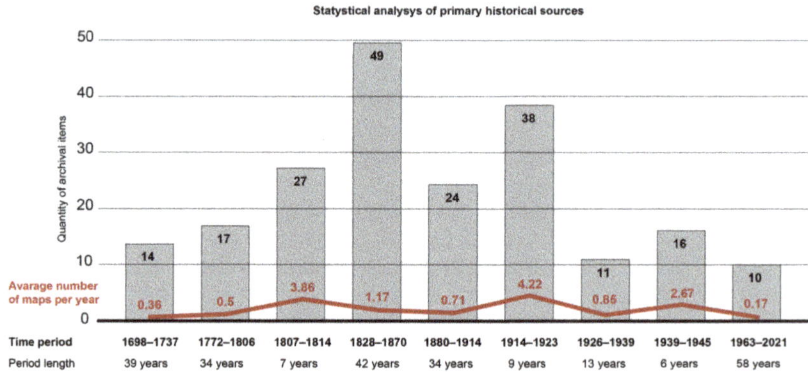

Chart 1. Statistical analysis of cartographical primary sources used in the research. The average number of analysed maps of Westerplatte per year.

Although history is a linear process, the landscape transformations of Westerplatte were taking place at different speeds in particular periods. Moreover, the differences in quantity and quality of historical sources depanding on the time of their creation (and other data useful for reconstruction of spatial changes) make the problem of their representativeness. Sometimes, we may analyse many nearly identical sources, other times we have to rely only on a unique one. Summarizing Chart 1, it should be emphasized that the most intense and detailed period of landscape transformation took place in the period from a few years before the Second World War to its first year. For this reason, aerial photographs were also used to capture these quick and subtle changes not sufficiently recorded by cartography. Due to the strategic role of Westerplatte as the main gateway to the port of Gdańsk, the peninsula has been recorded in an incomparably greater number of aerial photos than any other place in this part of Europe [44–49].

The main scientific method used in the described research is based on cartographic retrogression combined with the results of archaeological excavations [Figure 2]. Selected maps and plans were mutually aligned for maximum geographical precision. Comparing different maps required them to be mutually calibrated. The main problem was the geographical accuracy of older maps from before the 19th century. Due to the drawing techniques used at that time, they had many distortions in the mapping of the terrain. Moreover, they could have been further distorted later during the reprographic process. Therefore, the maps were always matched in reverse chronological order, taking the most recent and most accurate map as a reference. Each older map was scaled accordingly (after conversion from old units of measurement to metric) and adjusted to the newer one. In the absence of anoriginal scale or large discrepancies, the matching of the maps was improved by referring to the constant elements present on both sheets. It was usually buildings, fortifications, hydrotechnical structures, road layout or other characteristic terrain points, etc. The technical irregularities were then interpolated and compared in Auto CAD 2019 software for which the reference map is compatible and inscribed in a coordinate system compliant with the applicable standard. Moreover, recent field works and archaeological excavations conducted in 2016–2022 provided the required feedback to assess the quality of data provided from written and cartographic sources [16]. It is worth mentioning that this type of doubled procedure of research is seldom applied in similar projects. Thanks to this process, satisfactory accuracy was achieved for further analysis, meaning that all the information could be compared and cross-referenced in the interpolation process, and then physically verified in situ.

The maps were the basis for further analysis of the coastline transformation, the civilian and military facilities, changes in the transport network, forestation, etc. Consequently, the history of Westerplatte has been described and illustrated as a process of permanent transformation, which is crucial for conservation issues. Moreover, the collection of maps created is linked with the current geodesic map from the State Resources, allowing the presentation of the results of this research on more accessible platforms such as the Google Maps application.

The method is crucial for the verification and synthesizing of the multiple map sources as well as aerial photographs [50,51], with different levels of detail and cartographical accuracy [52]. It allows accurate reconstruction of the spatial transformation process, including its natural and anthropogenic character.

Although the authors implemented modern methods of documentation (including GIS and CAD modelling), the results of the research were illustrated in nine synthetic maps visually presenting the reconstructed process of Westerplatte's spatial transformations. To facilitate referring information from the text to the maps, the most important elements of the described development (landscape) are marked with letters in the text and on individual illustrations.

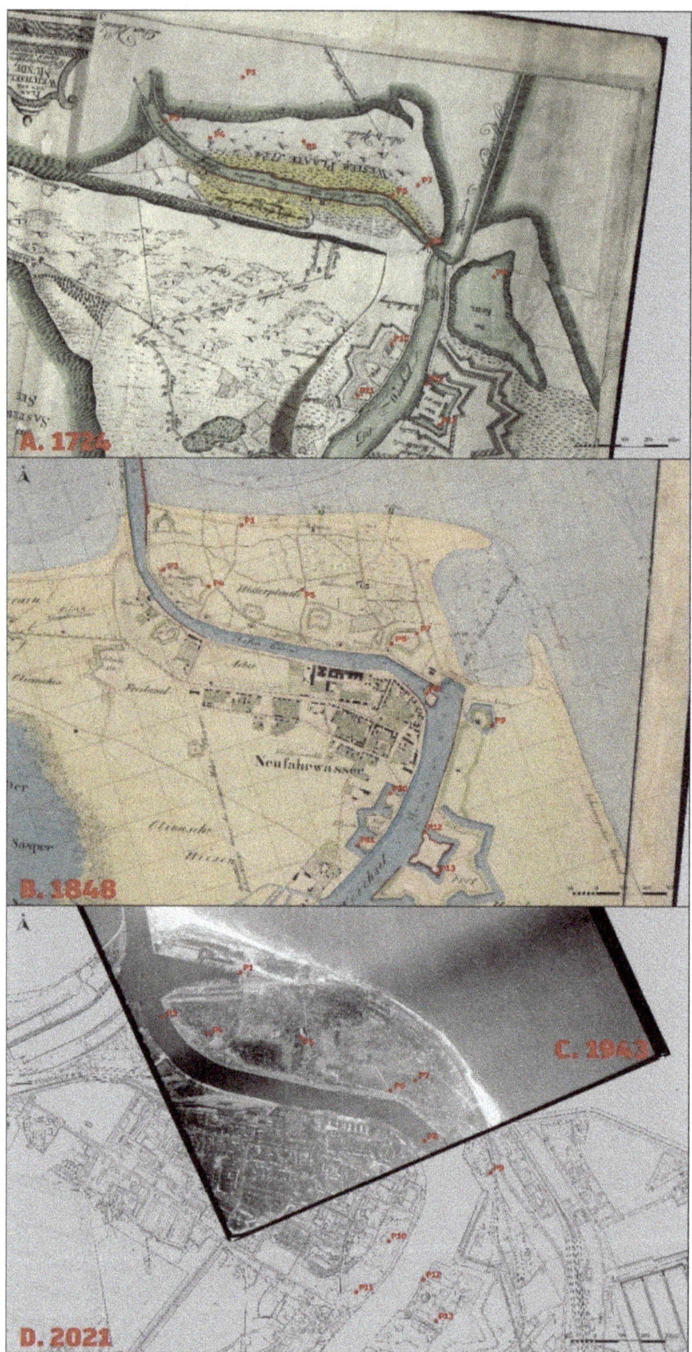

Figure 2. Methods of interpolating and calibrating the historical sources: (**A**) hand-drawn map from 1724; (**B**) printed geodesic map from 1848; (**C**) aerial photo from 1943; (**D**) modern CAD map.

3. Results—Spatial Transformations of Westerplatte

In the 14th century, Gdansk became the main harbour of the State of Teutonic Knights in Prussia (and later the Polish-Lithuanian Commonwealth). In the 1370s, harbour facilities were also developed outside the city and thus the area was inspected by the customs officials from the castle. As a result, an additional checkpoint was built by the Vistula mouth which gave origin to the Wisłoujście Fortress, guarding the entrance to the Gdańsk harbour for the next four hundred years [53]. Every spring since 1593, the waterway through uncertain waters of the river mouth was tracked by the special commission. The reports and sketches made in the course of those procedures established a solid base for further historical research [Figure 3].

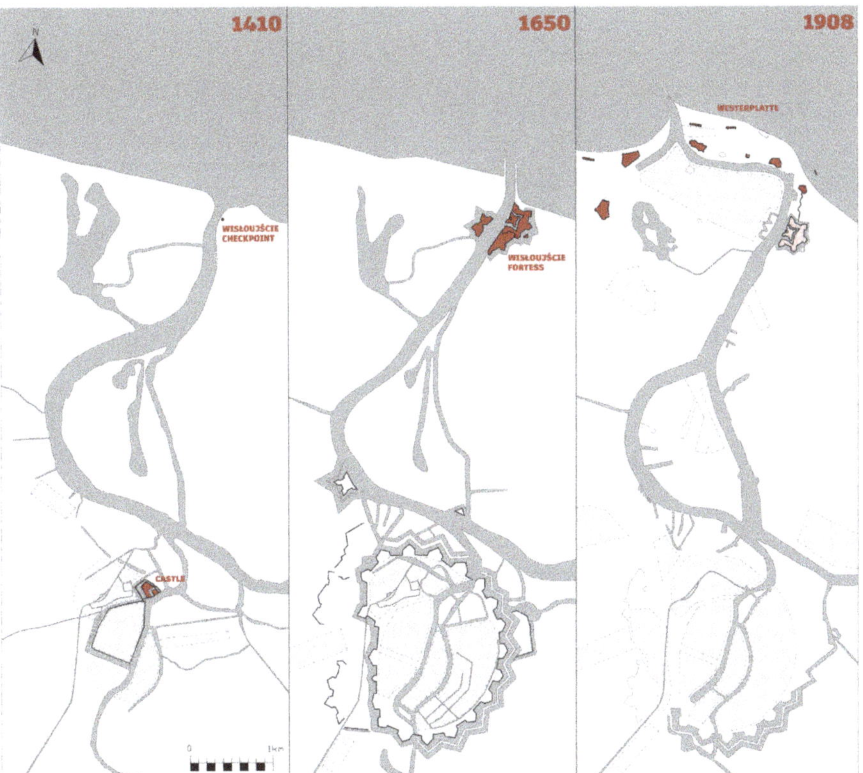

Figure 3. Scheme of the Gdansk city's development along with its direct protection of the port's entry.

3.1. Stage 1: 1698–1737

In the late 17th century, two isles appeared on both sides of the mouth—Oster- and Wester-Platte. The Osterplatte was an ephemeral site but its western neighbour survived. It was separated from the dry land by a shallow lagoon which was turned into the alternative entrance to the port, bypassing the problematic river mouth causing the nautical problems mentioned before. A special dam protecting the new canal from silting and linking the island with the mainland was built in 1686 and 1698 [54]. Between the canal and the Wisłoujście Fortress, there was the so-called Balastkrug—a place where the gravel carried as ballast by ships arriving in Gdańsk was left and subsequently used in hydro-technical works to reinforce the Westerplatte island and the banks of the lagoon. The first engineering works that allowed the water way through the lagoon (A) to be tracked were undertaken in 1716–1724 [55]. For a few subsequent decades, the city councillors did not anticipate, however, that the new strip of land might have such military importance [Figure 4]

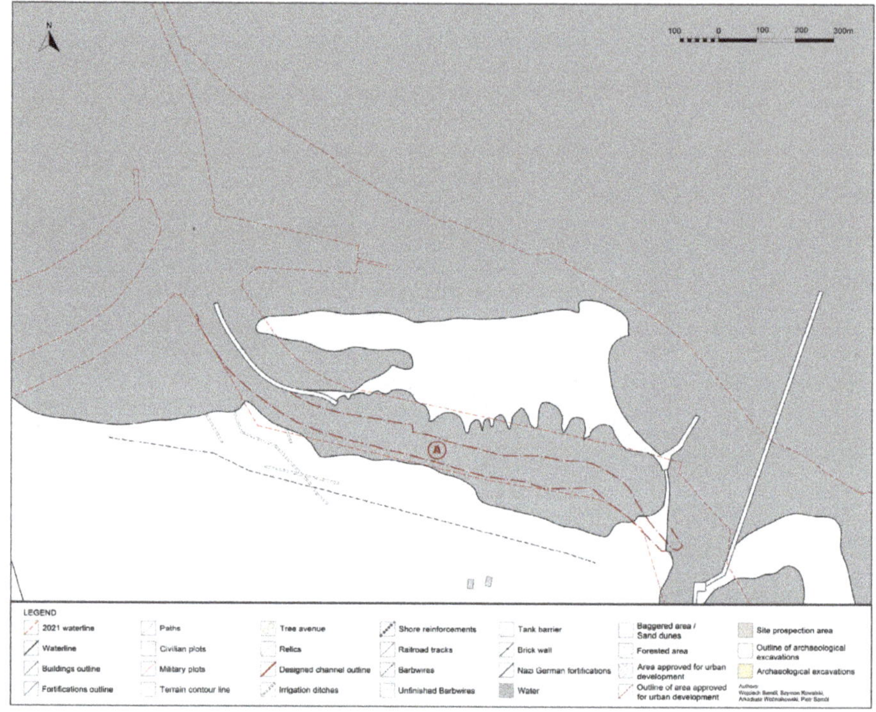

Figure 4. Westerplatte was a lagoon after 1698.

This neglect was laid bare during the Polish Succession War in 1734 when the French military expedition landed on the island of Westerplatte and tried to help Stanisław Leszczyński—the legal pretender to the Polish throne and personally the father-in-law of King Louis XV of France, who was besieged in Gdańsk. The 2000-man corps built a provisional fortification (B) on the lagoon's northern bank. The French units were too weak to break the Gdansk lockdown that had been set by the Saxo-Russian corps under the field-marshal Munich (ca. 18,000 soldiers) and could only occupy the isle [2]. After the capitulation of Gdańsk, those fortifications were abandoned, and quickly disappeared due to increasing erosion [Figure 5]. In 1737 a new waterway led through the lagoon, and a new dam (C) was erected [56]. It was probably then that the rest of the French camp was demolished; however, the presence of French soldiers was confirmed by archaeological surveys conducted in 2017–2019 [16]

3.2. *Stage 2: 1772–1806*

In the 18th century, the significance of the canal as an alternative entrance to the Gdańsk harbour increased. However, the true milestone in Westerplatte's history was the First Partition of Poland in 1772, when the whole Pomeranian province (except Gdańsk itself), including the western bank of the Vistula river, was taken over by Prussians. They were aware of the site's strategic role and thus located the tax chamber and warehouses there. The canal separated from the mainstream of the Vistula river by the mentioned dam might have been used as a harbour, similar to the docks by the Thames or Liver rivers [57–59]. This new urban area (D), called Neufahrwasser (new waterway), was joined with the former suburbs of Gdańsk (Chełm/Stoltzenberg and Stare Szkoty/AltesSchottland) to create a competitive administrative complex for the Polish city—the so-called combined municipality. Not later than 1783, the western part of Westerplatte was fortified by an earthen battery (E) with a guardhouse [60,61]. However, in 1788, the new Prussian King, Frederick Wilhelm

II, decided to change alliances and supported the Polish-Lithuanian Commonwealth in reforming the state. At that time, the Russian invasion was considered the most severe threat to the security of Neufahrwasser. In 1788–1790,it caused the Prussian government to build four new autonomic ramparts (F) on the southern side of Westerplatte [62]. The western bastion incorporated the older battery (E) by the entrance to the harbour [63].

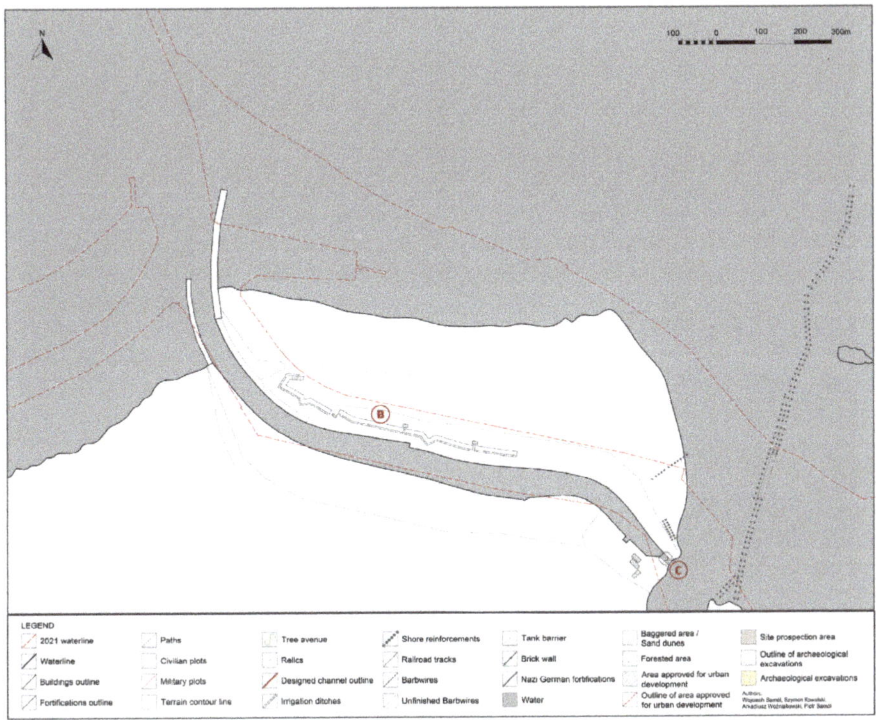

Figure 5. Westerplatte with French fortifications (after 1734).

Just three years after the bastions were completed, Prussia took over Gdańsk and there was no longer a reason for Westerplatte to be fortified. Moreover, Neufahrwasser became a part of Gdańsk harbour, and the improvement of traffic between all its features became a priority. In 1804, the port's administration widened the mouth, and an earth wall strengthened the eastern bank. Before 1802, a second lock was set, parallel to the old one (C), between the Vistula and the canal [64].

3.3. *Stage 3: 1807–1814*

The military and political situation changed once again in 1807 when Gdańsk was besieged and captured by Napoleon's Great Army. After the treaty of Tilsit, Gdańsk was declared a free city and a French protectorate, and thus it became one of the most important imperial military bases in Eastern Europe [65]. Undoubtedly, it was the reason why the French army administration decided to renew its fortifications (E, F), including those located on Westerplatte (e.g., rampart no. 4 was transformed). Due to increasing traffic in the external port of Neufahrwasser, the Westerplatte banks of the canal were adopted as harbour facilities, including two new kitchens (G) for sailors [66] [Figure 6].

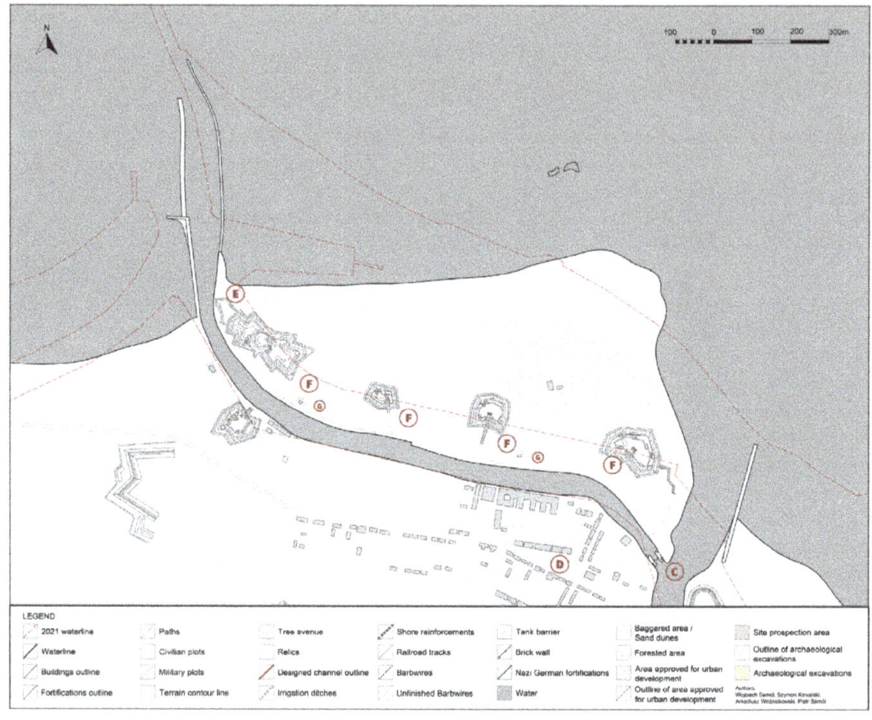

Figure 6. Westerplatte with its Prussian fortifications (in 1812).

The first attempts to stabilize the dunes around the mouth of the river with plantings were carried out in the mid-18th century by the Society of Nature, active in Gdańsk since 1743, and weresomeof the first examples of environmental engineering [67]. However, similar solutions were also being applied in Prussian coast fortresses in Pillau and Kolbergfromthe end of the 18th century [62]. A forest plantation established around 1803 in the eastern part of Westerplatte inspired future recomposition of the area in the following years.

After the failure of Napoleon's invasion of Russia, Gdańsk was besieged by Prussian-Russian forces. Westerplatte played an important military role and it was the place where one of the bloodiest battles occurring during the siege was fought (16 September 1813). In 1734, the city being cut off from the sea caused its capitulation [68].

3.4. Stage 4: 1828–1870

In 1828, the ice jam in the Vistula mouth brought the most catastrophic flood in the history of Gdańsk [69]. A similar tragedy happened in 1840 when the Vistula broke a narrow strip of land around 5 km east of Gdansk and found a new estuary to the Baltic Sea. During that catastrophic flood, the village of Gorki was wiped off the earth. However, this disastrous event allowed the closure of the old river mouth beside Westerplatte (H), turning the isle into a peninsula. The construction work took a few years and in 1840–1845 there was only a stone dyke across the old mouth of the Vistula [70]. The sea currents created a temporary pond (I) between the dyke and the sea (finally filled with earth at the turn of the 20th century) [71–73].

The linking of the former island with the mainland allowed the development of new facilities such as baths, which had been gaining popularity in Europe since the 1810s. Although Martin Kruger bought16 plots with the intention to build sea baths in Westerplatte in 1829 [74], the first installation for sea-bathing admirers (J) opened only in 1842 [70,75]. A decade later, the fortifications at Westerplatte were strengthened by two coastal field batteries (K) located in the baths' close neighbourhood, which did not facilitate the symbiosis between them [76–79] [Figure 7].

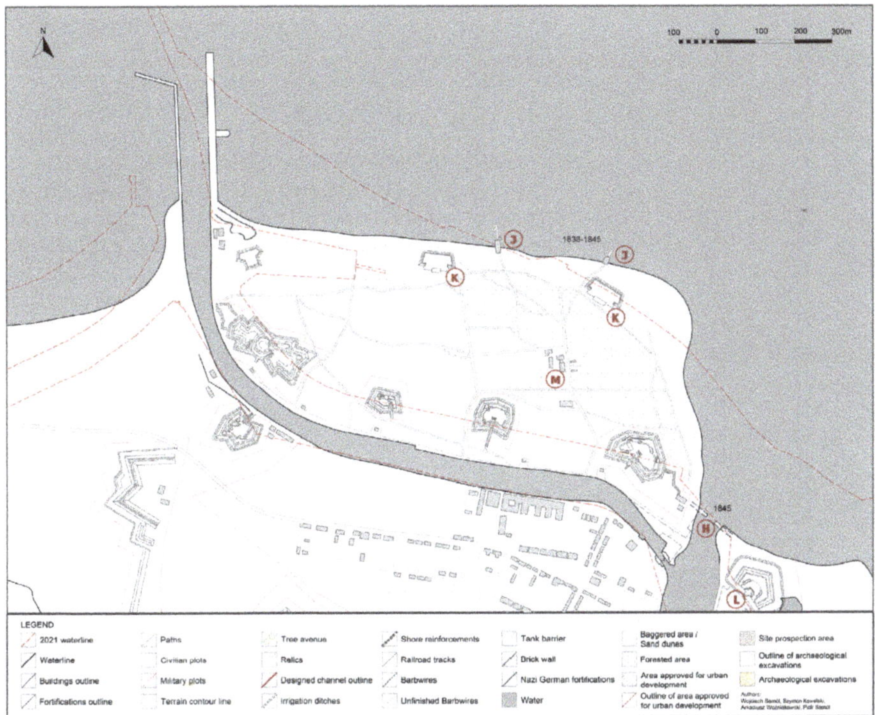

Figure 7. Westerplatte strengthened with the coastal batteries along with the beginnings of the resort facilities (in 1846).

Due to the change of the Vistula mouth, it was decided that the Westerplatte ramparts would be connected with the Wisłoujście Fortress through an intermediate fortification (L)—the Seagull Fort (Möwenschanze), which was completed in 1846 [70,76,80,81].

As technological advancements of artillery in the 1870s and onwards significantly increased effectiveness, range and salvo weight, as well as the accuracy of modern ordnance, a need to expand and modernize the old fortifications emerged [82]. Consequently, the Prussian military administration erected new fortifications in Brzeźno/Brösen and between Westerplatte and Wisłoujście Fortress, which caused the abandonment of two central (nos. 2 and 3) ramparts out of the four on the peninsula [83–85]. The new limitation of the restricted fortified zone was established, designating the central part of the peninsula as a less restrictive area (M) where the erection of civilian buildings of light construction was permitted [86,87].

3.5. Stage 5: 1880–1914

Although in 1890 the Prussian government permitted the building of a new Schichau's shipyard in the close vicinity of the Gdansk fortifications and, a few years later, it allowed the demolition of part of the inner circle of bastions surrounding the city [88], its policy on Westerplatte was not so liberal [85]. Moreover, the changes in the German strategic plans of 1905–1907 (the so-called Schlieffen's plan) caused Gdańsk to maintain its military potential, especially in terms of coastal fortifications as part of the eastern fortresses for the protection of the borderland against Russia [89–91]. Although some of the plans were not implemented, the new structures of 1905–1914 were stronger than those built before. On the eve of the First World War (1914–1918), a new stationary battery (N) and observation bunker were built as part of a modern coastal battery fire control system [79].

At the turn of the 19th century, Gdańsk became one of the Reich's main shipyard hubs. As a result, canal and harbour entrance enhancements were made in 1908–1912. During that modernization, some parts of rampart no. 4 of 1783/1788 were demolished [89,92]. Meanwhile, at the end of the 1880s, the "Weichsel Gesellschaft" (the company which took control over the swimming site at Westerplatte [Figure 8]) started the construction works of a villa settlement in the centre of the peninsula [93–95]. This caused Westerplatte to experience an extraordinary development boom, with more hotels, resorts, seaside swimming premises and parks (P). Such tendencies in urban planning were related to Howard's ideas of "Garden cities" [96]. A telephone network and electricity were also brought in. The baths of Westerplatte became quite a popular resort on the shores of the Baltic Sea. There were over 40 buildings erected, mostly made of wood and thin brick walls (up to 12 cm). Light construction resulted from the above-mentioned restrictions concerning buildings in fortified zones so that all civilian structures could be demolished within the first hours after a proclamation of war. According to the legal regulations of the act of Reichstag and directions of "Reichs-Rayon-Komission", one to three fortification zones around every fortification work had been established (each with a width of 600 to 1275 m). Inside these zones, every investment or change in the landscape (such as mason structures) was either prohibited or the subject of arrangements with the military administration of Gdansk Fortress [97].

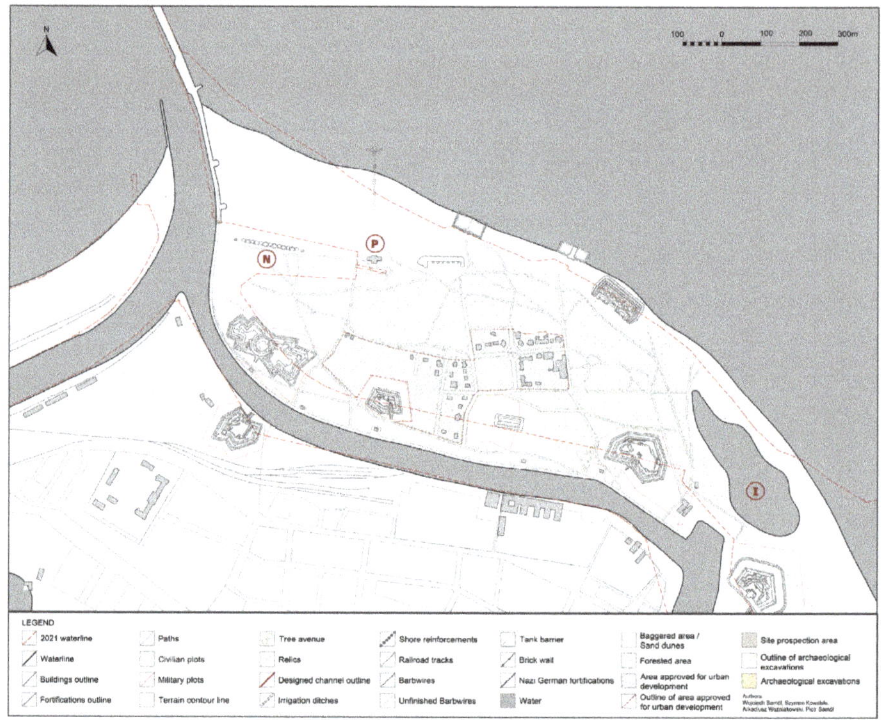

Figure 8. Westerplatte during extensive urban development with summer villas and an SPA resort (in 1895).

3.6. *Stage 6: 1914–1918/1923*

The First World War ended in Germany's failure. According to the Treaty of Versailles, Gdańsk was declared a demilitarized free city, dependent on the High Commissioner of the League of Nations. Still, Polish national interests were protected because the railways' and harbour's administration were subordinated to the Polish Government. The treaty's political and functional details were defined in the Polish-Gdańsk Convention signed in Paris in November 1920. As a result, all military structures on Westerplatte were abandoned. Because of the impediments to guaranteeing Polish privileges in Gdańsk, in the early 1920s, Poland started secretly purchasing plots (O) in the central part of the peninsula [98]. Finally, after many meetings of the League of Nations Council, as a result of the action, the exterritorial supply base for the Polish Army was established. Formally, any military installations in the Free City of Gdańsk/Danzig were banned, and thus the whole complex was officially named a restricted area of military warehouses [99]. It must be mentioned here that the idea of building a separate harbour for the Polish state initially occurred in 1917–1918 when the regency over the occupied Congress Poland was established by two emperors of Germany and Austria-Hungary [100]. What is more, the Prussian military administration considered building a new ammunition basin on Westerplatte in the final years before the First World War. Therefore, neither location of the Polish base nor its military facility was new, because the advantages of Westerplatte had been analysed before [Figure 9].

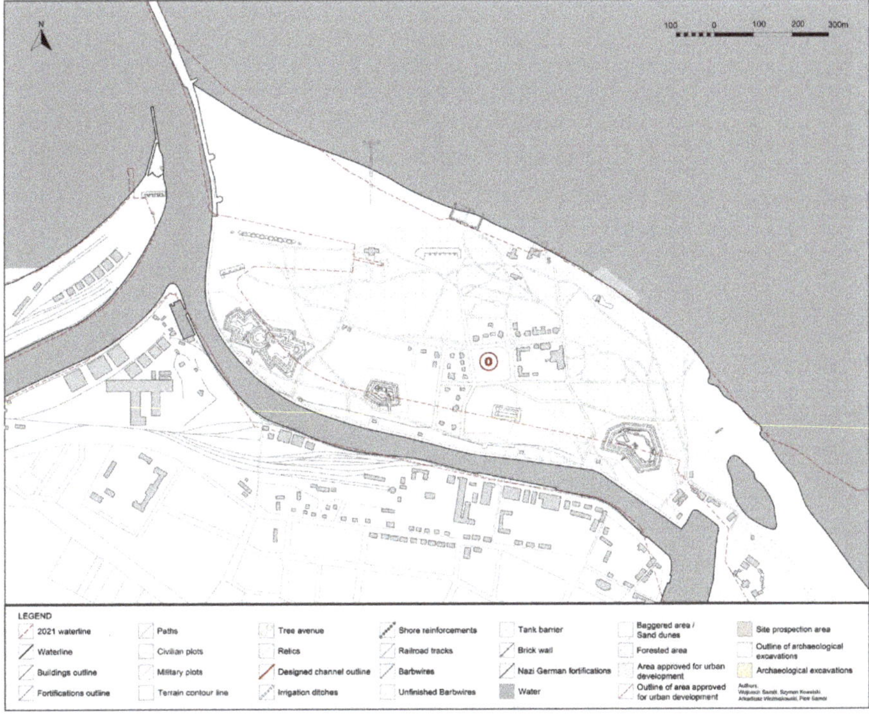

Figure 9. Westerplatte during the First World War (1914–1918).

3.7. *Stage 7: 1926–1939*

Following the Resolution of the League of Nations of 14 March 1924, Westerplatte was designated for storage, loading, unloading and transit of munitions for Poland [Figure 10].

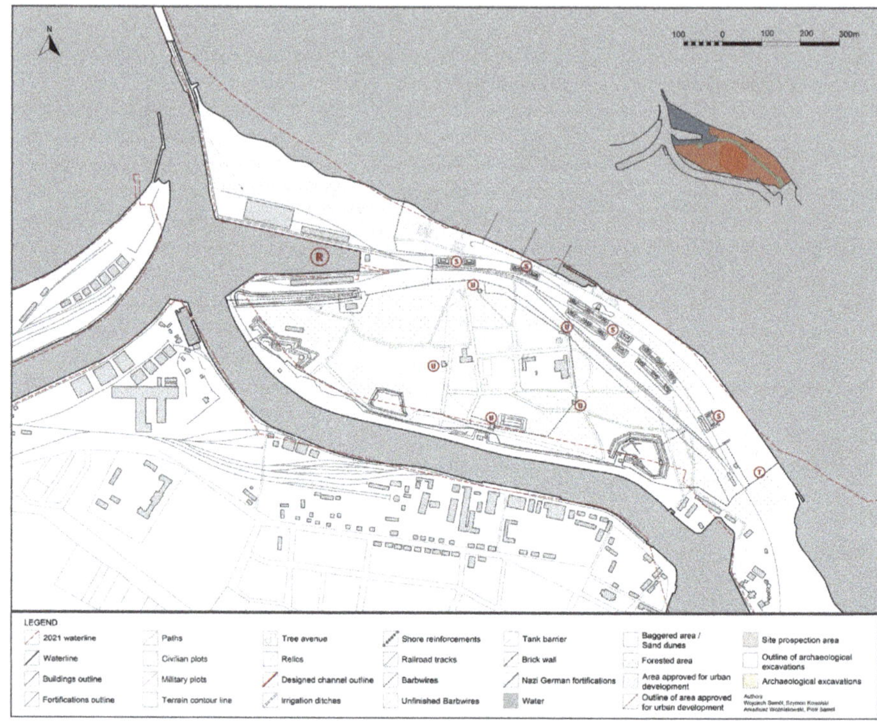

Figure 10. Westerplatte in 1939 as a fortified Polish military transit depot. The top scheme shows three different areas of defence. Blue—port facilities, green—train transit area, red—military area with inner circle perimeter.

The area was approximately 3.5 km in perimeter and covered 60 hectares of land. To provide the basic port infrastructure, the Polish government decided to dig an ammunition basin (about 950 m of coastline) along with installing six bayside cranes and building three ammunition warehouses (R), and nineteen ammunition depots (S) sheltered by earthen embankments [101]. In addition, a part of the earthen fortifications was dismantled and the remaining inhabitants of the peninsula were forced to relocate. An essential element was the location of the railway tracks forming the new port service infrastructure [102–104]. A high border wall (T) was built from the base of the peninsula to the dyke along the southern edge of the canal. The League of Nations limited crew size to 88 soldiers and issued an explicit ban on building new fortifications on the Westerplatte peninsula [105].

At first, the garrison on the peninsula occupied former spa buildings and villas (O). With the growing need to create better living conditions for the soldiers, after the political crisis of 1932, military authorities built new barracks for more than 100 crew members, and constructed five new guardhouses around them (U). This decision was induced by the public mood which in 1933 led to the National Socialists winning the elections [4]. Notification for the initial projects was also sent to the Gdańsk authorities, which initially strongly objected but eventually granted permission to build the facilities [106]. An interesting concept was the secret location of combat shelters under the former spa buildings and in the basements of the new guardhouses which the Germans attacking Westerplatte did not know about. In addition, shortly before the outbreak of World War II, Polish commanders built an additional chain of field fortifications in front of the main guardhouses.

3.8. Stage 8: 1939–1945

Preparations made in the last weeks of August 1939 resulted in a significant increase in the defensive value of the Polish outpost and its effectiveness during the fighting. Thanks to a seven-day defence of the Polish garrison, Westerplatte has gone down in history as a symbolic place regarding the outbreak of World War II. After the capitulation, a process of obliteration of the traces of Polish presence began. The slave labour of hundreds of Polish prisoners, who quickly dismantled the buildings, irreversibly changed this place [2,5]. Most of the buildings in Westerplatte were demolished in 1939/1940 and the building material was used to build the Stutthof concentration camp. Despite cleaning up the area, most of Westerplatte remained undeveloped until the end of the war. Most probably it was connected with unspecified and unrealized plans for a greater extension of the shipyard and the port of Gdansk. The area of almost one square kilometre at the entrance to the port with was certainly foreseen as a backup for the developing Kriegsmarine base. However, until spring 1945 the area of Westerplatte was used only within the ammunition pool. The fighting in March 1945 and the post-war discharge of ammunition and unexploded ordnance in the preserved buildings irreversibly devastated the last of the preserved (large-sized) buildings of the former Polish depot. It can be assumed that these years left their greatest mark on the historical buildings of the former resort and the later buildings of the Polish depot [Figure 11].

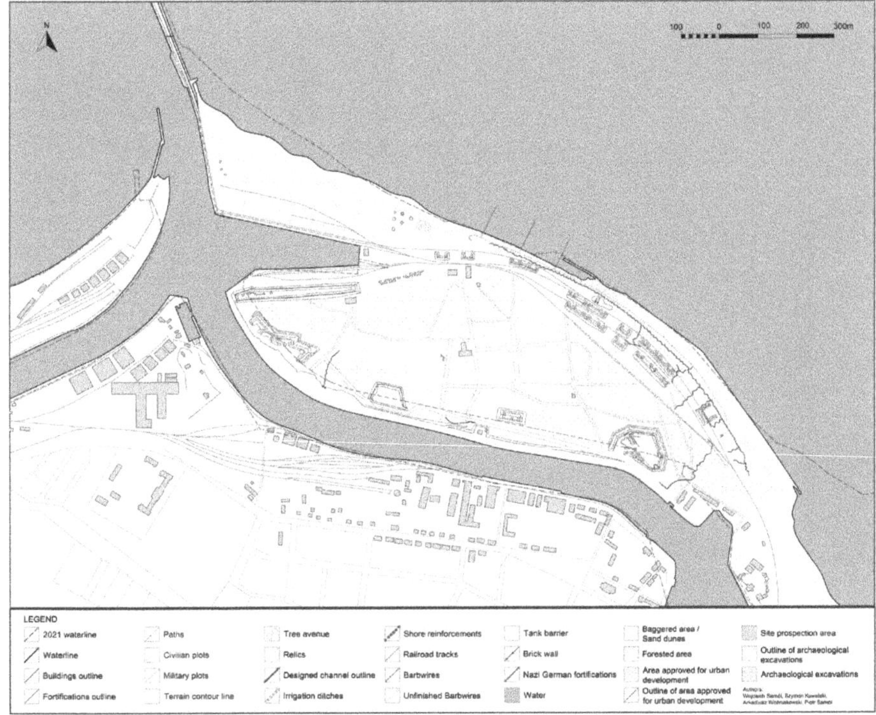

Figure 11. Westerplatte in 1945 with Nazi German Fortifications and dismantled resort facilities.

3.9. Stage 9: 1963–1966

After the end of the hostilities, as during World War II, the area was mainly subject to military investments such as a radio station and a navy observation post, as well as a shooting range and the observation tower of one of the new coastal batteries in the central part of the peninsula. Additionally, a branch of the Naval Shipyard was established within the ammunition basin and damaged quays. The whole area of Westerplatte was again fenced off and access became strictly limited. Despite this, as early as 1946, a symbolic cemetery for the fallen Polish soldiers (W) was created in place of the destroyed guardhouse no. 5 [107]. The subsequent years, during which an interest in the symbol of Westerplatte grew, coincided with the economic development of the People's Republic of Poland along with the city of Gdańsk. For logistical reasons, the port channel had to be widened so that the shipyard and port could handle larger sea vessels (V). That irreversibly destroyed traces of most of the historic structures on the southern and eastern parts of the peninsula, in particular those that were used for defence in 1939, which from today's perspective represented the greatest historical value. The expansion of port infrastructure also contributed much to the degradation of the remaining land. Along with the military area, only a strip in the central part of the peninsula remained partially original [108,109]. For those reasons, from the mid-1960s onwards, the question of how to honour this place and, on the other hand, how to emphasize its historical value began to arise. The answer was the 1963 nationwide architectural competition for the development of a public area in Westerplatte. Although the winning project was not fully implemented, its most characteristic element—a mound with a monument (Y)—was created, and since then has been towering over the entrance to the port. As part of this project, new paths (X) and parking lots (Z) were built, often without any respect to the historical plan of the garrison, which disturbed the image of this historic space, already preserved to a small extent. Moreover, as a result of the transformation most visitors could no longer recognize the values of authenticity and integrity that should have been crucial for a historical site [Figure 12].

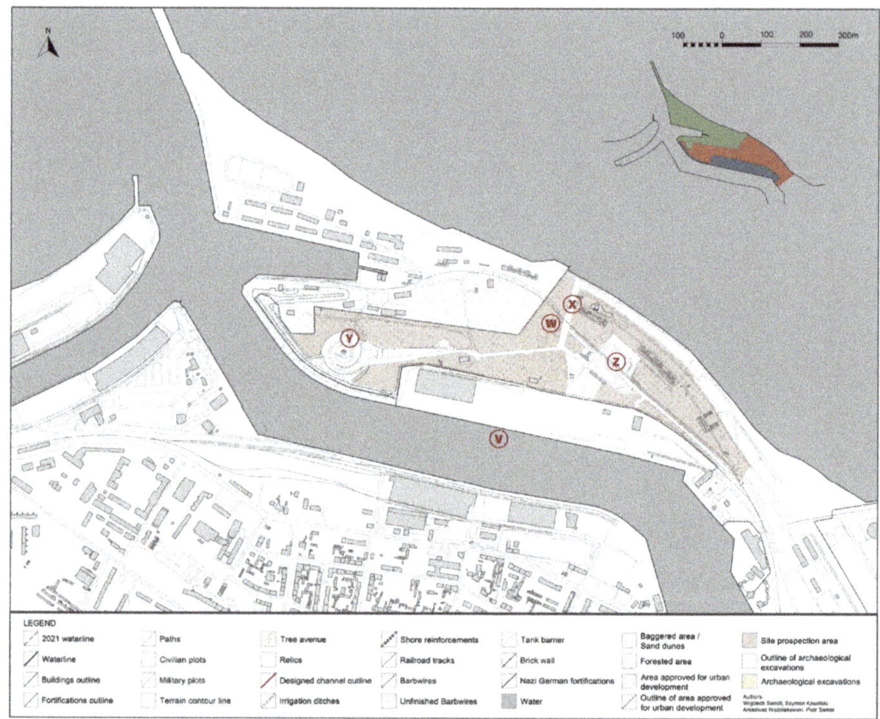

Figure 12. Modern-day Westerplatte with a monumental statue on the mound, container terminal and military area. The map also includes a prospection of the terrain and archaeological excavations. The top scheme shows the division problem—today's Westerplattehas three different land users. Green—Army, red—memorial site, blue—container terminal.

3.10. Perspectives

The future image of Westerplatte is still not determined—the main struggle is to set a strict agenda for the Museum of the Second World War conservation recommendations. However, the most genuine threat to its cultural landscape might come from the current harbour administration which aims to develop a new industrial area located on the waters of the gulf, directly north of the peninsula (including new containers and oil terminals). New visions of Westerplatte itself span from the restitution of the stage on the eve of the war in 1939, which means the reconstruction of the landscape of the military depot, to the modern implementation of new facilities on the peninsula. This article may help to solve the problem of the lack of full understanding of Westerplatte's multi-layered history, which has marked its space and cultural landscape.

4. Discussion

The reconstructed history of Westerplatte shows what a challenge it is to treat the historic landscape in the context of the changes taking place. On the one hand, HUL transformations are, as it were, written into their definition [21]. On the other hand, the peninsula is a special place due to the fact that a large part of it is occupied by a museum. As early as the second half of the 20th century, the concept of an open-air museum was superseded by the "Landschaftmuseum" [110,111]. However, this idea means protecting its current status and inviolability, which can be of limited use in urban areas. The awareness of multitemporal evolutions and the non-linear process of landscape formation should encourage researchers into interdisciplinary documentation of the evidence of its changes. Otherwise, the lack of the stakeholders' respect for the complex history might cause attempts at "musealization" of the space, focusing on one particular period and blurring the others.

The spatial transformation of Westerplatte, therefore, proves that any analysis of the historical landscape must also take into account the fourth dimension—time [19]. The basic problem in the current research concerning battlefields is the spot analysis of the landscape [12–15]. Meanwhile, the presentation of Westerplatte's history utilizing a set of traditional maps or ultimately a virtual map that allows the tracking of the changes throughout a period of over 300 years provides an ideal opportunity to understand the events of the past as a process. Obviously, the detail of this type of consideration depends on the available sources—historical and archaeological (also archaeobotanical) [24–30]. Nevertheless, the use of new technologies and the management of big data collections are always associated with the need to transfer dispersed historical sources to the digital language. The selection of historical primary sources (assessment of their credibility) is always a qualitative and not quantitative activity [41–43]. Thus, it cannot be easily automated because historical landscape research depends on the accessibility and representativeness of these sources. s. This is probably the reason that many studies on multitemporal landscape transformations have been focused on the last century [21,112,113].

Comparing the air photos, geodesic plans and hand-written maps brings with it the crucial problem of representativeness of evidence in such a research activity. Numerous layers and complicated archival material have showed the importance of in situ research for the verification and supplementing of sources. The stratification (stratigraphy) of the peninsula's plan (including individual architectural structures, e.g., barracks, casinos, officers' villas) is—due to so many transformations—not only a way of documenting the history of the place but also a supplementary research method [114]. Certainly, the next step in presenting knowledge about history (including new research findings) may be the use of virtual reality techniques. However, such actions must be based on properly selected, analysed and synthesized sources.

A separate problem is the apparent interdisciplinarity of scientific research. Most of the older literature on Westerplatte's history [2–6] shows only selected aspects of transformations of the historical landscape. For this reason, literature studies are only an introduction to the research and will not replace the laborious collection of archival data. Once developed, however, they become an open resource—enabling the development of comparative studies in the future. According to the authors, the key to the success of collecting archival data and using new methods (GIS, LIDAR) is consistency in tracking the entire historical process [17,18,23,24,26–28,37]. On a greater scale, the application of geomorphologic studies is crucial [9–12,31–34], but they must be compared with the traditional historical primary sources.

Following the modern tendencies in humanities, some geographers have proposed concepts of narratives and storytelling as a new approach towards comprehensive landscape studies [112,115,116]. They are usually focused on using and testing new methods of presenting the history of landscape (e.g., virtual reality, extended reality, etc.), instead of analysing HUL itself [115]. Moreover, the popularity of landscape research and the polyphony of methods used in various scientific disciplines often do not contribute to expanding the evidence base but only to the reinterpretation of the traditional historical narrative. On the other hand, the scale of research also matters. In micro-scale (e.g., Westerplatte) analysis, operating on the best possible reference material is essential, making it often only the interdisciplinary use of available historical maps that can make progress.

5. Conclusions

The results described above show how multifaceted the analysis of the historical landscape evolution is. Westerplatte—a natural sandbank at the mouth of the river, transformed into a peninsula; the subject of temporary and more permanent military investments for 150 years; one of the seaside resorts and an unfinished villa district—all these elements of the landscape were part of the reality that became the battlefield in 1939. Later, political and economic decisions gradually began to blur the symbolic space. Paradoxically, when large financial efforts were finally made to establish a cultural park themed around the battle, the ad hoc expansion of the port area made implementing these plans difficult and sometimes impossible. Moreover, part of Westerplatte is still an important, inaccessible military area, which is the continuation of this place's functions from the beginning of its existence.

The main impact of the detailed analyses presented above is to highlight the process of joining different types of data. Although archaeological, archaeobotanical or geomorphological data have to be elaborated by a specialist from the appropriate field of science, there should be no doubt that all of them—because of the purpose of use—should be analysed as the primary historical sources for the multitemporal landscape evolution.

Moreover, the role of time as a factor in shaping the landscape is surprisingly rarely addressed. At Westerplatte, each of the presented periods was characterized by different vegetation, development and surroundings. It should be emphasized that the scope of the study of four hundred years of transformation differs from geomorphological studies beyond historical cognition (written sources, archaeology of civilizations, etc.). Although historical politics influence the emphasis on one particular period—which was pointed out in the title of this article (the symbolic place where World War II broke out in Europe)—it does not mean, or at least should not, to cross out the four centuries of history of the Westerplatte peninsula.

Author Contributions: Conceptualization, W.S., S.K., A.W. and P.S.; methodology, W.S. and P.S.; software, A.W. and S.K.; investigation, W.S. and A.W.; resources, W.S., S.K. and A.W.; data curation, W.S.; writing—original draft preparation, W.S. and P.S.; writing—review and editing, W.S., S.K., A.W. and P.S.; visualization, S.K.; supervision, P.S. All authors have read and agreed to the published version of the manuscript.

Funding: This research received no external funding. The APC was partially funded by the Gdańsk University of Technology, Faculty of Architecture.

Data Availability Statement: Not applicable.

Conflicts of Interest: The authors declare no conflict of interest.

References

1. Davis, N. No Simple Victory. In *World War II in Europe 1939–1945*; Penguin Books: New York, NY, USA, 2008.
2. Biernat, C.; Cieślak, E. *History of Gdansk*; Wydawnictwo Morskie: Gdańsk, Poland, 1995.
3. Drzycimski, A. *Westerplatte 1939. Przed Szturmem*; Słowo/ObrazTerytoria: Gdańsk, Poland, 2009.
4. Drzycimski, A. *Westerplatte: Reduta w Budowie*; Fundacja Gdańska: Gdańsk, Poland, 2014.
5. Drzycimski, A. *Westerplatte: Reduta Wojenna*; Fundacja Gdańska: Gdańsk, Poland, 2014.
6. Drzycimski, A. *Westerplatte. Kulisy Niemieckiej Wojny Propagandowej*; Muzeum Gdańska: Gdańsk, Poland, 2019.
7. Błyskosz, T. Westerplatte—Przekształcenia Pola Bitwy Po 1945 r. (Do Początku XXI w.). In *Materialne Pozostałości Konfliktów i Zbrodni XX Wieku w Świetle Najnowszych Badań Archeologicznych*; Mik, H., Węglińska, W., Eds.; Muzeum Drugiej Wojny Światowej: Gdańsk, Poland, 2019; pp. 342–362.
8. Snyder, T. Poland vs. History. *The New York Review of Books*, 3 May 2016. Available online: https://www.nybooks.com/online/2016/05/03/poland-vs-history-museum-gdansk/ (accessed on 31 August 2022).
9. Brandolini, P.; Cappadonia, C.; Luberti, G.M.; Donadio, C.; Stamatopoulos, L.; Di Maggio, C.; Faccini, F.; Corrado, S.; Vergari, F.; Paliaga, G.; et al. Geomorphology of Anthropocene in Miditerranean urban areas. *Prog. Phys. Geogr. Earth Environ.* **2019**, *44*, 461–494. [CrossRef]
10. Del Monte, G.; D'Orefice, M.; Luberti, G.M.; Marini, G.; Pica, A.; Vergari, F. Geomorphological classification of urban landscapes: The case study of Rome (Italy). *J. Maps* **2016**, *12*, 178–189. [CrossRef]
11. Zwoliński, Z.; Hildebrandt-Radke, I.; Mazurek, M.; Makohonienko, M. Anthropomorphological metamorphosis of an urban area in the Postglacial landscape: A case study of Poznań city. In *Urban Geomorphology, Landforms and Process in the Cities*; Thornbushand, M.J., Allen, C.D., Eds.; Elsevier: Amsterdam, The Netherlands, 2018; pp. 55–77.
12. Šantrůčková, M.; Salašová, A.; Sokolová, K.; Sedlacek, J. Mapping military landscape as a cultural heritage: Case study of the Austerlitz/Slavkov battlefield site. *AUC Geogr.* **2020**, *55*, 66–76. [CrossRef]
13. De Matos-Machado, R.; Arnaud-Fassetta, G.; Batard, F.; Bilodeau, C.; Jacquemot, S.; Amat, J.-P. Warfare as a new field of study in archaeo-geomorphology: The case of the battlefield of Verdun (France). In Proceedings of the 6th Young Geomorphologists' Day the Geomorphology for Society from Risk Knowledge to Landscape Heritage, Cagiliari, Italy, 28–30 September 2015. Available online: https://www.academia.edu/32801630/Warfare_as_a_new_field_of_study_in_archaeo_geomorphology_The_case_of_the_battlefield_of_Verdun_France_ (accessed on 21 December 2022).
14. Uribe, P.; Angás, J.; Romeo, F.; Pérez-Cabello, F.; Santamaría, D. Mapping Ancient Battlefields in a multi-scalar approach combining Drone Imagery and Geophysical Surveys: The Roman siege of the oppidum of Cabezo de Alcalá (Azaila, Spain). *J. Cult. Herit.* **2021**, *48*, 11–23. [CrossRef]
15. Gheyle, W.; Stichelbaut, B.; Saey, T.; Note, N.; Van den Berghe, H.; Van Eetvelde, V.; Van Meirvenne, M.; Bourgeois, J. Scratching the surface of war. Airborne laser scans of the Great War conflict landscape in Flanders (Belgium). *Appl. Geogr.* **2018**, *90*, 55–68. [CrossRef]
16. Dziewanowski, A.; Kuczma, F.; Samól, W. BadaniaArcheologiczne. Pola Bitwy Na Westerplatte. WynikiPracProwadzonych w Latach 2016–2017. In *Materialne Pozostałości Konfliktów i Zbrodni XX Wieku w Świetle Najnowszych Badań Archeologicznych*; Mik, H., Węglińska, W., Eds.; Muzeum Drugiej Wojny Światowej: Gdańsk, Poland, 2019; pp. 235–274.
17. Spennemann, D.H.R.; Poynter, C. Using 3D Spatial Visualisation to Interpret the Coverage of Anti-Aircraft Batteries on a World War II Battlefield. *Heritage* **2019**, *2*, 2457–2479. [CrossRef]
18. Briaud, J.-L. Normandy cliff stability: Analysis and repair. In *Geomechanics and Geodynamics of Rock Masses, Proceedings of the 2018 European Rock Mechanics Symposium, St. Petersburg, Russia, 22–26 May 2018*; Litvinenko, V., Ed.; CRC Press: London, UK, 2019; pp. 611–616.
19. Altizer, W.E. Time Perspectivism, Temporal Dynamics, and Battlefield Archaeology: A Case Study from the Santiago Campaign of 1898. *Neb. Anthropol.* **2008**, *36*, 62–79.
20. Bandarin, F.; Van Oers, R. (Eds.) *Reconnecting the City: The Historic Urban Landscape Approach and the Future of Urban Heritage*; John Wiley and Sons: Oxford, UK, 2015.
21. Taylor, K. The Historic Urban Landscape Paradigm and Cities as Cultural Landscapes. Challenging Orthodoxy in Urban Conservation. *Landsc. Res.* **2016**, *41*, 471–480. [CrossRef]
22. Giliberto, F.; Appendino, F. Handling change in historic urban landscapes: An analysis of urban heritage conservation approaches in Bordeaux (France), Edinburgh (UK) and Florence (Italy). In *Landscape as Heritage. International Critical Perspectives*; Pettenati, G., Ed.; Routledge: Abingdon, UK; New York, NY, USA, 2023; pp. 216–230.
23. Balletti, C. Georeference in the Analysis of the Geometric Content of Early Maps. *E-Perimetron* **2006**, *1*, 32–42.
24. Huffman, K.L.; Giordano, A.; Bruzelius, C. (Eds.) *Visualizing Venice*; Routledge: New York, NY, USA, 2017.
25. Bruzelius, C.; Tronzo, W. *Medieval Naples*; Italica Press: New York, NY, USA, 2011.
26. Mapping Historical New York City A Collaboration to Map Immigration and Neighbourhood Change in New York City during the Late Nineteenth and Early Twentieth Centuries. 2022. Available online: https://c4sr.columbia.edu/Projects/Mapping-Historical-New-York-City (accessed on 20 November 2022).
27. Layers of London Mapping Project. Available online: https://www.layersoflondon.org/ (accessed on 20 November 2022).
28. Spatial History of San Francisco. Available online: https://web.stanford.edu/Group/Spatialhistory/Cgi-Bin/Site/Project.Php?Id=1141 (accessed on 21 November 2022).

29. Siebert, L. Using GIS to Document, Visualize, and Interpret Tokyo's Spatial History. *Soc. Sci. Hist.* **2000**, *24*, 537–574.
30. Opll, F. The European Atlas of Historic Towns. Project, Vision, Achievements. *Ler História* **2011**, *60*, 169–182. [CrossRef]
31. Bellotti, P.; Davoli, L. Landscape Diachronic Recontruction in the Tiber Delta during historical time: A holistic approach. *Geogr. Din. Quat.* **2018**, *2*, 3–21.
32. Hujizenveld, A.A.; Bellotti, P.; Gisotti, G. Alle foci del Tevere: Territorio, Storia, Attualita. *Geol. Dell'Ambiente* **2019**, *3*, 1–48.
33. Jegliński, W. Rozwój wybrzeża Zatoki Gdańskiej w rejonieujścia Wisły Martwej. *Przegląd Geol.* **2013**, *61*, 587–595.
34. Przezdziecki, P.; Uścinowicz, S.; Jegliński, W. The geological structure and evolution of the area around the Copper Ship. In *The Copper Ship, a Medieval Shipwreck and Its Cargo*; Centralne Muzeum Morskie: Gdańsk, Poland, 2014; pp. 161–176.
35. Earley-Spadoni, T. Spatial History, Deep Mapping and Digital Storytelling: Archaeology's Future Imagined through an Engagement with the Digital Humanities. *J. Archaeol. Sci.* **2017**, *84*, 95–102. [CrossRef]
36. Riding, T. "Making Bombay Island": Land Reclamation and Geographical Conceptions of Bombay, 1661–1728. *J. Hist. Geogr.* **2018**, *59*, 27–39. [CrossRef]
37. Spatial History Project. 2022. Available online: https://web.stanford.edu/Group/Spatialhistory/Cgi-Bin/Site/Project.Php (accessed on 18 August 2022).
38. Vegari, F.; Luberti, G.M.; Pica, A.; Del Monte, M. Geomorphology of the historic centre of the Urbs (Rome, Italy). *J. Maps* **2021**, *17*, 6–17. [CrossRef]
39. Logemann, D. On "Polish History": Disputes over the Museum of the Second World War in Gdańsk. *Cultures of History Forum*, 21 March 2017. Available online: https://digital.herder-institut.de/publications/frontdoor/index/index/searchtype/collection/id/16274/start/6/rows/100/docId/95 (accessed on 20 October 2022).
40. Siddi, M.; Gaweda, B. The National Agents of Transnational Memory and Their Limits: The Case of the Museum of the Second World War in Gdańsk. *J. Contemp. Eur. Stud.* **2019**, *27*, 258–271. [CrossRef]
41. Hershberg, T. The New Urban History. Toward an Interdisciplinary History of the City. *J. Urban Hist.* **1978**, *5*, 3–40. [CrossRef]
42. Elias, A.J. Defining Spatial History in Postmodernist Historical Novels. In *Narrative Turns in Minor Genres in Postmodernist*; D'haen, T., Bertens, H., Eds.; Rodopi: Amsterdam, Atlanta, GA, USA, 1995; pp. 105–114.
43. White, R. What Is Spatial History, Spatial History Lab. 1 February 2010. Available online: https://web.stanford.edu/group/spatialhistory/cgi-bin/site/pub.php?id=29 (accessed on 1 September 2022).
44. Signature RG373 N/452 F:2023. National Archives and Records Administration: College Park, MD, USA.
45. Signature RG373 N/859 F:5103. National Archives and Records Administration: College Park, MD, USA.
46. Signature RG373 106W/24 F: 4084. National Archives and Records Administration: College Park, MD, USA.
47. Signature RG373 106G/962 F:4044. National Archives and Records Administration: College Park, MD, USA.
48. Signature RG373 106G/5095 F:3063. National Archives and Records Administration: College Park, MD, USA.
49. Zapłata, R.; Różnicki, S. Historic Aerial Photographs in the Analysis of Cultural Landscape—Case Studies from Poland. In *Recovering. Lost Landscapes*; Ivanisevic, V., Veljanowski, T., Cowley, D., Kiarszys, G., Bugarski, I., Eds.; Institute of Archaeology: Belgrade, Serbia, 2015; pp. 107–115.
50. Różycki, S.; Osińska-Skotak, K.; Świątek, A. *Zdjęcia Lotnicze Polski z Okresu II Wojny Światowej*; Oficyna Wydawnicza Politechniki Warszawskiej: Warszawa, Poland, 2020.
51. Kuna, J. The Orthophotomap of Lublin 1944: From Luftwaffe photographs to map application—idea, methods, contemporary challenges of processing and publishing archival aerial photographs. *Pol. Cartogr. Rev.* **2022**, *54*, 123–142. [CrossRef]
52. El-Hussainy, M.; Moustafa, A.; Baraka, A.; El-Hallaq, M.A. A Methodology for Image Matching of Historical Maps. *E-Perimetron* **2011**, *6*, 7795.
53. Samól, P.; Hirsch, R.; Woźniakowski, A. History of the Lighthouse of the Wisłoujście Fortress in Light of a 2018 Architectural Study. *Wiad. Konserw. J. Herit. Coservat.* **2021**, *66*, 21–36.
54. Signature 8109,16. Architectural Museum of the Technical University of Berlin: Berlin, Germany.
55. Signature 300 MP/266. State Archive: Gdansk, Poland.
56. Signature 300 MP/1248. State Archive: Gdansk, Poland.
57. Paul, D. *Liverpool Docks: A Short History*; Fonthill Media: Stroud, UK, 2016.
58. Rule, F. *London's Docklands, A History of the Lost Quarter*; Ian Allan Publishing: Shepperton, UK, 2012.
59. Behrendt, S.D.; Hurley, R.B. Liverpool as a Trading Port: Sailors' Residences, African Migrants, Occupational Change and Probated Wealth. *Int. J. Marit. Hist.* **2017**, *29*, 875–910. [CrossRef]
60. Signature 300 MP/898. State Archive: Gdansk, Poland.
61. Bukal, G. *Fortyfikacje Gdańska i Ujścia Wisły 1454–1793*; Bukal: Sopot, Poland, 2012.
62. Podruczny, G. *Twierdze z Papieru, Fortyfikacje Pruskie w Latach 1786–1807*; Avalon: Warszawa, Poland, 2020.
63. Signature300 MP/613. State Archive: Gdansk, Poland.
64. Signature 9, 2/914. State Archive: Gdansk, Poland.
65. Signature 300 MP/3. State Archive: Gdansk, Poland.
66. Signature 9, 2/720. State Archive: Gdansk, Poland.
67. Signature 9, 2/60. State Archive: Gdansk, Poland.
68. Signature 9, 2/1641. State Archive: Gdansk, Poland.
69. Signature 9, 2/745. State Archive: Gdansk, Poland.

70. Signature XI. HA, FpK, B 70101. Geheimes Staatsarchiv Preussischer Kulturbesitz: Berlin, Germany.
71. Signature, 1121/1. State Archive: Gdansk, Poland.
72. Signature, 1121/199. State Archive: Gdansk, Poland.
73. Signature, 1121/202. State Archive: Gdansk, Poland.
74. Dargacz, J. *Od Sopotu Po Stogi. Początki Kąpielisk Morskich w Okolicach Gdańska (1800–1870)*; Muzeum Gdańska: Gdańsk, Poland, 2020.
75. Signature 1121/177. State Archive: Gdansk, Poland.
76. Signature 9,2/917. State Archive: Gdansk, Poland.
77. Signature XI. HA, FpK, C 70718. Geheimes Staatsarchiv Preussischer Kulturbesitz: Berlin, Germany.
78. Signature XI. HA, FpK, E 71251, Bl. 8-9. Geheimes Staatsarchiv Preussischer Kulturbesitz: Berlin, Germany.
79. Woźniakowski, A. The History and Material Relics of the Westerplatte Fortifications from before 1920. In *MaterialnePozostałościKonfliktówiZbrodni XX Wieku w ŚwietleNajnowszychBadańArcheologicznych*; Mik, H., Węglińska, W., Eds.; Muzeum Drugiej Wojny Światowej: Gdańsk, Poland, 2019; pp. 276–302.
80. Signature 1121/182. State Archive: Gdansk, Poland.
81. Signature 1121/178. State Archive: Gdansk, Poland.
82. Rolf, R. Die Entwicklung des Deutschen Festungssystems Seit 1870. In *Vollständige und Bearbeitete Ausgabe des Manuskriptes*; Fortress Books: Borger-Odoorn, The Netherlands, 2000.
83. Signature XI. HA, FpK, B 70826. Geheimes Staatsarchiv Preussischer Kulturbesitz: Berlin, Germany.
84. Signature XI. HA, FpK, E 71218. Geheimes Staatsarchiv Preussischer Kulturbesitz: Berlin, Germany.
85. Signature 1121/242. State Archive: Gdansk, Poland.
86. Signature 1121/162. State Archive: Gdansk, Poland.
87. Signature 1121/172. State Archive: Gdansk, Poland.
88. Omilanowska, M. DefortyfikacjaGdańska Na TlePrzekształceńMiastNiemieckich w XIX Wieku. *Biul. Hist. Szt.* **2010**, *72*, 293–334.
89. Signature 1121/249. State Archive: Gdansk, Poland.
90. Signature XI. HA, FpK, G 70471. Geheimes Staatsarchiv Preussischer Kulturbesitz: Berlin, Germany.
91. Signature XI. HA, FpK, G 70447. Geheimes Staatsarchiv Preussischer Kulturbesitz: Berlin, Germany.
92. Signature XI. HA, FpK, A 70834. Geheimes Staatsarchiv Preussischer Kulturbesitz: Berlin, Germany.
93. Signature XI. HA, FpK, A 70836. Geheimes Staatsarchiv Preussischer Kulturbesitz: Berlin, Germany.
94. Signature BZ-I 06,039. Architectural Museum of the Technical University of Berlin: Berlin, Germany.
95. Signature BZ-I 07,043. Architectural Museum of the Technical University of Berlin: Berlin, Germany.
96. Herdy, D. *From Garden Cities to New Towns. Campaigning for Town and Country Planning 1899–1946*; Chapman and Hall: London, UK, 1991.
97. Rolf, R. *Die Entwicklung des Deutschen Festungssystems*; Fortress Books: Borger-Odoorn, The Netherlands, 2000.
98. Jałowiecki, M. *WolneMiasto*; SpółdzielniaWydawniczaCzytelnik: Warszawa, Poland, 2002.
99. Signature 1027/411. State Archive: Gdansk, Poland.
100. Daniluk, J. Danzig, Fiume, Memel—Concept of the Free Cities after World War Iand the Principle of National Self-Determination and Protection of National Minorities. *Historie. Jahrb. Zent. Hist. Forsch. Berl. Pol. Akad. Wiss.* **2020**, *13*, 92–114.
101. Signature 1126/433. State Archive: Gdansk, Poland.
102. Signature 1081/3256. State Archive: Gdansk, Poland.
103. Signature 1081/1377. State Archive: Gdansk, Poland.
104. Signature 1081/1475. State Archive: Gdansk, Poland.
105. Signature I.300.63.236. Central Military Archive: Warsaw, Poland.
106. RM 20/1967 p.8. Federal Archive—Military Archive: Freiburg im Breisgau, Germany.
107. Zajączkowski, K. *Westerplatte Jako Miejsce Pamięci 1945–1989*; Instytut Pamięci Narodowej: Warszawa, Poland, 2015.
108. Signature 1153/5216. State Archive: Gdansk, Poland.
109. Signature 1153/5217. State Archive: Gdansk, Poland.
110. Howard, P. The eco-museums: Innovation that Risks the Future. *Int. J. Herit. Stud.* **2002**, *8*, 63–72. [CrossRef]
111. Sturani, M.L. Landscape as heritage in museums. A critical Appraisal of past and present experiences. In *Landscape as Heritage. International Critical Perspectives*; Pettenati, G., Ed.; Routledge: Abingdon, UK; New York, NY, USA, 2023; pp. 289–300.
112. Davies, D.K. Reading Landscapes and Telling Stories: Geography, the Humanities and the Environmental History. In *Envisioning Landscapes, Making Worlds: Geography and the Humanities*; Daniels, S., DeLyser, D., Entrikin, J.N., Richardson, D., Eds.; Routledge: Abingdon, UK, 2011; pp. 170–176.
113. Mendoza, J.E.; Etter, A. Multitemporal Analysis (1940–1996) of land cover changes in the southwestern Bogota highplain (Comobia). *Landsc. Urban Plan.* **2002**, *59*, 147–158. [CrossRef]
114. San-Antonio-Gómez, C.; Velilla, C.; Manzano-Agugliaro, F. Urban and landscape changes through historical maps: The Real Sitio of Aranjuez (1775–2005), a case study. *Comput. Environ. Urban Syst.* **2014**, *44*, 47–58. [CrossRef]

115. Goldstein, B.E.; Wessells, A.T.; Lejano, R.; Butler, W. Narrating Resilience: Transforming Urban Systems Though Collaborative Storrytelling. *Urban Stud.* **2015**, *52*, 1285–1303. [CrossRef]
116. Albuquerque, A.R.; Betelho, M.L.; Crozat, D. Storytelling and online media as narrative practices for engaging with Historical Urban Landscapes (HUL). The case study of Porto, Portugal. In *Landscape as Heritage. International Critical Perspectives*; Pettenati, G., Ed.; Routledge: Abingdon, UK; New York, NY, USA, 2023; pp. 67–79.

 land

Communication

Erosion Modelling Indicates a Decrease in Erosion Susceptibility of Historic Ridge and Furrow Fields near Albershausen, Southern Germany

Johannes Schmidt [1,2,*,†], Nik Usmar [1,†], Leon Westphal [1,†], Max Werner [1,†], Stephan Roller [3], Reinhard Rademacher [4], Peter Kühn [5], Lukas Werther [3] and Aline Kottmann [6]

1 Institute of Geography, Leipzig University, 04103 Leipzig, Germany
2 Historic Anthropospheres Working Group, Leipzig Lab, Leipzig University, 04107 Leipzig, Germany
3 Institute of Prehistory, Early History and Medieval Archaeology, Eberhard-Karls-Universität, 72070 Tübingen, Germany
4 County Archaeology, District Administration, 73008 Göppingen, Germany
5 Research Area Geography, Soil Science and Geomorphology, Eberhard-Karls-Universität, 72070 Tübingen, Germany
6 Archaeology, State Office for Cultural Heritage Baden-Wuerttemberg, 73728 Esslingen, Germany
* Correspondence: j.schmidt@uni-leipzig.de
† These authors contributed equally to this work.

Abstract: Ridge and furrow fields are land-use-related surface structures that are widespread in Europe and represent a geomorphological key signature of the Anthropocene. Previous research has identified various reasons for the intentional and unintentional formation of these structures, such as the use of a mouldboard plough, soil improvement and drainage. We used GIS-based quantitative erosion modelling according to the Universal Soil Loss Equation (USLE) to calculate the erosion susceptibility of a selected study area in Southern Germany. We compared the calculated erosion susceptibility for two scenarios: (1) the present topography with ridges and furrows and (2) the smoothed topography without ridges and furrows. The ridges and furrows for the studied site reduce the erosion susceptibility by more than 50% compared to the smoothed surface. Thus, for the first time, we were able to identify lower soil erosion susceptibility as one of the possible causes for the formation of ridge and furrow fields. Finally, our communication paper points to future perspectives of quantitative analyses of historical soil erosion.

Keywords: historic soil erosion; USLE; Anthropocene; Archaeology; Wölbäcker; ridge and furrow; historic land use; GIS; erosion management; historic anthroposphere

Citation: Schmidt, J.; Usmar, N.; Westphal, L.; Werner, M.; Roller, S.; Rademacher, R.; Kühn, P.; Werther, L.; Kottmann, A. Erosion Modelling Indicates a Decrease in Erosion Susceptibility of Historic Ridge and Furrow Fields near Albershausen, Southern Germany. *Land* **2023**, *12*, 544. https://doi.org/10.3390/land12030544

Academic Editors: Bellotti Piero and Alessia Pica

Received: 25 January 2023
Revised: 15 February 2023
Accepted: 20 February 2023
Published: 23 February 2023

1. Introduction

Erosion caused by surface runoff leads to a serious loss of soil based ecosystem services. Areas of lower historical soil erosion are valuable areas of landscape and biodiversity resilience [1]. In the course of the current climate change, especially due to changing precipitation patterns and intensive land use, the average soil loss in Central Europe is expected to increase significantly [2,3]. The question of historical cultivation forms and their effects on soil erosion has so far been answered mainly qualitatively [4–7]. Models that simulate erosion quantitatively have existed for some time with the introduction of the USLE (Universal Soil Loss Equation) by Wischmeier and Smith [8] and the German adaption ABAG (*Allgemeine Bodenabtragsgleichung*) [9]. Beside actualistic studies [10–12], an application of such a concept to historical geomorphological conditions are rare [13,14].

As land use and topographic properties are crucial factors estimating soil erosion [15], the imprint of historical cultivation techniques in topographic properties are still important factors. In several areas of Germany [16–19], but also in other parts of Europe and beyond [6,18,20–22], large parts of former ridge and furrow field systems are still preserved. The most evident structures are ridge and furrow fields, which consist of several parallel ridge structures of considerable length (up to 700 m) and relatively small width (about 5–20 m), separated from each other by rather small furrows. They are clearly visible in LiDAR (Light Detection and Ranging) images [16,18] and characterise large parts of present-day meadow landscapes, also on low mountain slopes, where they have been well preserved through extensive use or afforestation after these fields were abandoned [23]. Due to their longevity and enormous spatial extent, they can be considered as a geomorphological key signature of the Anthropocene [24].

Origins and reasons for the establishment of ridge and furrow fields in low mountain ranges as well as path dependencies to the present appearance of the landscape are the focus of a collaborative research project in the district of Göppingen [25]. So far, various explanations for their establishment have been discussed in the literature: A common view sees a causal link between the mouldboard plough with fixed, non-turning ploughshare and long strip fields: it is argued that the fact that this type of plough led to an accumulation of clods of soil in the middle of the plots and the fact that the turning manoeuvres at the end of the field required a lot of energy, led to an elongated shape of the plots that required fewer trips for each owner [26,27]. Apart from the fact that ridge and furrow fields can also be formed by other types of soil tillage, additional causes are also discussed such as the function as boundary markers within open field systems [26], the improvement of water management or at least the increase of internal hydrological variability to secure crop yields [27] and the accumulation of organic fertiliser [21,28].

Therefore, we present a case study, where historic ridge and furrow fields are still visible in a loess landscape with high soil fertility. We focused on a field system with well-preserved ridge and furrow fields in the municipality of Albershausen, southern Germany (in total, c. 13 % of the area is still covered with remains of ridge and furrow structures). Many of the ridge and furrow fields there were oriented perpendicular to the slope, raising the question of whether this orientation was deliberately chosen to reduce erosion of the vital soil and humus cover and/or to support the run-off of excess rainwater. To answer this question, we used a GIS-based USLE approach to model and compare the erosion susceptibility using a digital elevation model of the present topography and a smoothed topography without ridge and furrow structures.

2. Materials and Methods

2.1. Study Area

The plot *Höfelbett* is located northwest to the village of Albershausen on the fringe of the Swabian Karst. It has been mentioned in 1476/77 for the first time in a stock book [29] and is nowadays used as an extensive orchard, which helped the old field system to survive in a well preserved and visible way up to today (Figure 1). Albershausen was first mentioned in 1275 in the *liber decimationis* [29] but may be of older, presumably early medieval origin due to its characteristic place name suffix-*hausen* [30]. One of the questions asked was how long has the field system been in use for. Today the region is characterised by a Cfb climate with 1058 mm mean annual precipitation and a mean annual temperature of 9.3 °C. The wettest months are in spring and summer. Heavy rainfall is also most frequent during this period [31,32]. The geology of the *Höfelbett* consists of a pisolith-rich claystone (Obtusus Clay Formation) on the upper slope and sandstone (Angulate Sandstone Formation, Lower Jurassic) on the middle to lower slope [33]. Both formations are covered by a lower periglacial layer (mainly clay loam with intercalated sandstone fragments) and an upper periglacial layer consisting mainly of loess loam with a few small sandstone fragments and pisoliths. Eroded Luvisols frequently occur in areas with thicker loess loam, while stagnic Luvisols and Vertisols are developed in more clayey and Cambisols in more sandy substrates. The boundaries of the individual fields and today's parcels are identical to the field boundaries from the original cadastre of 1828 [34]. Their size varies considerably with lengths between 86 and nearly 400 m and widths between 8 and 20 m.

The individual ridges are still between 25 and 80 cm high (Figure 1). The plot *Höfelbett* is situated at a North-East facing slope with a mean slope angle of c. 5° and cut off on the north by a narrow valley leading east. Between 2020 and 2022 five archaeological trenches were opened as accompanying measures of the development of the site. The soils were described according to WRB [35] and samples were taken for dating, soil chemical and soil physical analyses as well as for archaeobotanical and ancient DNA analyses [25].

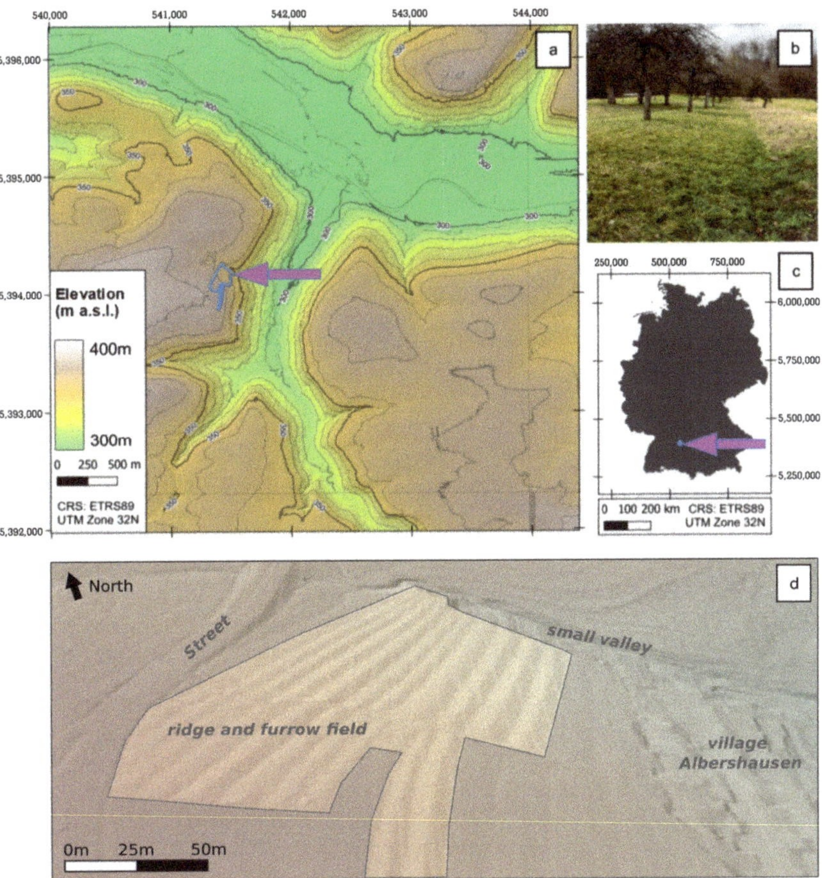

Figure 1. (**a**) Topographical map of the area surrounding the plot "Höfelbett" in Albershausen based on LiDAR DEM data (©www.lgl-bw.de (accessed on 24 January 2023)) [36]. The blue line frames the plot *Höfelbett*. (**b**) Picture of the plot with visible ridge- and furrow structures. (**c**) Location of the study area within Germany. (**d**) 3D view of the study area (elevation and hillshade) based on LiDAR DEM data (©www.lgl-bw.de (accessed on 24 January 2023)).

2.2. Data

To model soil erosion, the USLE requires four factors to be multiplied for any given point in the study area (Table 1): slope gradient and length (LS-factor), soil erodibility (K-factor), rainfall erosivity (R-factor) and land cover (C-factor) [8]. To calculate the LS-factor, we used a LiDAR DEM provided by LGL (*Landesamt für Geoinformation und Landesentwicklung Baden-Württemberg*) [36]. This DEM has a spatial resolution of 0.5 m and is therefore suitable to accurately depict the ridge and furrow structures in the study area [16,18]. For the second factor of the USLE, R-factor, we used a data set from the German Weather Service (DWD - *Deutscher Wetterdienst*), which was derived from radar measurements of precipitation. This dataset has a spatial resolution of 1000 m due to the low variability of rainfall erosivity on a local scale [37,38]. The complete study area lies within a single cell of this data set, which appoints a value of $R = 104.497864915$ kJ/m^2/mm. The K-factor data set has been created by the German Federal Institute for Geosciences and Natural Resources (BGR - *Bundesanstalt für Geowissenschaften und Rohstoffe*) based on their land use stratified soil map of Germany at scale 1:1,000,000 and according to *DIN 19708:2005-02* [39]. This data set has a spatial resolution of 250 m and offers one value for the complete study area which is $K = 0.22912$ (t/ha)/(N/ha). Concerning C-Factor, there are various possible approaches to calculate modern day values depending on the data available [40–43]. As stated earlier, we assume that historical land use was different from the meadow orchard found there today. In absence of historical data accurately describing early medieval farming practices and arable crops in the area, we assigned $C = 0.2$. This value is meant to represent wheat farming with standard tillage practices as implemented in the Europe-wide RUSLE 2015 soil erosion study [42].

Table 1. Basic USLE factor data product acquisition and description.

Data Product	Data Source	Availability	Spatial Resolution (m)
LiDAR DEM	LGL [36]	upon request	0.5
R-factor	DWD [37]	open access	1000
K-factor	BGR [39]	open access	250
C-factor	ESDAC [42]	open access	-

2.3. Modelling Routine

To approximate the influence of ridge and furrow structures on soil erosion for our study area, we chose to simulate soil erosion rates under two scenarios (procedure in Figure 2): (1) The first is the *present surface* scenario, which represents the relief of the study area exactly as it is now, including the ridge and furrow structures. (2) The second scenario is called *smoothed surface*, which describes soil erosion rates for the study area using a virtually smoothed surface that does not include the ridge and furrow structures, while at the same time preserving the larger scale topographical properties of the study area. Therefore, we assume this smoothed surface to serve as an approximate model for the pre-ridge and furrow palaeosurface. The first step after acquiring all necessary data from the sources listed in Table 1 was to create the virtually smoothed surface in Quantum-GIS (QGIS). To do this, we applied the *Gaussian Filter* function [44] to the DEM using the following settings: *Standard Deviation = 50; Kernel Type = Circle; Radius = 20.* The *Gaussian Filter* and specific settings were chosen by a trial and error approach. Methods mentioned in other publications [45,46] for slightly different objectives, e.g., the *Low pass Filter* changed the underlying relief more than necessary in order to smooth out only the ridge and furrow structures.

After the virtually smoothed surface had been created using the Gaussian Filter, we proceeded to use the USLE, for estimating soil erosion rates. In addition to the factors (K, R and C), described above, the slope gradient and length (LS) is necessary for the calculation of the USLE [8]. K, R and C were gained from external data sets as shown in Table 1 and are used in both scenarios without alteration. As the K, R and C factors are independent from the shape of the surface and equal in both scenarios, LS must be the crucial factor in this study. For its calculation, we applied the exact same procedure to the present surface and the virtually smoothed surface. We used the function *LS-factor, field based* with the method of calculation set to *Wischmeier and Smith 1978* [8]. Hrabalikova et al. 2017 [47] found with empirical measurements that USLE models using this method for LS calculation predicted erosion most accurately in a manual and GIS environment. The settings were as follows: *Type of Slope = [0] local slope; Specific Catchment Area = [1] contour length dependent on aspect; Rill/Interrill Erosivity = 1; Stability = [0] stable.*

We then multiplied each of the two resulting LS-factor Layers with all other factors using *Raster Calculator* in QGIS (Figure 2). The extent was set to match the DEM-based catchment area of the plot *Höfelbett*. This provided two erosion models representing the *present surface* and *smoothed surface* scenarios.

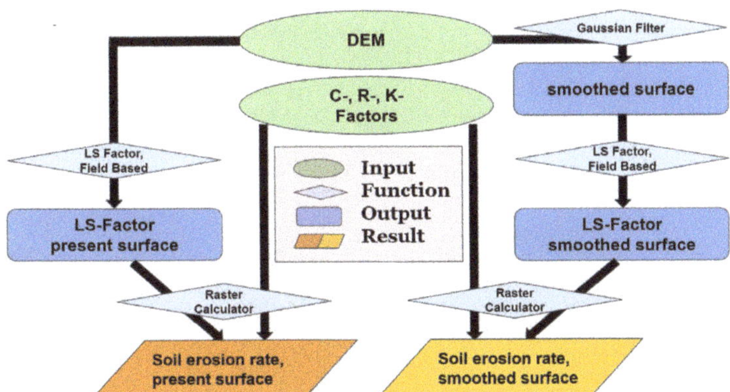

Figure 2. Flowchart of the GIS-based modelling procedure for both scenarios for calculating the potential soil erosion.

3. Results

The difference between our erosion model scenarios is the topography. In the calculation of the USLE, only the calculated LS factor differs. The smoothed surface model no longer shows ridges and furrows (Figure 3). However, the general topography of the slope is maintained. The drainage direction is therefore also preserved. In addition to the linear structures on the agricultural land, further anthropogenic structures such as buildings (visible on the western margin of Figure 3) have also been cleared away.

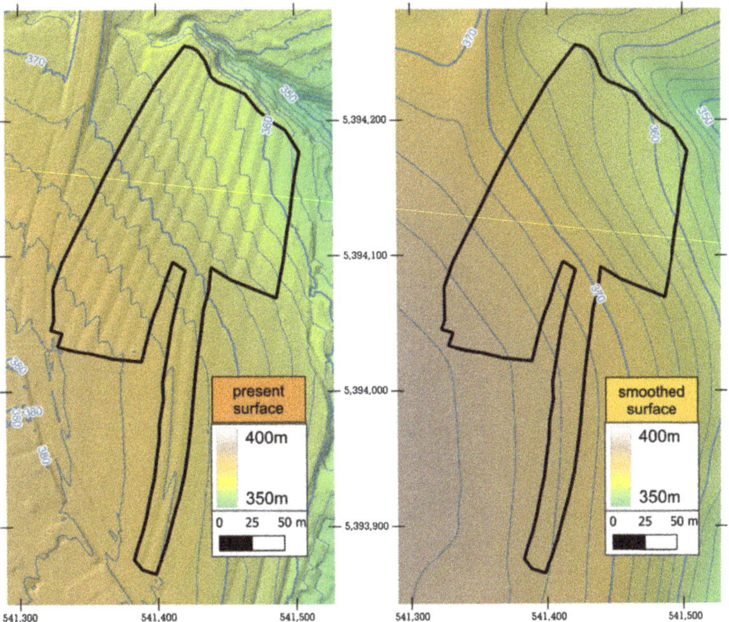

Figure 3. Comparison of the present (**left**) and smoothed (**right**) surface topography based on the LiDAR DEM (©www.lgl-bw.de (accessed on 24 January 2023)) [36]. The black line frames the classified cultural site within the plot *Höfelbett*.

The *present surface* scenario model (Figure 4 left) shows the highest erosion rates on top of the ridges of the ridge and furrow structures, with very low erosion rates in the furrows. Sharper edges of the ridges create extreme values in potential soil loss. In contrast, under the *smoothed surface* scenario, the rates are evenly low at the western, higher upper slope of the plot and gradually increase towards the east. The slope gradient is essentially constant across the plot (see Figure 3), so these differences must be a result of the increased slope length and connectivity. Maximum values, comparable with extreme values from the *smoothed surface* scenario only appear in the northernmost part in context of the steeps slopes of the small valley at the northern margin (Figure 4).

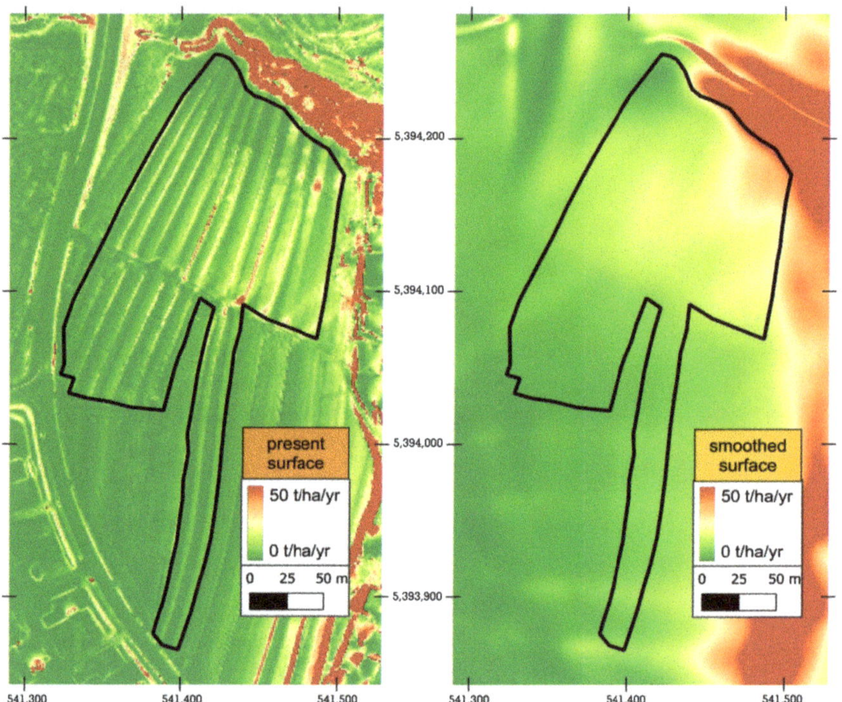

Figure 4. Maps of the two (**left**: present surface, **right**: smoothed surface) erosion models for the study area. The red-to-green colour range shows the mean annual soil loss. The black line frames the cultural site within the plot *Höfelbett*.

4. Discussion

4.1. Erosion Susceptibility of Ridge and Furrow Structures

The reconstruction of a palaeosurface, based only on a deductive filtering approach, such as the Gaussian filter, can produce unintentional effects. The surface smoothing in the area of the bridge (at the northern margin of the map, Figure 3) shows insufficient removal. However, since the area of the bridge lies in the periphery of the catchment area of the *Höfelbett*, no bias of the results of the erosion model is to be expected. A coupled approach in building up a (ridge-and-furrow-free) palaeosurface should consider large linear structures and remove them accordingly [45,48,49].

Summary statistics for all raster cells (0.5 m spatial resolution as given by the LiDAR data) of the modelled potential soil loss show clear differences between both scenarios (Figure 5). The results suggest that the ridge and furrow structures reduce erosion by approximately 50% according to the arithmetic mean. However, the median is more meaningful because extreme values at the edges of the ridges are small scale and and may have been artificial outliers. The median shows a 3-fold increase for potential erosion for the smoothed surface scenario and a 3-fold decrease for the present surface with ridge and furrow features.

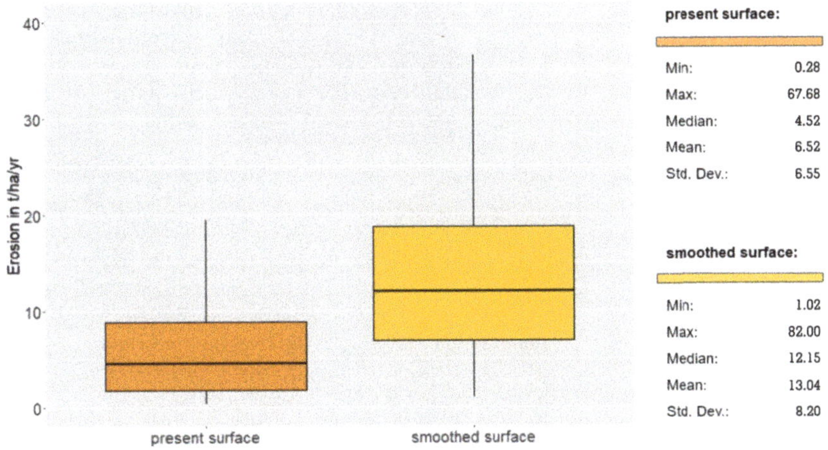

present surface:	
Min:	0.28
Max:	67.68
Median:	4.52
Mean:	6.52
Std. Dev.:	6.55

smoothed surface:	
Min:	1.02
Max:	82.00
Median:	12.15
Mean:	13.04
Std. Dev.:	8.20

Figure 5. Boxplot and summary statistics of raster cell values of both erosion model scenarios.

It is clear that a topography structured by ridges and furrows significantly reduces erosion. Even today, the altered erosion rates through old ridges and furrows are noticeable [50,51]. Ridges and furrows have also been linked to changed erosion patterns [52], e.g., they were attributed to a potentially increased erosion through high landscape connectivity in times of increased cultivation during the 9th and 10th century AD on landscape scale [6]. In general, changing erosion rates can be quantified based on scenarios of historical land uses and their effects [14].

Compared to RUSLE calculations by ESDAC [53], both our *smoothed surface* and our *present surface* scenario models predict soil erosion rates above the European average for arable land of 2.46 t/ha/yr (see Figure 5). This difference of maximum values cannot be attributed to a single factor of our model. Generally, it seems that the combination of the USLE factors in our study leads to such high erosion rates compared to the European scale RUSLE calculations, which are more sensitive for surface runoff distribution [37,54,55].

Today, we still observe large areas with ridge and furrow fields in the district. There has been a conversion of arable land, including many ridge and furrow fields to grassland, orchards and also forests, which are of different origins and periods [56]. The late medieval desertification phase [57,58], but also the intensification and increased productivity of agriculture in the last century has significantly changed the structure of the rural areas and the extent of cultivated land. This conversion often occurred on slopes that were labour-intensive for arable farming [59] and resulted in large areas of ridge and furrow fields still being preserved in the study area.

4.2. Future Perspectives

1. We see a high potential to quantify landscape-altering processes such as erosion (as well as their geo-ecological consequences and yield changes) in a historical context using erosion models with different topographic scenarios. This will allow for a quantification of anthropogenic impacts, e.g., on hydrology and soil preservation, on a local scale and therefore contribute to global models of pathways to the Anthropocene.
2. In order to refine the validity of erosion models, it is crucial to adjust the USLE factors to the historical context. Information of the historical land use (C-factor) and land use techniques (P-factor) could be derived from local palaeoecological archives (e.g., pollen, phytoliths, sedimentary DNA, geo-biomarkers) as well as from archaeological and written sources [60,61]. The K-factor can be adjusted by

local, empirical sedimentological analyses of soil properties, especially because of the exposure of different soil horizons (with varying erodibility) during erosion phases. The reconstruction of a historical R-factor can be achieved using model scenarios of rainfall erosivity [62] but also accompanied by regional proxy-based palaeoclimatic reconstructions [63,64] and (semi-) quantitative data from written sources [65].

3. As our approach is based on the comparison of the present surface with a ridge and furrow-free smoothed surface, precise palaeosurface reconstructions are essential to improve the results. We encourage the use and adaptation of existing deductive, inductive and coupled surface modelling (see review in [45]) approaches to create supervised (cross-checked with historical and land use information) and reproducible *pre-modern* Digital Terrain Models with high spatial resolution.
4. One further hypothesis about the functions of ridge and furrow fields may be addressed using a similar approach: the assumption of a positive effect on the hydro management of soils. The furrows are meant to drain surface runoff from precipitation and the humus-rich ridges increased water storage capacities. Case modelling can generate quantitative data about such anthropogenic impacts on the soil moisture and proof historic *geo-engineering* on a local level.
5. The coupling of empirical sedimentological-geoarchaeological data of soil erosion and erosion models offers the chance of (i) a model verification and (ii) the examination to what extent parameters of the erosion model have to be adjusted. Therefore, we see field-based geomorphological, sedimentological and archaeological mappings and information gain indispensable. The link to soil geochemical laboratory results (e.g., soil organic matter) also allows the distribution of carbon stocks to be modelled using e.g., machine learning approaches in digital soil mapping [66,67].
6. To clarify the question of the extent to which erosion reduction contributes to the installation of ridges and furrows, a systematic mapping of ridge and furrow fields at (supra-) regional level is important. The comparison with geomorphological indices, such as slope, aspect, angle of the ridges and furrow to the slope, as well as geological-sedimentological characteristics allows the identification of common patterns.

5. Conclusions

For the first time, we have shown how valuable a quantitative erosion modelling approach can be for assessing what effects historic ridge and furrow fields can have on soil erosion. With a GIS-based erosion modelling approach based on the Universal Soil Loss Equation (USLE), we compared two scenarios with surface topographies in a small study area in Southern Germany: (1) The present topography with ridges and furrows and (2) the smoothed topography without ridges and furrows. This allowed us to show that historical land uses with ridges and furrows led to a strongly reduced erosion susceptibility. From this, we drew six key perspectives for future research on the topic of historical anthropospheres regarding ridge and furrow fields.

Author Contributions: Conceptualization: J.S., N.U., L.W. (Leon Westphal), M.W. and L.W. (Lukas Werther); methodology: J.S., N.U., L.W. (Leon Westphal) and M.W.; software: J.S., N.U., L.W. (Leon Westphal) and M.W.; validation: J.S., N.U., L.W. (Leon Westphal), M.W., P.K., A.K. and L.W. (Lukas Werther); formal analysis: J.S., N.U., L.W. (Leon Westphal) and M.W.; investigation: J.S., N.U., L.W. (Leon Westphal) and M.W.; resources: J.S., N.U., L.W. (Leon Westphal), M.W. and S.R.; data curation: J.S., N.U., L.W. (Leon Westphal), S.R. and M.W.; writing—original draft preparation: J.S., N.U., A.K, S.R. and P.K.; writing—review and editing: J.S., N.U., L.W. (Leon Westphal), M.W., A.K., L.W. (Lukas Werther) and P.K.; visualization: N.U., L.W. (Leon Westphal) and M.W.; supervision: J.S., A.K., L.W. (Lukas Werther), P.K. and R.R.; project administration: A.K., P.K., L.W. (Lukas Werther) and R.R.; funding acquisition: A.K. and R.R. All authors have read and agreed to the published version of the manuscript.

Funding: This research received no external funding. The Article Processing Charge (APC) was funded by Leipzig University within the program of Open Access publishing.

Data Availability Statement: The data presented in this study are available on request from the corresponding author. The data are not publicly available due to legal reasons.

Acknowledgments: We would like to thank Michael Weidenbacher (County Archaeology, District Administration, Göppingen) and Ralf Hesse (State Office for Cultural Heritage Baden-Wuerttemberg, Esslingen) for their support in the field and concerning data compilation. Furthermore, we thank Alexander Bolland for English spell checking. We thank four anonymous reviewers for their helpful comments.

Conflicts of Interest: The authors declare no conflict of interest.

References

1. Baude, M.; Meyer, B.C.; Schindewolf, M. Land use change in an agricultural landscape causing degradation of soil based ecosystem services. *Sci. Total Environ.* **2019**, *659*, 1526–1536. [CrossRef]
2. Uber, M.; Rössler, O.; Astor, B.; Hoffmann, T.; van Oost, K.; Hillebrand, G. Climate Change Impacts on Soil Erosion and Sediment Delivery to German Federal Waterways: A Case Study of the Elbe Basin. *Atmosphere* **2022**, *13*, 1752. [CrossRef]
3. Marcinkowski, P.; Szporak-Wasilewska, S.; Kardel, I. Assessment of soil erosion under long-term projections of climate change in Poland. *J. Hydrol.* **2022**, *607*, 127468. [CrossRef]
4. Larsen, A.; Robin, V.; Heckmann, T.; Fülling, A.; Larsen, J.R.; Bork, H.R. The influence of historic land-use changes on hillslope erosion and sediment redistribution. *Holocene* **2016**, *26*, 1248–1261. [CrossRef]
5. Dreibrodt, S.; Lubos, C.; Terhorst, B.; Damm, B.; Bork, H.R. Historical soil erosion by water in Germany: Scales and archives, chronology, research perspectives. *Quat. Int.* **2010**, *222*, 80–95. [CrossRef]
6. Brown, A.G. Colluvial and alluvial response to land use change in Midland England: An integrated geoarchaeological approach. *Geomorphology* **2009**, *108*, 92–106. [CrossRef]
7. Dotterweich, M. The history of soil erosion and fluvial deposits in small catchments of central Europe: Deciphering the long-term interaction between humans and the environment—A review. *Geomorphology* **2008**, *101*, 192–208. [CrossRef]
8. Wischmeier, W.H.; Smith, D.D. *Predicting Rainfall Erosion Losses—A Guide to Conservation Planning*; Agriculture Handbook; U.S. Department of Agriculture: Washington, DC, USA, 1978; Volume 537.
9. Schwertmann, U.; Vogl, W.; Kainz, M. *Bodenerosion durch Wasser: Vorhersage des Abtrags und Bewertung von Gegenmaßnahmen*; Eugen-Ulmer Verlag: Stuttgart, Germany, 1987.
10. Tetzlaff, B.; Friedrich, K.; Vorderbrügge, T.; Vereecken, H.; Wendland, F. Distributed modelling of mean annual soil erosion and sediment delivery rates to surface waters. *CATENA* **2013**, *102*, 13–20. [CrossRef]
11. Schmidt, S.; Alewell, C.; Meusburger, K. Monthly RUSLE soil erosion risk of Swiss grasslands. *J. Maps* **2019**, *15*, 247–256. [CrossRef]
12. Gericke, A.; Kiesel, J.; Deumlich, D.; Venohr, M. Recent and Future Changes in Rainfall Erosivity and Implications for the Soil Erosion Risk in Brandenburg, NE Germany. *Water* **2019**, *11*, 904. [CrossRef]
13. Favis-Mortlock, D.; Boardman, J.; Bell, M. Modelling long-term anthropogenic erosion of a loess cover: South Downs, UK. *Holocene* **1997**, *7*, 79–89. [CrossRef]
14. Ayala, G.; French, C. Erosion modeling of past land-use practices in the Fiume di Sotto di Troina river valley, north-central Sicily. *Geoarchaeology* **2005**, *20*, 149–167. [CrossRef]
15. Sun, W.; Shao, Q.; Liu, J.; Zhai, J. Assessing the effects of land use and topography on soil erosion on the Loess Plateau in China. *CATENA* **2014**, *121*, 151–163. [CrossRef]
16. Sittler, B. Revealing historical landscapes by using Airborne Laser Scanning: A 3D-Model of ridge and furrow in forests near Rastatt (Germany). *Int. Arch. Photogramm. Remote Sens. Spat. Inf. Sci.* **2004**, *8*, 258–261.
17. Schmoock, I.; Gehrt, E. Verbreitung und Charakterisierung der Wölbackerböden in Niedersachsen. Jahrestagung der DGB, Kom V. 2017. Available online: https://eprints.dbges.de/1271/ (accessed on 5 January 2023).
18. van Gils, H.; Kasielke, T. Historical parcellation and ridge-and-furrow relics of open strip-fields in the north-west European lowlands. *Landsc. Hist.* **2022**, *43*, 77–102. [CrossRef]
19. Langewitz, T.; Fülling, A.; Klamm, M.; Wiedner, K. Historical classification of ridge and furrow cultivation at selected locations in Northern and central Germany using a multi-dating approach and historical sources. *J. Archaeol. Sci.* **2020**, *123*, 105248. [CrossRef]
20. Catsadorakis, G.; Mougiakou, E.; Kizos, T. Ridge-and-Furrow Agriculture around Lake Mikri Prespa, Greece, in a European perspective. *J. Eur. Landsc.* **2021**, *2*, 7–20. [CrossRef]
21. Langewitz, T.; Wiedner, K.; Polifka, S.; Eckmeier, E. Pedological properties related to formation and functions of ancient ridge and furrow cultivation in Central and Northern Germany. *CATENA* **2021**, *198*, 105049. [CrossRef]
22. Palmer, R. A further case for the preservation of earthwork ridge-and-furrow. *Antiquity* **1996**, *70*, 436–440. [CrossRef]
23. Schreg, R. Mittelalterliche Feldstrukturen in deutschen Mittelgebirgslandschaften—Forschungsfragen, Methoden und Herausforderungen für Archäologie und Geographie. In *Agrarian Technology in the Medieval Landscape*; Klápště, J., Ed.; Brepols Publishers n.v: Turnhout, Belgium, 2016; pp. 351–370.

24. Brown, A.G.; Tooth, S.; Bullard, J.E.; Thomas, D.S.G.; Chiverrell, R.C.; Plater, A.J.; Murton, J.; Thorndycraft, V.R.; Tarolli, P.; Rose, J.; et al. The geomorphology of the Anthropocene: Emergence, status and implications. *Earth Surf. Process. Landf.* **2017**, *42*, 71–90. [CrossRef]
25. Kottmann, A.; Harding, S.; Kühn, P.; Marinova-Wolff, E.; Miller, C.E.; Nelle, O.; Rademacher, R.; Schreg, R.; Vogt, R.; Werther, L. *Untersuchungen der Wölbäcker im Gewann "Höfelbett" in Albershausen*; Archäologische Ausgrabungen in Baden-Württemberg: Darmstadt, Germany , 2020; pp. 322–326.
26. Ewald, K.C. Agrarmorphologische Untersuchungen im Sundgau (Oberelsass) unter besonderer Berücksichtigung der Wölbäcker. *Tätigkeitsbericht Der Nat. Ges. Baselland* **1968**, *2*, 9–178.
27. Bartussek, I. Die Gewölbten Ackerbeete in der historischen Landwirtschaft—Ihr Relikte und Nachwirkungen in der Gegenwart: Diploma Thesis. Ph.D. Thesis, Georg-August-Universität Göttingen, Göttingen, Germany, 1982.
28. Langewitz, T.; Wiedner, K.; Fritzsch, D.; Eckmeier, E. Improvement of soil fertility in historical ridge and furrow cultivation. *Geoarchaeology* **2022**, *37*, 750–767. [CrossRef]
29. Hänßler, E. *700 Jahre Albershausen*; Angebotene Zahlungsarten: Albershausen, Germany, 1975.
30. Ziegler, W. (Ed.) Vom frühen Mittelalter bis zur Neuzeit. In *Der Kreis Göppingen*; 1985; pp. 83–103.
31. Krähenmann, S.; Walter, A.; Brienen, S.; Imbery, F.; Matzarakis, A. Monthly, Daily and Hourly Grids of 12 Commonly Used Meteorological Variables for Germany Estimated by the Project TRY Advancement. Available online: https://doi.org/10.5676/DWD_CDC/TRY_BASIS_V001 (accessed on 5 January 2023).
32. Climate data. Station Göppingen. Daily Observations. 2021. Available online: https://opendata.dwd.de/climate_environment/CDC/observations_germany/climate/ (accessed on 5 January 2023).
33. Franz, M.; Arp, G.; Niebuhr, B. Schwarzjura-Gruppe. 2020. Available online: https://litholex.bgr.de/pages/Einheit.aspx?ID=10000049 (accessed on 5 January 2023).
34. Grams, G. Württembergische Landesvermessung in der Zeit von 1818-1840. In *200 Jahre Landesvermessung Baden-Württemberg*; Ministerium für Ländlichen Raum und Verbraucherschutz Baden-Württemberg, Ed.; 2018; pp. 19–32. Available online: https://www.lgl-bw.de/Ueber-Uns/Geschichte/200-Jahre-Landesvermessung/ (accessed on 5 January 2023).
35. IUSS Working Group WRB. *World Reference Base for Soil Resources: International Soil Classification System for Naming Soils and Creating Legends for Soil Maps*, 4th ed.; International Union of Soil Sciences (IUSS): Vienna, Austria, 2022.
36. Landesamt für Geoinformation und Landesentwicklung (LGL). Laserscandaten "ALS_2". Available online: https://www.lgl-bw.de/Produkte/3D-Produkte/Laserscandaten/ALS_2/ (accessed on 5 January 2023).
37. Fischer, F.; Winterrath, T.; Junghänel, T.; Walawender, E.; Auerswald, K. Mean Annual Precipitation Erosivity (R Factor) Based on RADKLIM Version 2017.002. 2019. Available online: https://doi.org/10.5676/DWD/RADKLIM_Rfct_V2017.002 (accessed on 5 January 2023).
38. Sauerborn, P. Die Erosivität der Niederschläge in Deutschland: Ein Beitrag zur Quantitativen Prognose der Bodenerosion Durch Wasser in Mitteleuropa. In *Bonner Bodenkundliche Abhandlungen*; Institut für Bodenkunde: Hanover, Germany, 1994; Volume 13.
39. Volker Hennings. *Erodierbarkeit der Böden in Deutschland*; Bundesanstalt für Geowissenschaften und Rohstoffe (BGR): Hannover, Germany, 2021.
40. Renard, K.G. *Predicting Soil Erosion by Water: A Guide to Conservation Planning with the Revised Universal Soil Loss Equation (RUSLE)*; United States Government Printing: Washington, DC, USA, 1997.
41. Efthimiou, N.; Psomiadis, E.; Papanikolaou, I.; Soulis, K.X.; Borrelli, P.; Panagos, P. A new high resolution object-oriented approach to define the spatiotemporal dynamics of the cover-management factor in soil erosion modelling. *CATENA* **2022**, *213*, 106149. [CrossRef]
42. Panagos, P.; Borrelli, P.; Meusburger, K.; Alewell, C.; Lugato, E.; Montanarella, L. Estimating the soil erosion cover-management factor at the European scale. *Land Use Policy* **2015**, *48*, 38–50. [CrossRef]
43. Auerswald, K.; Ebertseder, F.; Levin, K.; Yuan, Y.; Prasuhn, V.; Plambeck, N.O.; Menzel, A.; Kainz, M. Summable C factors for contemporary soil use. *Soil Tillage Res.* **2021**, *213*, 105155. [CrossRef]
44. Kokalj, Ž.; Hesse, R. *Airborne Laser Scanning Raster Data Visualization: A Guide to Good Practice*, 1st ed.; ZRC SAZU: Ljubljana, Slovenia, 2017; Volume 14.
45. Schmidt, J.; Werther, L.; Zielhofer, C. Shaping pre-modern digital terrain models: The former topography at Charlemagne's canal construction site. *PLoS ONE* **2018**, *13*, e0200167. [CrossRef]
46. Hesse, R. LiDAR-derived Local Relief Models—A new tool for archaeological prospection. *Archaeol. Prospect.* **2010**, *17*, 67–72. [CrossRef]
47. Hrabalikova, M.; Janecek, M. Comparison of different approaches to LS factor calculations based on a measured soil loss under simulated rainfall. *Soil Water Res.* **2017**, *12*, 69–77. [CrossRef]
48. van der Meulen, B.; Cohen, K.M.; Pierik, H.J.; Zinsmeister, J.J.; Middelkoop, H. LiDAR-derived high-resolution palaeo-DEM construction workflow and application to the early medieval Lower Rhine valley and upper delta. *Geomorphology* **2020**, *370*, 107370. [CrossRef]
49. Henselowsky, F.; Rölkens, J.; Kelterbaum, D.; Bubenzer, O. Anthropogenic relief changes in a long–lasting lignite mining area ("Ville", Germany) derived from historic maps and Digital Elevation Models. *Earth Surf. Process. Landf.* **2021**, *46*, 1725–1738. [CrossRef]

50. Deumlich, D. Beeinflusst Historische Landnutzung Die Aktuelle Bodenerosion? 2009. Available online: https://www.researchgate.net/publication/279416046_Beeinflusst_historische_Landnutzung_die_aktuelle_Bodenerosion (accessed on 5 January 2023).
51. Eyre, S.R. The curving Plough-strip and its Historical Implications. *Archaeol. Hist. Rev.* **1955**, *3*, 80–94.
52. Møller, P.G. Ridge and Furrow Fields: Field Systems ca. 1000–1800 as a Stabilising Factor in an Agricultural Society—A Danish Example. In *Agricultural and Pastoral Landscapes in Pre-Industrial Society*; Retamero, F., Schjellerup, I., Davies, A., Eds.; Earth Ser.; Oxbow Books: Philadelphia, PA, USA, 2015; pp. 159–171.
53. Panagos, P.; Borrelli, P.; Poesen, J.; Ballabio, C.; Lugato, E.; Meusburger, K.; Montanarella, L.; Alewell, C. The new assessment of soil loss by water erosion in Europe. *Environ. Sci. Policy* **2015**, *54*, 438–447. [CrossRef]
54. Panagos, P.; Borrelli, P.; Meusburger, K. A new European slope length and steepness factor (LS-Factor) for modeling soil erosion by water. *Geosciences* **2015**, *5*, 117–126. [CrossRef]
55. Panagos, P.; Ballabio, C.; Borrelli, P.; Meusburger, K.; Klik, A.; Rousseva, S.; Tadić, M.P.; Michaelides, S.; Hrabalíková, M.; Olsen, P.; et al. Rainfall erosivity in Europe. *Sci. Total Environ.* **2015**, *511*, 801–814. [CrossRef] [PubMed]
56. Niedertscheider, M.; Kuemmerle, T.; Müller, D.; Erb, K.H. Exploring the effects of drastic institutional and socio-economic changes on land system dynamics in Germany between 1883 and 2007. *Glob. Environ. Chang.* **2014**, *28*, 98–108. [CrossRef]
57. Grees, H. Zur Siedlungs- und Landschaftsentwicklung der Ostalb. Die Wüstungsvorgänge des ausgehenden Mittelalters und ihre Folgewirkung. *Karst Höhle* **1993**, 363–378.
58. Schreg, R. 15. Human Impact on Hydrology. In *The Power of Urban Water*; Chiarenza, N., Haug, A., Müller, U., Eds.; De Gruyter: Berlin, Germany, 2020; pp. 249–264. [CrossRef]
59. Landesanstalt für Landwirtschaft, Ernährung und Ländlichen Raum. Ertragsmesszahlen der Gemarkungen in Baden-Württemberg. Available online: https://www.lel-web.de/app/ds/lel/a3/Online_Kartendienst_extern/Karten/33279/index.html (accessed on 5 January 2023).
60. Scherer, S.; Deckers, K.; Dietel, J.; Fuchs, M.; Henkner, J.; Höpfer, B.; Junge, A.; Kandeler, E.; Lehndorff, E.; Leinweber, P.; et al. What's in a colluvial deposit? Perspectives from archaeopedology. *CATENA* **2021**, *198*, 105040. [CrossRef]
61. Scherer, S.; Höpfer, B.; Deckers, K.; Fischer, E.; Fuchs, M.; Kandeler, E.; Lechterbeck, J.; Lehndorff, E.; Lomax, J.; Marhan, S.; et al. Middle Bronze Age land use practices in the northwestern Alpine foreland—A multi-proxy study of colluvial deposits, archaeological features and peat bogs. *Soil* **2021**, *7*, 269–304. [CrossRef]
62. Diodato, N.; Borrelli, P.; Fiener, P.; Bellocchi, G.; Romano, N. Discovering historical rainfall erosivity with a parsimonious approach: A case study in Western Germany. *J. Hydrol.* **2017**, *544*, 1–9. [CrossRef]
63. Büntgen, U.; Franke, J.; Frank, D.; Wilson, R.; González-Rouco, F.; Esper, J. Assessing the spatial signature of European climate reconstructions. *Clim. Res.* **2010**, *41*, 125–130. [CrossRef]
64. Büntgen, U.; Brázdil, R.; Heussner, K.U.; Hofmann, J.; Kontic, R.; Kyncl, T.; Pfister, C.; Chromá, K.; Tegel, W. Combined dendro-documentary evidence of Central European hydroclimatic springtime extremes over the last millennium. *Quat. Sci. Rev.* **2011**, *30*, 3947–3959. [CrossRef]
65. Esper, J.; Büntgen, U.; Denzer, S.; Krusic, P.J.; Luterbacher, J.; Schäfer, R.; Schreg, R.; Werner, J. Environmental drivers of historical grain price variations in Europe. *Clim. Res.* **2017**, *72*, 39–52. [CrossRef]
66. Ließ, M.; Schmidt, J.; Glaser, B. Improving the Spatial Prediction of Soil Organic Carbon Stocks in a Complex Tropical Mountain Landscape by Methodological Specifications in Machine Learning Approaches. *PLoS ONE* **2016**, *11*, e0153673. [CrossRef] [PubMed]
67. Behrens, T.; Schmidt, K.; MacMillan, R.A.; Viscarra Rossel, R.A. Multi-scale digital soil mapping with deep learning. *Sci. Rep.* **2018**, *8*, 15244. [CrossRef]

 land

Article

Evolving Cultural and Historical Landscapes of Northwestern Colchis during the Medieval Period: Physical Environment and Urban Decline Causes

Galina Trebeleva [1,*], Andrey Kizilov [2], Vasiliy Lobkovskiy [3] and Gleb Yurkov [4]

1 Institute of Archaeology of the Russian Academy of Sciences, 117292 Moscow, Russia
2 Federal Research Centre the Subtropical Scientific Centre of the Russian Academy of Sciences, 354002 Sochi, Russia
3 Institute of Geography of the Russian Academy of Sciences, 119017 Moscow, Russia
4 N.N. Semenov Federal Research Center of Chemical Physics of the Russian Academy of Sciences, 119334 Moscow, Russia
* Correspondence: g_gis@mail.ru

Abstract: In Late Antiquity and the Early Middle Ages, both coastal and sub-mountainous parts of Colchis underwent rapid urbanization. In the 12th century, the processes of decline began: Large settlements were replaced by small farmsteads with light wooden buildings, and the economy transformed from commodity-based to subsistence-based. What caused this decline? Was it the social and political events linked to the decline of the Byzantine Empire and changes to world trade routes, or were there other reasons? This article provides the answer. The synergy of archaeological, folkloristic, historical cartographic, climatological, seismological, and hydrological data depicts a strong link between these processes and climate change, which occurred at the turn of the 12th–13th centuries. The beginning of cooling led to a crisis in agriculture. A decline in both farming and cattle breeding could not fail to affect demography. Seismic activity, noted in the same period, led to the destruction of many buildings, including temples, and fortresses, and changes in hydrological networks, which were directly linked to climate change and caused water logging, led to a loss of the functions of coastal areas and their disappearance.

Keywords: Colchis; archaeology; GIS; settlement patterns; urbanization; physical environment; seismic activity; hydrological networks; climate change

Citation: Trebeleva, G.; Kizilov, A.; Lobkovskiy, V.; Yurkov, G. Evolving Cultural and Historical Landscapes of Northwestern Colchis during the Medieval Period: Physical Environment and Urban Decline Causes. *Land* **2022**, *11*, 2202. https://doi.org/10.3390/land11122202

Academic Editors: Bellotti Piero, Alessia Pica and Hannes Palang

Received: 23 October 2022
Accepted: 1 December 2022
Published: 4 December 2022

1. Introduction

Northwestern Colchis is partly located in the present-day Russian Federation (Greater Sochi Area, Krasnodar Region) and the Republic of Abkhazia. Nevertheless, from ancient times to the end of the 20th century, Colchis was a single historical and cultural area (Figure 1). Throughout history, this area was of major importance for various cultural interactions, and in Late Antiquity and the early Middle Ages, the area experienced urban growth and transformation of the settlement structure, resulting in the temple stone architecture marking densely populated areas. Ancient towns and settlements in the area have been insufficiently studied by archaeologists because the buildings were traditionally constructed from sun-dried brick, hiding such structures in the landscape. However, analysis of the temple size and the distribution of the temples gives a clear idea of demographic processes. Other important markers are the remains of stone fortresses, which reveal the main caravan routes connecting the Caucasus Range mountains and coastal trading posts. Thus, archaeological evidence depicts vivid signs of the rise of urbanization, both in the coastal part and in the mountainous areas during the period of Late Antiquity and the early Middle Ages. By the 12th century, archaeological findings confirm that the process was reversed, and settlements fell into disrepair. During this process, temples collapsed, and large settlements with stone buildings were replaced by small farmsteads with light wooden buildings [1]. What brought about the decline? Several factors may have been involved, including social and political events associated with the decline of the Byzantine Empire and changes in world trade routes, as well as other causes of a global or local nature. The impact of climatic factors on the level of socioeconomic development in society during the entire period of mankind can be traced clearly enough [2–7]. This article presents a synthesis of archaeological and paleogeographic data showing the effects of climatic, hydrogeological, and seismic factors on the development of settlement structure, together representing an attempt to recreate the historical period being analyzed.

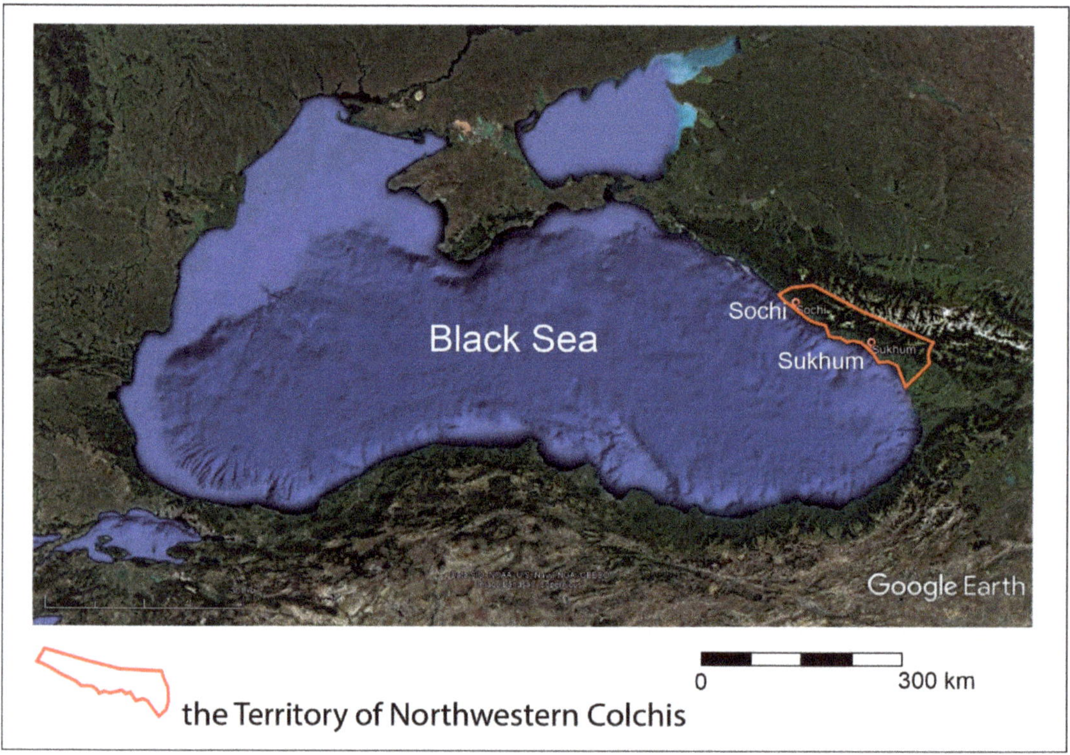

Figure 1. The territory of northwestern Colchis on a satellite image.

2. Materials and Methods

The main research method in this study is a comprehensive analysis of the settlement structure in GIS, applying relevant data from other natural science disciplines.

Geo-information systems have been a traditional research tool for many years, not only in geographical but also in historical research around the world [8–16]. Our team has performed field research since 2001 [17] to find the sites of northwestern Colchis, and currently the database includes 1780 sites, including dolmens, temples, fortresses, settlements, individual sites, and burial grounds, without any architectural remains above the surface. Data on the location of the object, submitted as geographical coordinates, were recorded using GPS.

The archaeological data set was presented in the form of point objects together with an associated primary database containing the following key positions:

1. Title
2. Type of site
3. Ancient (antique or local) toponym (if any)
4. Link to reference in written sources (if available)
5. Names of researchers of the site (if available)
6. Years of research of the site (if available)
7. Description of the site
8. Bibliography (if available)
9. Level of preservation
10. Modern use of the territory
11. Area (if defined)
12. Dating
13. Interpretation

Due to certain objective reasons, i.e., different states of preservation and the study of sites, the primary database is being constantly updated and modified correspondingly by applying layering and hyperlinks. Via hyperlinks, raster and media data are connected to point objects, including plans, photographs, and 3D models. As object data in point-layers format do not carry complete information about the location of the site and its outline in the landscape, the GIS is supplemented with polygonal data obtained from orthophotomaps and digital terrain models (DTM) of sites included in the GIS [18].

The study of temples is crucial for the analysis of settlement structures in the periods of Late Antiquity and the Middle Ages. According to W. Christaller [19], temples are regarded as "central places", each representing a type of political and administrative center. Temples, therefore, mark not merely the direct spread of Christian religion but also the administrative and political division of the territory. Moreover, taking into account the size of churches, which depended on the number of parishioners, it is possible to estimate the demographic situation in the area in order to trace the dynamics of population settlements based on different terrain altitudes, to identify the most favorable geographic zones in each chronological period.

A critical issue for the reconstruction of settlement structure is dating. The vast majority of temples have not been studied and cannot be allocated to a certain period. However, prior assumptions about dating can be made based on several factors:

(1) A detailed description of the architectural features of temples, followed by finding similarities to certain architectural details and features between a number of temples. Recording the frequency of the occurrence of such features in the studied area enables one to obtain the required data to date and characterize different architectural schools [20];
(2) Analysis of mortar in the masonry of the site [21];
(3) Analysis of lifting materials, including plinthite analysis [22];
(4) Data obtained during excavation works.

Therefore, dating is based on an integrated approach that involves the organization of archaeological surveys, work with archival data and literature, analysis of architectural features, and the use of natural science methods. Simultaneously, it is important to find two main dates: the date of construction of the temple (Figure 2) and the date that the temple ceased to function (Figure 3).

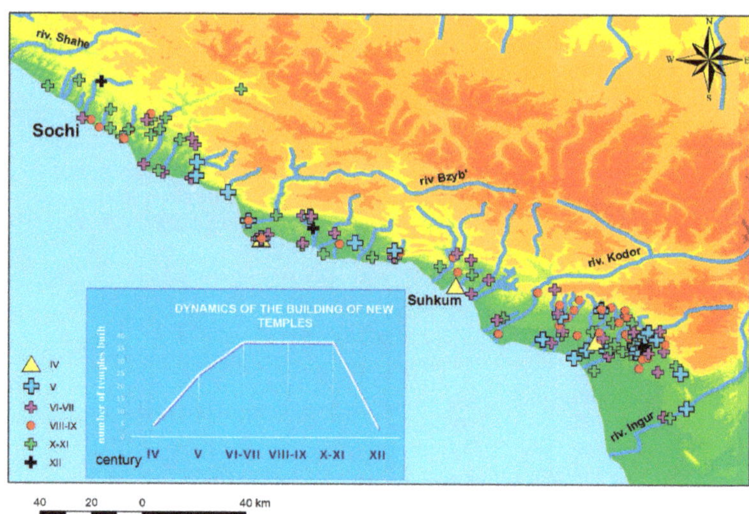

Figure 2. Map of temples with a chronological classification according to the date of construction of the temple.

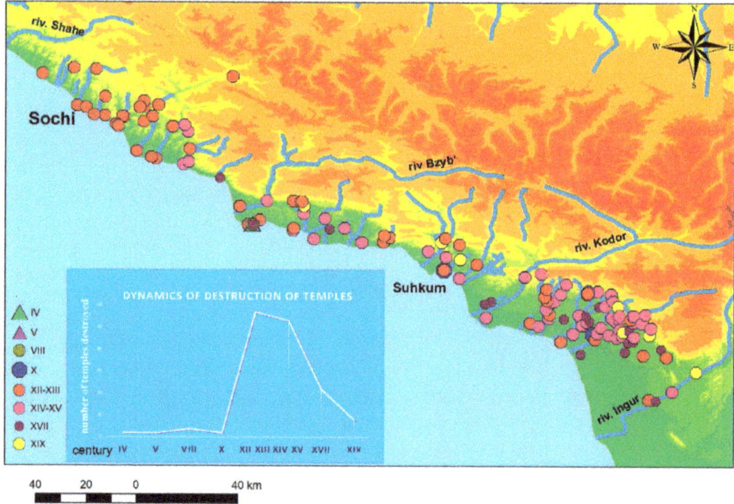

Figure 3. Map of temples with a chronological classification according to the date of destruction of the temple.

It should be noted that the former date is generally more accurate than the latter since it is very problematic to identify and precisely determine the date that a temple ceased to function. However, in some cases, it remains possible to acquire such a date. Excavation and narrative data represented by chronicles and church literature, with references to existing temples and monasteries, Genoese portals, where certain sites are marked, and maps from the modern era, can be used to confirm the date that a temple ceases to function; meanwhile, reference to the temple may indicate its functioning during a certain period. For example, by analyzing travelers' descriptions and finding reference to 'ruins', one can assume that the temple had ceased to function by the time of that evidence. However, for certain sites, a simplified method must be used: interpolation. In this scenario, if a temple in the village Vesyoloe was ruined by an earthquake [23], and next to it (in about 3 km), the ruins of another temple ("Sovkhoz Russia") were found, then it can be assumed that the reason for the destruction of both temples was undoubtedly the same. Thus, under various assumptions, we can obtain maps of temples for various chronological periods in addition to graphs of the dynamics of the building of temples and their destruction (Figures 2 and 3).

Another important aspect relates to data on hydrographic networks, including rivers, reservoirs (lagoons and lakes), and changes in coastline development. Notably, direct research was not conducted by the present authors; all of the information was borrowed from the research of hydrologists and palaeogeographers published in the second half of the 20th century, since modern studies in this area have not been carried out in the territory of the region of interest [24–28]. Data from the literature were entered into the existing GIS, partly by overlaying various layers of maps presented in the literature and geocoding based on reference points, i.e., recording a river's shoreline or lines of the road and other objects easily distinguishable in the landscape.

Climate curves represented the next set of data. For this area, data could only be found in the works of Soviet scientists [29].

Maps of tectonic faults and seismic zones were the most crucial layer in GIS. These materials are well-represented in the works of modern researchers of the 21st century [30–32]. It should be noted though that the available data refer only to the northwestern part of the area. Comprehensive studies on the area east of the city of Sukhum have not been carried out. The findings of our archaeological team revealed evidence of earthquakes and landslides on archaeological sites in the Ochamchira region of the Republic of Abkhazia, in the Mokva River basin (Figure 4).

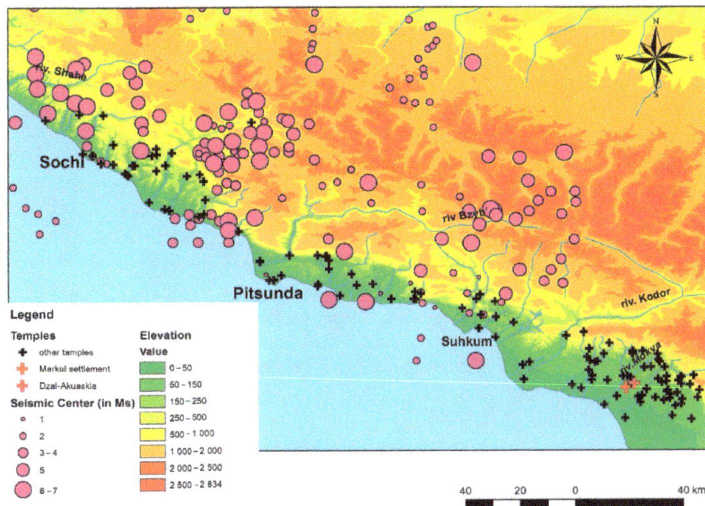

Figure 4. A map of earthquake centers (pink circles) recorded by our team (data from the centers are taken from [32]) and the Markul settlement (red cross), where multiple traces of earthquakes and landslides were recorded after our expedition's excavations.

Seismic zones, in contrast to climatic and hydrographic ones, are not usually taken into account in archeology. Studies related to the influence of the physical environment on human evolution usually focus on climate as the main external driving force of evolution and cultural changes. However, recently, the attention of archaeologists has increasingly focused on the effects of tectonic factors, including theoretical works [33–36] and research on the impacts of those factors on individual sites, e.g., in the European Bosporus (Crimea Peninsula) [37], Greece [38–42], Turkey [43], Jordan [44], and many other parts of the globe [45–49]. However, no similar investigation in northwest Colchis has yet been conducted. Meanwhile, these factors are crucial for the analysis of all settlement structures since archeologists must detect evidence for the occurrence of tectonic factors. Therefore, the available maps of tectonic fault zones were included as a separate layer in the GIS to determine the number of sites within areas of high seismic activity. Only a minor portion of sites (no more than 10%) have been thoroughly studied. Nevertheless, by knowing about those zones, we can determine the areas where settlements (and, accordingly, traces of their architectural structures) could have been destroyed due to seismic factors. For the parts of the territory where no seismic maps are available, we can obtain information about the traces of earthquakes after archaeological research and then add these zones to the areas of high seismic activity (Figure 5).

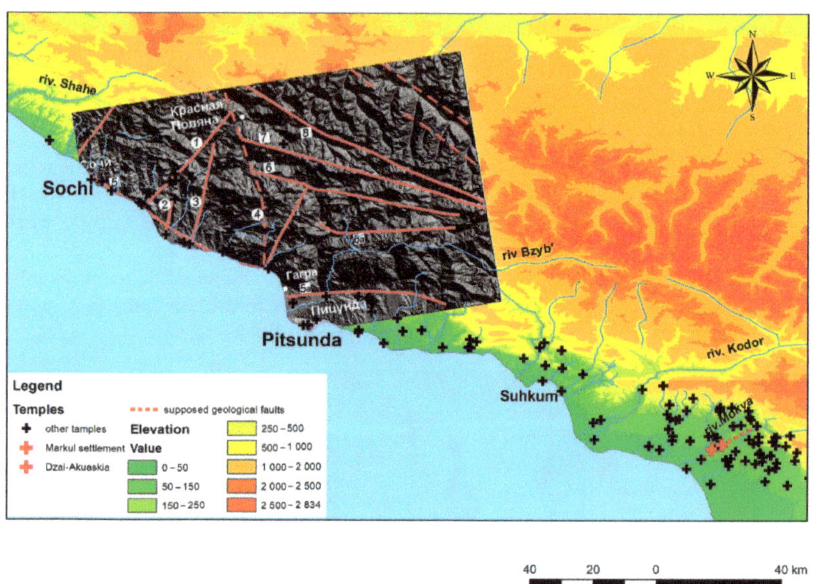

Figure 5. Synthesis of tectonic data in GIS according to the data from seismologists: The raster with red lines and numbers indicates longitudinal (5, 5a, 6, 6a, 7, 8) and transverse (1, 2, 3, 4) tectonic faults [31]; selection of faults according to our field studies (red dotted line).

The following GIS analytical functions were used: Thiessen (or Voronoi) Map and Distance Cost Weighted. Thiessen (or Voronoi) Map functions of GIS analytical tools made it possible to determine the areas that belonged to a particular temple as "central places". The adjustment of polygons, taking into account hydro networks, was performed in manual mode. Distance Cost Weighted was used to reconstruct the road networks and, on the basis of these, temple clusters were identified. The DEM used to create the cost raster was obtained from the open source ASTER Global Digital Elevation Map. Standard tools of ArcGis such as map overlay, selection, data sorting, and others were used.

Another important source of information, not directly included in the GIS, is folklore. Folklore is a traditional source for historical research. In the local people's memory, events of the distant past are preserved, transformed, and then reflected in legends. For example, the Nart saga mentions "climatic cataclysms", containing lines about cold weather: "One year heavy, deep snow covered the Narts' dwelling, so none of them could get the cattle out", "It was a hard winter in Nart", "A terrible year was for the Narts, there was nothing left for their cattle to eat", "A bitterly cold winter and hunger was there for the Narts, and they despaired: 'What else can we do, where can we find food?'", "A harsh winter was there for the Narts. There was no food, the starvation killed their herds", "A terrible winter was for the Narts: there was a heavy snowfall (zalty). Their cattle had nothing to eat. They fell into despair: 'how shall we save our animals, if our horses die out, then a horseless man is no different from a wingless bird'", "Once, in early spring, the heaviest snowfall ever fell over the village of Narts. Farm fields were of no use, they were in despair as they did not know what to do with the cattle", and "The tough times were in the village of the Narts: a harsh winter, a winter of misery. And the winter was long, the spring was late, and their cattle were in a desperate state" [50].

3. Results

When analyzing the reasons underlying the system of placing temple buildings, it was noted that the main chains of temples are located along the rivers that reach a certain port on the coast and diverge inland in fan-shaped routes. Thus, certain clusters can be distinguished, each of them requiring a separate analysis and synthesis of the results from the analysis of each cluster. For the present study, we carried out a detailed analysis [51] of a cluster of sites along the Tamysh–Dgamsh and Mokva rivers (Figure 6):

(1) During the Late Antique and early Medieval periods, the most populated, and, therefore, the most favorable zone for living, was the zone from 50 to 150 m above sea level (i.e., flat but not flooded or swampy);
(2) During the period of the united Abkhazian Kingdom (8th–12th centuries), a significant increase in the population of at least 2.5 times was noted;
(3) After the 12th–13th centuries, a sharp decline began: New temples were built, and in the 14th–15th centuries, those temples were abandoned and destroyed.

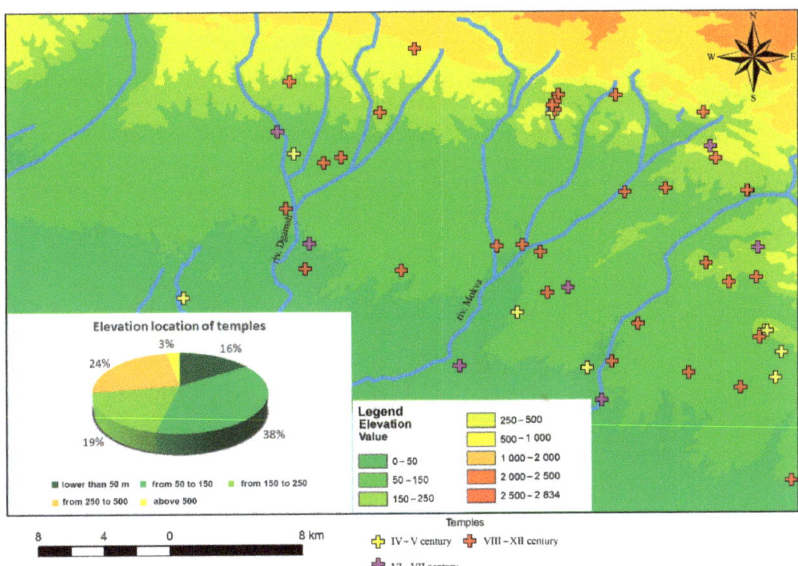

Figure 6. Temples in the areas of the Dgamsh and Mokva rivers.

An indirect but objective indicator of population growth during the period of the Abkhazian Kingdom (8th–12th centuries) is a sharp increase in the number of workshops to create pottery and containers for storing food. During this period, the majority of pithoi, which are large storage containers used primarily for the bulk storage of fluids and grains, were found [52].

Moreover, stamps of unified workshops were found. A meshed circle stamp was found in the Mamai-Kale fortress near Sochi [53], as well as during the excavations of the Markul settlement, meaning that pottery was delivered to different parts of northwestern Colchis [54] (Figure 7).

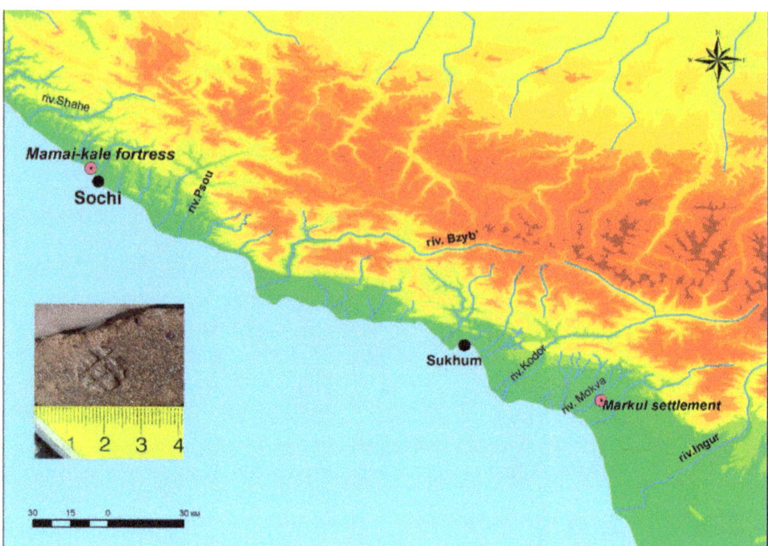

Figure 7. The stamp on the pithoi in the form of a meshed circle and the places where they were found marked on the map.

When discussing agriculture, the economic foundation of the Late Antique and Medieval periods, it is important to take into account the results of analyses by related scientific disciplines. Indeed, the basis of nutrition was determined to be agricultural products, after an analysis of the isotopic indicators of nitrogen and carbon in the teeth of buried people. This analysis was carried out on bodies buried at the Markul temple and in the village of Vesyoloe, yielding similar outcomes [17] (Figure 8).

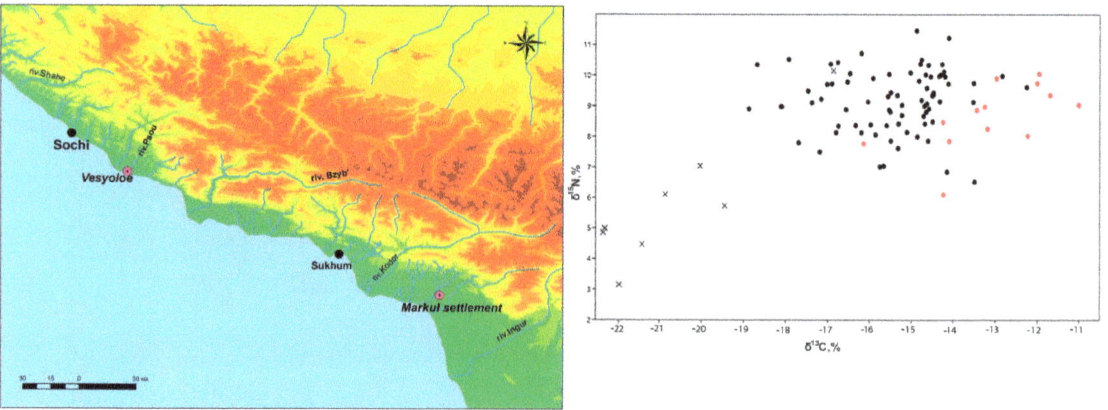

Figure 8. Maps and isotopic indicators of nitrogen and carbon: comparison of data from the village of Vesyoloe (black dots) and the Markul settlement (red dots), and samples of animals (crosses).

In terms of paleoclimatic studies, attention was drawn to the results of the analysis of the soil of the watershed surface between the Bzugu and Sochapa rivers close to the Experimental Gardens of the Institute of Floriculture and Subtropical Crops (Sochi). The scientists of the Laboratory of Agrochemistry and Soil Science identified a mineral horizon atypical of this soil lying at a depth of 12–30 cm. In contrast to the contemporary natural soil of similar landscapes, the newly formed surface organic horizon, formed on the artificially created medieval carbonate claystone, is characterized by a low content of humus. Taking into account the fact that this horizon was formed to a fully mature state, this lower humus content was likely caused by changes in climatic conditions during the Holocene (the time of the formation of modern soils). The modern Subatlantic period of the Holocene, which continues to this day, is referred to as a regressive climatic stage. Average annual temperatures in the Subatlantic period were, on average, below the levels of the Middle Holocene Atlantic climatic optimum. The low content of humus in the newly formed surface of the organogenic horizon, formed on artificially created medieval masonry of carbonate claystone, served as the basis for the signal point located on this massif and indicated that this site has not been used since the Holocene climatic optimum [55], which, as shown by the climatic curve derived by A.S. Monin and Yu. A. Shishkov (Figure 9), occurred approximately in the 11th–12th centuries.

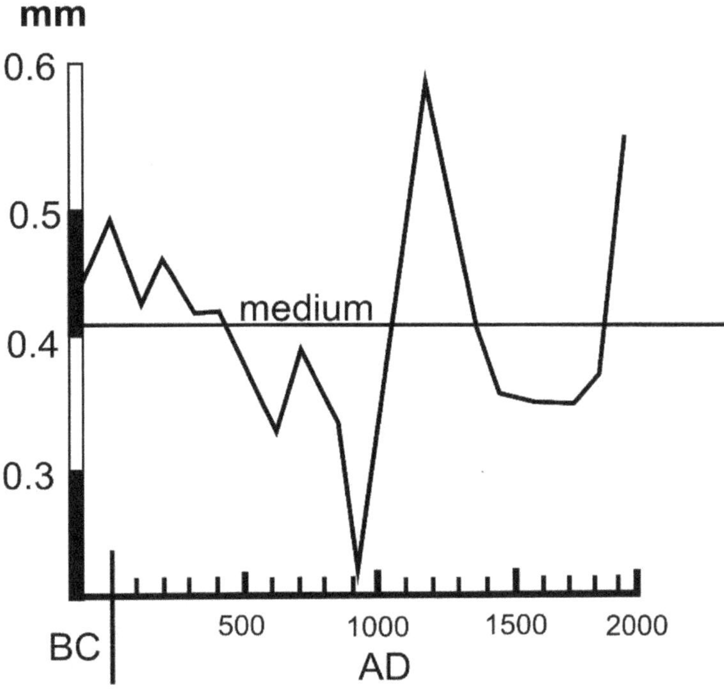

Figure 9. Climatic curve for the last two thousand years based on data on the width of the rings for one of the pine species (in mm); the curve was created based on the data given in [29].

Second only to agriculture in the Late Antique and Medieval periods was distant pastoralism. An important factor for the efficiency of this method was the length of the grazing period in alpine meadows. During the period of the climatic optimum, "the upper boundary of the vertical distribution of forests in the Alps, in the mountains of Central Europe and in Scandinavia increased by 100–200 m" [29]. Thus, it can be presumed that the period of summer vegetation of alpine meadows was longer. However, after the onset of cooling, this pattern changed. An indicative factor in the study of atsanguars (pastoral stock farming areas) is the presence of arrowheads typical for the Middle Ages and the complete absence of any material associated with firearms. Furthermore, the folk epos depicts atsanguars as the houses of Atsan dwarves, and such homes appear in a mythological, fairy-tale context; this suggests that the true purpose of these structures and their role in the economy of the region have been wiped from people's memories in recent centuries. Climatic variability was also described in greater detail in the medieval epos. As mentioned above, the Nart epos actively reflects severe cooling as a climatic cataclysm [50].

Thus, on the basis of the above facts, it can be assumed that the end of Early Middle Ages coincides with the deterioration of agriculture in the entire region. Comparing the above conditions with the borders of European territories, the objective correlation becomes obvious. During the Little Ice Age, the Caspian Sea level increased, indicating a rise in the amount of precipitation in the basin of the river Volga and, possibly, reduced evaporation. Chronicles and other historical documents confirm that cold snowy winters and rainy summer seasons became more frequent, and floods occurred more often [29].

Hydrographic networks are also an important factor in the reconstruction of paleogeographic patterns, since the physical and geographical data associated with such networks are extremely important for an objective assessment of the historical and cultural landscape and its development. Let us take the Imereti Lowlands as an example (Figure 10). The hydrographic network in the territory of the Imereti Lowlands area is a part of the Black Sea basin and represented by the Mzymta and Psou rivers, bounding the lowlands, from the west and east, respectively, as well as inland water bodies of natural (sea retreats) and artificial (excavated ponds) origin [56]. The Imereti Lowlands is a territory featuring shallow groundwater (1–2 m), which emerges during the rainy season. Nearly the entire territory of the lowlands is flooded and swampy. The central part is flooded especially intensively during periods of river floods [56]. The effects of flooding are now considerably mitigated by a drainage system designed in the lowlands and consisting of a network of drainage channels. In the period preceding construction, a significant part of the territory was occupied by malarial swamps, namely, lakes filled with silt and peat [57].

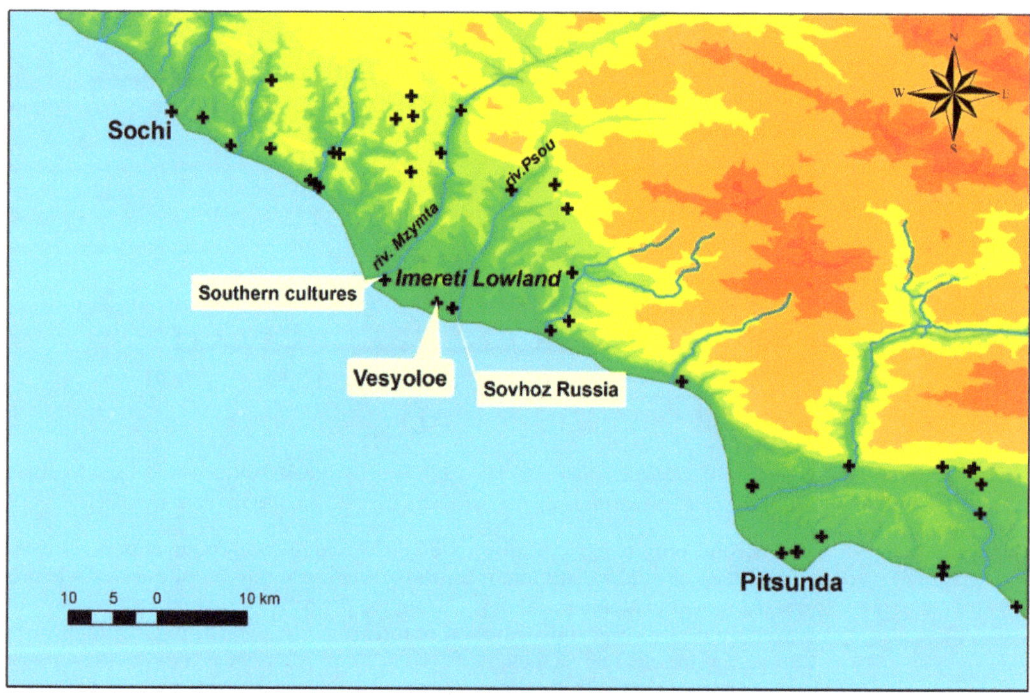

Figure 10. Temples in the Imereti Lowlands.

Archaeological excavations in the Imereti Lowlands during the construction of the Olympiad facilities in 2010–2014 revealed a number of new sites and made it possible to carry out complete excavations of previously identified archaeological sites. Archaeologists examined almost the entire Imeretiskaya lowlands for two years. All the sites identified by archaeologists and dating from the most ancient sites to the late Middle Ages were located on the outskirts of the valley, while nothing was found in the center where the stadiums were built [57]. Based on the results of the research, it was concluded that there used to be a large freshwater lake on the territory of the Olympic Park, and people settled along its shores.

Another important point is the identification of historical sites along the coast. Particular attention was given to the temple in the park "Southern Cultures", which is marked on historical navigation maps under the toponym "Saint Sophia". The fact that this location is marked on navigation maps is of crucial importance and may indicate the presence of a seaport in this area in Antiquity. Indirect evidence for this seaport is the presence of a "bay" mentioned in the reports of I.K. Nedolya, who examined this basilica in the middle of the 20th century [58]. Today, this bay does not exist.

A seaport with a large temple, which, most likely, was located inside a fortress that has not survived (a lake in the center of the lowlands), allows us to draw an analogy with a similar region on the coast that has been fairly well explored by paleogeographers and for which a detailed reconstruction has already been made: the area of Pitsunda Cape and the ancient Roman fortress Great Pitiunt [59]. Here, on the modern area of swampy lakes, there was a large inland lake open to the sea in the Ancient and early Medieval periods. Therefore, it is possible that the lake in the Imereti Lowlands could have been open to the sea. On the site of the temple in the park "Southern Cultures" on the territory of modern Sochi, there was a fortress that was one of the points of the Roman–Byzantine Pontic Limes [60]. The presence of such a fortress in that area appears logical if we analyze the settlement structure, as there are traces of caravan routes directed toward the Krasnaya Polyana region and Aibga Pass [61].

A natural question arises: Why are there no traces of this fortress? The answer lies in the study of seismic activity maps (Figures 4 and 5).

The Imereti Lowlands is a part of the new Sochi–Adler Depression and is separated from Sochi by discontinuities of the Adler flexure–fault zone, which is a seismogenic structure. This territory belongs to the 8–9 point earthquake zone [56]. Traces of damage after earthquakes, in particular, can be seen in the temples of the villages Vesyoloe and Loo [62].

Based on an analysis of the available seismological databases, a consolidated unified-magnitude MS catalogue of historical and instrumentally registered earthquakes was compiled, covering the entire territory of the Western Caucasus (Figure 4). Comparison of the results showed an obvious predominance of seismic activity in the Sochi, Adler, Sukhumi and Ochamchira regions. However, in the Gudauta region, especially in the area of Pitsundsky and Bombor capes, Gali regions, and the the Ingur river basin, such activities were not observed. This result explains the different preservation levels of objects of historic and cultural heritage. Temples similar in time of construction, architecture, and dimensions in the village Vesyoloe and Lykhny have a radically different appearance from each other. According to archaeological research, the temple in Vesyoloe was destroyed between the 11th and 13th centuries, and the temple in Lykhny stands undamaged to this day (Figure 11).

Nothing remains of the fortress in the area of the Southern Cultures Park, and the temple is located at the level of the foundation, while the magnificent medieval cathedral in Pitsunda has survived to this day almost without serious damage; additionally, the walls of the fortress from the ancient period have been preserved.

Figure 11. Lykhny temple (a), Pitsundsky cathedral (b), and fortress (orthophotoplan) (c).

4. Discussion

The crucial point of the study was to identify the impact of natural factors upon the transformation of the settlement system, which led to deurbanization in northwestern Colchis in the 12th century. A retrospective analysis of literary references revealed the absence of previous studies of the area during this certain period. However, the correlation between changes in the natural environment and the development of human society in earlier eras has long been noted by numerous researchers. Global climate changes, transgressive–regressive developments in ocean affairs, and the glacial–interglacial cycles of the Russian plains and mountain zones have been outlined as the most significant natural factors for the Black Sea area [63].

Historically, the fall of the Byzantine Empire was considered the main factor in the decline and deurbanization of the Black Sea region in the Middle Ages [1] and the changes in trade routes. The present study demonstrated that, for this particular historical period, climate and coastline alterations were the primary causes. Natural factors made the bays (previously convenient for parking ships) shallow, thus leading to the formation of subaerial gravel–pebble deposits, which completely separated the bay from the sea. Steadily, former bays without access to the sea turned into swamps, and the fortresses that guarded those ports lost their importance and trade froze. This picture can be observed in Cape Pitsunda [26]. Recent geo-archaeological examinations initiated by our expedition [64] showed that, according to an analysis of geological deposits in Cape Bombora, the paleogeographic reconstruction is comparable to that of Cape Pitsunda. This has been confirmed, in particular, by the results of studies on the chemical and granulometric composition of deposits, their carbonate content, and the constructed lithological column (Figure 12). In the near future, in order to verify these findings, a wider range of studies, including isotopic, spore pollen, faunal analysis, etc., are planned. In addition, geological research of the Imereti Lowlands is essential. Although archaeological data mostly confirm our assumptions, it is impossible to prove the influence of natural factors on the development of the settlements without conducting comprehensive geological studies at the present level. Therefore, the conclusions of our article, based on an integrated approach, also present new challenges.

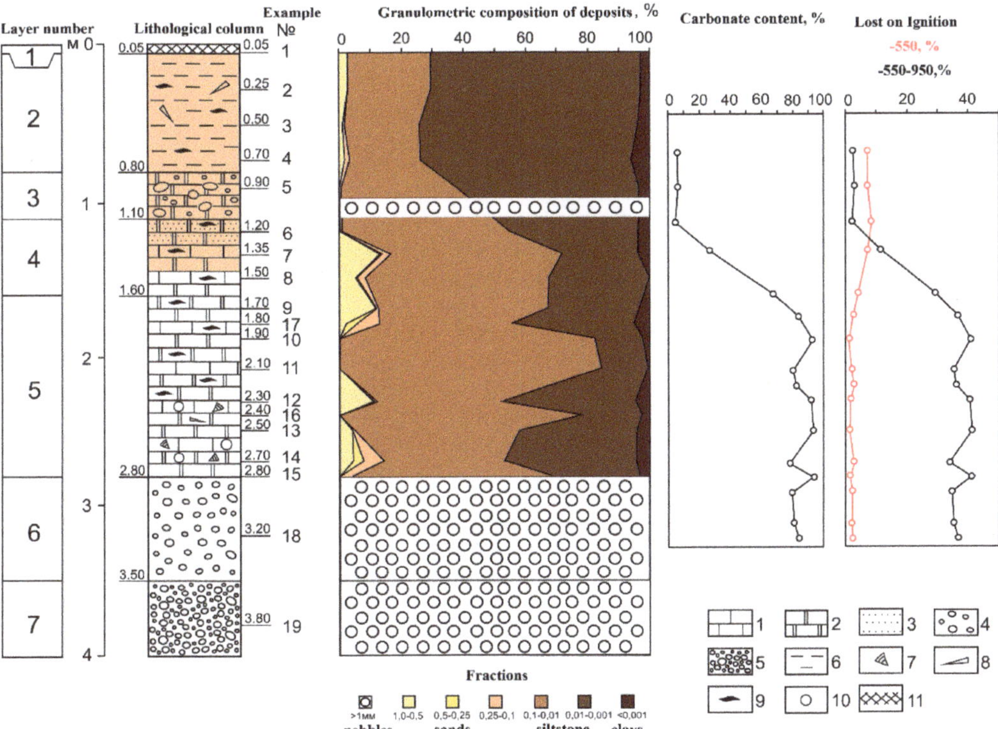

Figure 12. The structure and composition of the section of sedimentary rocks found in Cape Bombora: 1. calcite; 2. dolomite; 3. sand; 4. loose pebbles; 5. dense pebbles; 6. siltstone; 7. shell detritus; 8. fragments of lime microorganisms; 9. fragments of plant tissue; 10. carbonate concretions; 11. sod layer. Loss on Ignition (LOI) curve No. 550 indicates the amount of combustible organics, No. 950 shows the amount of carbonate minerals that release carbon dioxide at specified temperatures.

It should be noted that the participation of hydrologists and geologists in assessing changes in the channels of large rivers at their confluence with the Black Sea highlights important areas for further research. As can be seen from the corresponding analysis that many coastal have been insufficiently studied by archeologists. Wide estuary valleys were, and remain, the areas at the center of urbanization, so the imbalance between the number of archeological findings and modern settlements cannot be merely accidental. The transgressive–regressive developments that caused active changes in the Black Sea coastline and the contours of the riverbed, in addition to the subsequent modern urbanization of the area, could hide ancient settlements and fortresses vital for study. Moreover, estuarine valleys are the most active zones in the formation of sewage sediments from the entire drainage of large rivers.

Other factors associated with uneven precipitation on the coast are the area of Holocene terraces and their distance from the adjacent mountain slopes. These factors affect the thickness and rate of deposit formation. When studying and identifying architectural sites that have survived seismic impacts, these concerns are not as critical. However, for completely ruined objects, deposit formation is of extreme importance. A completely ruined temple in the zone, featuring the slow formation of soil influxes, is usually a low hill two to three meters in height. However, in the zone of active influx of soils on the foothill terrace, such hills can be completely buried at a maximum height to the horizon of the adjacent territory, which makes them particularly difficult to notice when applying conventional methods. Therefore, the application of an interdisciplinary approach is crucial and will assist in further clarifying and supplementing the layered composition of spatial information in the developed archaeological GIS.

5. Conclusions

This study was based primarily on archaeological data, which served as the basis for analyzing the settlement structures of historical and cultural heritage sites via the GIS. For territories that offer data left by previous inhabitants, analyses were carried out using the methods of various sciences, both natural (climatology, seismology, hydrology, etc.) and humanities-based (folklore, historical cartography, etc.). Synthesis of these data revealed a clear correlation between various factors. In terms of the impact of the physical environment on human evolution, data confirmed that climate change that occurred at the turn of the 12th century led to a crisis in agriculture, producing a decline in both farming and cattle breeding, which certainly affected demography. Seismic activity, noted in the same period, led to the destruction of many buildings (temples and fortresses) and a change in the nature of hydrological networks, which are directly linked to climate change. These changes yielded the formation of swamp lakes on the ground, and, as a result, led to the loss of a number of coastal sites and their abandonment.

Thus, it can be concluded that natural, rather than political, factors played the key role in changing the settlement structure (in terms of both demographic decline and deurbanization). The political changes that took place were the consequence of changes in the physical environment.

Author Contributions: Conceptualization, G.T. and A.K.; methodology, A.K. and V.L.; investigation, G.T. and A.K.; analysis archival materials, V.L. and G.Y.; writing—original draft preparation, G.T. and A.K.; writing—review and editing, G.Y.; visualization, V.L.; supervision, G.T.; project administration, G.Y.; funding acquisition, G.Y. All authors have read and agreed to the published version of the manuscript.

Funding: This research was funded by the Russian Science Foundation, grant number 22-18-00466.

Data Availability Statement: The data presented in this study are available on request from the corresponding author.

Conflicts of Interest: The authors declare no conflict of interest.

References

1. Voronov, Y.N. *Antiquities of Sochi and Its Environs;* Prince Publishing House: Krasnodar, Russia, 1979; p. 103.
2. Hager, B.H.; Clayton, R.W.; Richards, M.A.; Comer, R.P.; Dziewonski, A.M. Lower mantle heterogeneity, dynamic topography and the geoid. *Nature* **1985**, *313*, 541–545. [CrossRef]
3. King, G.C.P.; Bailey, G.N. *Dynamic Landscapes and Human Evolution;* GSA Special Paper 471; Geological Society of America: Boulder, CO, USA, 2010.
4. Bolikhovskaya, N.S.; Gorlov, Y.V.; Kaitamba, M.D.; Muller, K.; Porotov, A.V.; Parunin, O.B.; Fuash, E. Changes in the landscape and climatic conditions of the Taman Peninsula over the past 6 thousand years. *PIFK* **2002**, *12*, 257–271.
5. Gorlov, Y.V.; Trebeleva, G.V. Integrated paleoecological and archaeological research on the Taman Peninsula. In *Archaeological Discoveries 1991–2004. European Russia;* Makarov, A.N., Ed.; Institute of Archeology of the Russian Academy of Sciences: Moscow, Russia, 2009; pp. 167–173.
6. Koshelenko, G.A.; Gaibov, V.A.; Trebeleva, G.V. The main channel of the Murgab River in the Parthian time. In *GAUDEAMUS IGITUR;* Collection of Articles Dedicated to the 60th Anniversary of A.V. Podosinov; Djakson, T.N., Konovalova, I.G., Tsetskhladze, G.R., Eds.; Foundation for the Promotion of Education and Science: Moscow, Russia, 2010; pp. 225–234.
7. Bailey, G.N.; King, G.C.P. Dynamic landscapes and human dispersal patterns: Tectonics, coastlines, and the reconstruction of human habitats. *Quat. Sci. Rev.* **2011**, *30*, 1533–1553. [CrossRef]

8. Savage, S.H. *GIS in Archaeological Research, Interpreting Space: GIS and Archaeology*; Allen, K.M.S., Green, S.W., Zubrow, E.B.W., Eds.; Taylor & Francis: Abingdon, UK, 1990; pp. 22–32.
9. Verhagen, P. Spatial Analysis in Archaeology: Moving into New Territories. In *Digital Geoarchaeology*; Natural Science in Archaeology; Springer: Cham, Switzerland, 2018; pp. 11–25. [CrossRef]
10. Smekalov, S.L.; Fedorov, D.L. *Geoinformation Technologies in Archaeological Research*; Baltic State Technical University: St. Petersburg, Russia, 2004; 104p.
11. Trebeleva, G.V. GIS-technologies in Archaeological Conservation in Moscow Region (based on the experience of the Moscow Region Expedition of the IA RAS). In *Archeology of the Moscow Region*; Materials of the Scientific Seminar; IA RAN: Moscow, Russia, 2004; pp. 216–220.
12. Koshelenko, G.A.; Gaibov, V.A.; Trebeleva, G.V. Archaeological Geoinformation System of Margiana. *Archeol. Geoinform.* **2007**, *4*. Available online: https://www.archaeolog.ru/media/periodicals/agis/AGIS-4/Koshelenko/page1.html (accessed on 19 June 2022).
13. Trebeleva, G.V. Defense of borders of the Bosporian Kingdom during the Roman time: Historical modeling on the basis of GIS-technology. *Eirene* **2006**, *42*, 159–166.
14. Mlekuž, D. Time geography, GIS and archaeology. In Proceedings of the 38th Conference on Computer Applications and Quantitative Methods in Archaeology, Granada, Spain, 6–9 April 2010; pp. 1–7.
15. De Gruchy, M.; Jotheri, J.; Alqaragholi, H.; Al-Janabi, J.; Alabdan, R.; Al-Talaqani, H.; Almamouri, G.; Al-Rubaye, H. The Khandaq Shapur: Defense, Irrigation, Boundary, Frontier. *Land* **2021**, *10*, 1017. [CrossRef]
16. Lee, T.W.; Walker, J.H. Forests and Farmers: GIS Analysis of Forest Islands and Large Raised Fields in the Bolivian Amazon. *Land* **2022**, *11*, 678. [CrossRef]
17. Trebeleva, G.V.; Yurkov, G.Y.; Kizilov, A.S.; Glazov, K.A.; Shvedchikova, T.Y. Complex Investigation (GIS, Photogrammetry, and Natural-Scientific Methods) of the Northwestern Colchis Historical and Cultural Landscape in the Late Antique and Medieval Times. In Proceedings of the Geoarchaeology and Archaeological Mineralogy: 7th Geoarchaeological Conference, Miass, Russia, 19–23 October 2020; Springer Proceedings in Earth and Environmental Sciences. Springer: Berlin/Heidelberg, Germany, 2020; pp. 365–382. [CrossRef]
18. Trebeleva, G.; Glazov, K.; Kizilov, A.; Kizilova, A.; Yurkov, V.; Yurkov, G. Advanced technologies used in digitizing the cultural heritage of northwestern Colchis: The experience of the Markul Expedition. *Appl. Sci.* **2022**, *12*, 2052. [CrossRef]
19. Christaller, W. *Die Zentralen Orte in Süddeutschland eine Ökonomisch-Geographische Untersuchung über die Gesetzmässigkeit der Verbreitung und Entwicklung der Siedlungen mit Städtischen Funktionen*; Wissenschaftliche Buchgesellschaft Publ.: Darmstadt, Germany, 1980; 331p.
20. Trebeleva, G.V.; Sakania, S.M.; Glazov, K.A.; Kizilov, A.S.; Khondzia, Z.G.; Yurkov, G.Y. Project for Cataloging Late Antique and Medieval Temples in Abkhazia: Findings and Future Perspectives. *Archit. Archeol.* **2020**, *2*, 54–63. Available online: https://www.archaeolog.ru/media/periodicals/arch-arch/AA-2.pdf (accessed on 9 September 2022).
21. Trebeleva, G.V.; Yurkov, G.Y.; Gorlov, Y.V.; Tzvinaria, I.I.; Agumaa, A.S.; Sangulia, G.A.; Kaitan, S.G. Abkhaz medieval monuments: Analysis of construction technology and dating issues. *Probl. Hist. Philol. Cult. J. Hist. Philol. Cult. Stud.* **2013**, *2*, 297–303. Available online: https://www.pifk.magtu.ru/doc/pifk-02-2013.pdf (accessed on 9 September 2022).
22. Klemeshova, M.E.; Trebeleva, G.V.; Kizilov, A.S.; Glazov, K.A.; Sokolov, S.V. The experience of using the technique of A.A. Bobrinsky for the study of ceramic vessels in the study of the composition of the molding masses of plinths of medieval temples and fortresses of Eastern Abkhazia. In Proceedings of the VIII All-Russian Scientific Conference "Geoarchaeology and Archaeological Mineralogy", Miass-Chelyabinsk, Russia, 20–23 September 2021; pp. 97–101. Available online: https://meetings.chelscience.ru/geoarcheology/collection-geoarcheology/ (accessed on 9 September 2022).
23. Armarchuk, E.A.; Mimokhod, R.A.; Sedov, V.V. Christian church near the village Vesyoloe: Preliminary publication of the results of excavations in 2010. *Russ. Archeol.* **2012**, *3*, 78–90.
24. Shmidt, T.S. Hydrographic Description of Rivers, Lakes and Reservoirs. Transcaucasia and Dagestan. Western Transcaucasia. In *Resources of Surface Waters of the USSR*; Gidrometioizdat: Leningrad, Russia, 1974; Volume 9, 424p.
25. Khmaladze, G.N. *Silts Discharged by the Rivers of the Black Sea Coast of the Caucasus*; Gidrometioizdat: Leningrad, Russia, 1978; 168p.
26. Balabanov, I.P. *Paleographic Prerequisites for the Formation of Modern Natural Conditions and a Long-Term Forecast for the Development of the Holocene Terraces of the Black Sea Coast of the Caucasus*; Dalnauka: Vladivostok, Russia, 2009; p. 120.
27. Gerashchenko, I.N. Analysis of Geographical Features and Thermal Regime of the Rivers of the Russian Black Sea Region. Ph.D. Dissertation, Belgorod State University, Belgorod, Russia, 2002; 257p.
28. Gerashchenko, I.N. Hydrological and Hydrographic Characteristics of Some River Basins in the Humid Subtropics of the Black Sea region of the Krasnodar Territory. *Int. Sci. J. Symb. Sci.* **2017**, *3*, 230–232. Available online: https://cyberleninka.ru/article/n/gidrologo-gidrograficheskaya-harakteristika-nekotoryh-bassenov-rek-vlazhnyh-subtropikov-prichernomorya-krasnodarskogo-kraya/viewer (accessed on 19 June 2022).
29. Monin, A.S.; Shishkov, Y.A. *Climate History*; Gidrometeoizdat: Leningrad, Russia, 1979; 406p.
30. Nikonov, A.A. A New Approach to Assessing the Seismic Potential and Hazard on the Black Sea Coast of the Caucasus (Based on Archeoseismic Data). Available online: https://www.researchgate.net/publication/301370692 (accessed on 19 June 2022).
31. Balabanov, I.P.; Nikiforov, S.P. *Gagra Bay. Recreational Potential of Natural-Geological Conditions of the Coastal-Marine Zone*; Publishing House "Author's Book": Moscow, Russia, 2016; 288p.

32. Akimov, V.A.; Zaitsev, V.A.; Larkov, A.S.; Lutikov, A.I.; Ovsyuchenko, A.N.; Panina, L.V.; Rogozhin, E.A.; Rodina, S.N.; Sysolin, A.I. Seismic hazard maps of the Northwestern and Central Caucasus in a detailed scale. *Issues Eng. Seismol.* **2019**, *46*, 57–74.
33. Stiros, S. Identification of earthquakes from archaeological data: Methodology, criteria and limitation. In *Archaeoseimology*; Fitch Laboratory Occasional Papers No. 7; Stiros, S., Jones, R., Eds.; British School at Athens: Athens, Greece, 1996; pp. 129–152.
34. Galadini, F.; Hinzen, K.-G.; Stiros, S. Archaeoseismology: Methodological issues and procedure. *J. Seismol.* **2006**, *10*, 395–414. [CrossRef]
35. Zhuravlev, A.P. *Seismic Archeology—The Science of the Third Millennium*; Zemlia Zaonezhia: Petrozavodsk, Russia, 1999; 85p.
36. Martín-González, F. Earthquake damage orientation to infer seismic parameters in archaeological sites and historical earthquakes. *Tectonophysics* **2018**, *724–725*, 137–145. [CrossRef]
37. Vinokurov, N.I.; Nikonov, A.A. Traces of an earthquake of the Antic time in the west of the European Bosporus. *Russ. Archeol.* **1998**, *4*, 98–114.
38. Kazmer, M.; Kolaiti, E. Earthquake-induced deformations at the Lion Gate, Mycenae, Greece. In Proceedings of the 6th International INQUA Meeting on Paleoseismology, Active Tectonics and Archaeoseismology, Pescina, Fucino Basin, Italy, 19–24 April 2015; Volume 27, pp. 248–251. Available online: https://www.researchgate.net/publication/292139936_Earthquake-induced_deformations_at_the_Lion_Gate_Mycenae_Greece (accessed on 9 September 2022).
39. Ambraseys, N.; Psycharis, I. Assessment of the long-term seismicity of Athens from two classical columns. *Bull. Earthq. Eng.* **2012**, *10*, 1635–1666. [CrossRef]
40. Papantoniou, G.; Vionis, A.K. Landscape Archaeology and Sacred Space in the Eastern Mediterranean: A Glimpse from Cyprus. *Land* **2017**, *6*, 40. [CrossRef]
41. Tourloukis, V. The Early and Middle Pleistocene Archaeology Record of Greece: Current Status and Future Prospects. Ph.D. Thesis, Leiden University, Leiden, The Netherlands, 2010. Available online: https://scholarlypublications.universiteitleiden.nl/handle/1887/16150 (accessed on 9 September 2022).
42. Buck, V.A. *Archeoseismology in Atlanta Region, Central Mainland Greece: Theory, Method, and Practice*; BAR Publishing: Oxford, UK, 2006; 120p. Available online: https://www.barpublishing.com/archaeoseismology-in-atalanti-region-central-mainland-greece.html (accessed on 9 September 2022).
43. Altunel, E. Evidence for damaging historical earthquakes at Priene, western Turkey. *Turk. J. Earth Sci.* **1998**, *7*, 25–35.
44. Al-Tarazi, E.A.; Korjenkov, A.M. Archaeoseismological investigation of the ancient Ayla site in the city of Aqaba, Jordan. *Nat. Hazards* **2007**, *42*, 47–66. [CrossRef]
45. Saucier, R.T. Geoarchaeological evidence of strong prehistoric earthquakes in the New Madrid (Missouri) seismic zone. *Geology* **1991**, *19*, 296–298. [CrossRef]
46. Ambraseys, N.N.; Melville, C.; Adams, R. *The Seismicity of the Egypt, Arabia, and the Red Sea, a Historical Review*; Cambridge University Press: Cambridge, UK, 1994; 181p.
47. Galli, P.; Galadini, F. Surface faulting of archaeological relics. A review of case histories from the Dead Sea to the Alps. *Tectonophysics* **2001**, *335*, 291–312. [CrossRef]
48. Satuluri, S.; Gadhavi, M.S.; Malik, J.N.; Vikrama, B. Quantifying seismic induced damage at ancient site Manjal located in Kachchh Mainland region of Gujarat, India. *J. Archaeol. Sci. Rep.* **2020**, *33*, 102486. [CrossRef]
49. Karakhanian, A.S.; Trifonov, V.G.; Ivanova, T.P.; Avagyan, A.; Rukieh, M.; Minini, H.; Dodonov, A.E.; Bachmanov, D.M. Seismic deformation in the St. Simeon Monasteries (Qal'at Sim'an), Northwestern Syria. *Tectonophysics* **2008**, *453*, 122–147. [CrossRef]
50. Rakhno, K.Y. Cooling in the Nart epic of Ossetians and Avesta. In Proceedings of the Nartology in the 21st Century: Modern Paradigms and Interpretations, Collection of Scientific Papers, Vladikavkaz, Russia, 6–7 October 2019; pp. 54–55.
51. Trebeleva, G.V.; Sakania, S.M.; Kizilov, A.S.; Glazov, K.A. Abzhui Abkhazia, the area of the Tamysh (Toumysh) and Dgamsh river basins: Reconstruction of the settlement structure based on a spatial analysis of temple architecture monuments. *Probl. Reg. Ecol.* **2020**, *6*, 72–85.
52. Kizilov, A.S.; Platonov, A.P. Agriculture of Greater Sochi from the Eneolithic to the Middle Ages based on materials from archaeological and historical sources. *Subtrop. Ornam. Gard.* **2021**, *77*, 15–25.
53. Sizov, V.I. Eastern coast of the Black Sea. In *Archaeological Excursions*; Uvarova, P., Ed.; Imperial Moscow Archaeological Society: Moscow, Russia, 1889; Volume 2, p. 7.
54. Yurkov, V.G.; Klemeshova, M.E.; Trebeleva, G.V. Issues of the chronology of the Markul settlement in the light of new discoveries in 2021. In Proceedings of the Actual Archeology 6—Proceedings of the International Scientific Conference of Young Scientists, St. Petersburg, Russia, 4–7 April 2022; pp. 194–197.
55. Kizilov, A.S.; Zakharikhina, L.V. New Medieval objects of the on the left bank of the Sochapa River as evidence of the peculiarities of the life of a medieval person and the spatial and temporal variability of soil formation factors. In Proceedings of the IX "Anfimov Readings" on the Archeology of the Western Caucasus, Anapa, Russia, 29–31 May 2019; pp. 158–161.
56. Osipov, V.I.; Vadachkoriya, O.A.; Mamaev, Y.A.; Yastrebov, A.A. Geotechnical conditions and protection of the construction site of the Olympic Park in Sochi. *Geoecol. Eng. Geol. Hydrogeol. Geocriol.* **2010**, *6*, 483–493.
57. Kryukov, A.V. Objects of cultural heritage within the boundaries of the natural ornithological park in the Imereti lowland: Composition and problems of conservation. In Proceedings of the Sustainable Development of Specially Protected Natural Territories, Collection of Articles of the V All-Russian Scientific and Practical Conference, Sochi, Russia, 10–12 October 2018; pp. 169–178.

58. Nedolya, I.K. *Report on the Survey of Medieval Sites in the Greater Sochi Region in 1954–1968*; Archive of the Sochi Branch of the Russian Geographical Society: Adler, Russia, 1969.
59. Trebeleva, G.; Glazov, K.; Kizilov, A.; Sakania, S.; Yurkov, V.; Yurkov, G. Roman fortress Pitiunt: 3D-reconstruction of the monument based on the materials of archaeological research and geological paleoreconstructions. *Appl. Sci.* **2021**, *11*, 4814. [CrossRef]
60. Trebeleva, G.V.; Kizilov, A.S. Once again to the question of the "Pontic Limes", or on the geographical and geopolitical principles of the location of the ancient fortifications of the Black Sea coast in ancient times. In Proceedings of the Ancient and Traditional Cultures in Interaction with the Environment: Problems of Historical Reconstruction: Materials of the I International Interdisciplinary Conference, Chelyabinsk, Russia, 13–15 April 2021; Kupriyanova, E.V., Ed.; Chelyabinsk Publishing House: Chelyabinsk, Russia, 2021; pp. 51–60.
61. Kizilov, A.S.; Trebeleva, G.V. To the question of the geographical and geopolitical principles of the location of the ancient fortifications of the Black Sea coast: Between the Bosporus and Abazgia. *Probl. Reg. Ecol.* **2020**, 3, 98–111. [CrossRef]
62. Ovchinnikova, B.B. Ten years of the Loo Archaeological Expedition (1987–1997). In *News of the Ural State University*; Ural State University: Ekaterinburg, Russia, 1997; Volume 3, pp. 119–123.
63. Lisitsyn, A.P. (Ed.) *The Black Sea System*; Scientific World: Moscow, Russia, 2018; 808p.
64. Trebeleva, G.V.; Chepalyga, A.L.; Sakania, S.M.; Sadchikova, T.A.; Yurkov, G.Y. *Geoarchaeological Studies of the Coast of NORTH-WESTERN Colchis Geoarchaeology and Archaeological Mineralogy—2022*; Scientific Publication; YuUrGGPU Publishing House: Miass-Chelyabinsk, Russia, 2022; pp. 14–19. Available online: https://www.academia.edu/89025738/Geoarkheologia_2022_Bombora (accessed on 9 November 2022).

 land

Article

The Evolutionary Process and Mechanism of Cultural Landscapes: An Integrated Perspective of Landscape Ecology and Evolutionary Economic Geography

Zhiqiang Gong [1,2], Zhuting Zhang [1], Jianqin Zhou [1], Jiami Zhou [1] and Wenhui Wang [1,3,*]

1 School of Tourism, Nanchang University, Nanchang 330031, China
2 Wanli Administration Bureau, Nanchang 330004, China
3 Tourism Research Institute, Nanchang University, Nanchang 330031, China
* Correspondence: whwang@ncu.edu.cn

Abstract: Cultural landscapes are joint masterpieces of man and nature with outstanding universal value. Adequate knowledge of their evolutionary process and mechanism is crucial to their development, protection, and management. However, theoretical understanding about such has been limited as existing studies tend to focus on the descriptive and interpretative analysis of the evolutionary process and pay less attention to the underlying mechanism of the process. Integrating the traditional perspective of landscape ecology in cultural landscape research and theories of path dependence and path creation in evolutionary economic geography, this paper constructs a triple-layered integrated analytical framework of cultural landscape evolution and applies the framework to empirically examine the cultural landscape evolution of Mount Lushan. To grasp an accurate and full picture of the process, field observation and historical data collection were carried out, and a combination of thematic analysis and chronological organization was conducted. The research finds that the cultural landscape evolution of Mount Lushan has experienced three stages, i.e., coexistence and mutual influence of multiple cultures, conflict and integration of Chinese and Western cultures, as well as landscape transformation, revival, and expansion. Such evolution is a non-linear, dynamic, and complex process across which the elements, functions, and patterns of landscapes were constantly constructed and reconstructed. Fundamentally, it is the result of the synergistic effect of path dependence and path creation, and is driven by the interplay of the behavior of associated actors and the change of contextuality. The findings of this study can provide some strategic references for the management practice of cultural landscape heritage sites.

Keywords: cultural landscapes; landscape ecology; evolutionary economic geography; spatiotemporal evolution; Mount Lushan

Citation: Gong, Z.; Zhang, Z.; Zhou, J.; Zhou, J.; Wang, W. The Evolutionary Process and Mechanism of Cultural Landscapes: An Integrated Perspective of Landscape Ecology and Evolutionary Economic Geography. *Land* **2022**, *11*, 2062. https://doi.org/10.3390/land11112062

Academic Editors: Bellotti Piero and Alessia Pica

Received: 24 October 2022
Accepted: 16 November 2022
Published: 17 November 2022

1. Introduction

As a joint masterpiece of man and nature with outstanding universal value, cultural landscapes mingle stunning natural landscapes and unique cultural connotations [1]. In 1992, the 16th General Assembly of the World Heritage Committee formally introduced "cultural landscape heritage" and included it as a subcategory of cultural heritage. This initiative has further stimulated scholars in fields including geography, history, anthropology, and sociology to pay increasing attention to the nomination, development, conservation, and management of cultural landscapes. The utilization of cultural landscape heritage resources, such as the development of heritage tourism, is a necessary means to make their outstanding and unique value known to the public and to protect them sustainably. However, in many heritage sites, development actions failed to adequately acknowledge the uniqueness of cultural landscapes such that irreversible damage to the associated cultural landscapes [2], and issues of fragmentation, isolation, artificialization, and over-commercialization was induced [3–5].

An adequate ontological understanding of cultural landscapes is a prerequisite for their effective conservation and sustainable development [6]. Cultural landscapes do not emerge from a vacuum, but are evolved through a dynamic, complex, and multi-faceted historical process [7]. The systematic analysis of the historical evolution of cultural landscapes, the changes in the forms, elements, structures, and values of the landscapes through the evolutionary process, as well as the driving mechanism and influencing factors of the evolution, can provide strategic reference for the development, conservation, and management of respective cultural landscapes in terms of identifying heritage values and functions, establishing heritage development directions and methods, and formulating adaptive management tools [8]. However, the existing studies and management practices of cultural landscapes are mostly based on the status quo of particular heritage sites, imperfectly mapping the dynamic nature of the cultural landscapes and offering insufficient knowledge about their value identification, conservation, and management [9].

Landscape ecology, especially the landscape ecosystem theory, is the mainstream and traditional perspective of cultural landscape research [10–13]. From a holistic and integrated standpoint, it regards human activities and their surrounding natural environment as a whole, while emphasizing the subjectivity of humans [14,15]. It goes beyond the dualistic view of nature and humanity and offers a panoramic, multi-dimensional, and multi-scale perspective toward the evolution of cultural landscapes. The key notions of landscape ecology, such as pattern, function, elements, and process, provide a comprehensive analytical framework for explaining the spatial and temporal evolution of cultural landscapes. Regarding the driving mechanisms of cultural landscape evolution, landscape ecology highlights the joint impacts of economic, social, political, and technological factors [14,15]. Overall, it considers the influence of the exogenous contextual factors on landscape evolution, but pays little attention to the impacts of the intrinsic factors, such as heritage site resources and historical evolutionary paths, as well as the associated agents, such as governments, tourists, and heritage site residents, and their actions. As a result, more theoretical attempts have been investigated by cultural landscape researchers in order to reveal the underlying mechanism of cultural landscape evolution [16,17].

Path dependence and path creation, as the two major theories of economic geography to examine the spatio-temporal evolution of economic landscapes, may provide further inspirations in that regard. Path dependence theory considers economic landscapes as an open, non-unitary equilibrium system. It emphasizes the important role of time and history in analyzing the evolution of economic landscapes, taking it as a process governed by the past development paths of landscapes [18–20]. Path creation theory, on the other hand, focuses on the creativity and novelty in the evolution of economic landscapes and emphasizes the important role of the dynamic actions of relative agents through the evolutionary process [21]. Their emphasis on the historicity and subjectivity of the evolution of economic landscapes and their attention to the synergistic interaction of endogenous and exogenous factors through the spatio-temporal process provide a comprehensive analytical framework for understanding the deep underlying mechanisms of cultural landscape evolution [22–25]. In view of the above background, based on the notions of pattern, function, elements, and process in landscape ecology, and drawing inspirations from path dependence theory and path creation theory, this paper constructs an analytical framework for the evolution process and mechanism of cultural landscapes. It further takes the world cultural landscape heritage of Mount Lushan in China as an example to conduct an in-depth analysis on the process and mechanism of the spatio-temporal evolution of its elements, pattern, and function. The purpose of such is to enrich the theoretical understanding of cultural landscape evolution and to provide strategic reference for the conservation, development, and management of cultural landscapes.

2. Literature Review and Conceptual Framework

The concept of "cultural landscape" emerged with the rise of human geography. In the second half of the 19th century, German scholar F. Ratzel first proposed the concept of "cultural landscape", and O. Schluter (1906) further proposed the "cultural landscape theory", which clarified the important relevance of landscape to human society [26]. Thereafter, American scholar Carl. O. Sauer (1927) [26] founded the landscape school, and German scholar C. Troll (1939) founded landscape ecology, indicating an increasing academic focus on the landscape patterns and processes under human actions. The increasing academic research has promoted the currency and acceptance of the concept of cultural landscape, marked by that world cultural landscape heritage became a subcategory of world heritage in December 1992. Since then, a growing number of studies have been conducted surrounding different types of cultural landscapes, such as industrial landscapes [27], agricultural landscapes [28,29], landscapes of traditional villages [16], historic towns [30], and linear landscapes [11], and on diverse topics such as value identification, conservation of the authenticity and integrity [31], sustainable development (of heritage tourism) [32], adaptive management [33], the impacts of world heritage inscription on heritage conservation and management [34], and so forth. Overall, the existing research of cultural landscapes predominantly focus on the question of "what cultural landscapes should be?", while there is less concern about "why cultural landscapes have evolved into what they are" [6].

Landscape ecology, especially the landscape ecosystem theory, is the mainstream and traditional perspective of cultural landscape research. As a comprehensive interdisciplinary subject, landscape ecology concerns the interplay between a landscape's biophysical and socioeconomic components across spatial and temporal scales, emphasizing the heterogeneity of patterns throughout the process [35]. The major research object of landscape ecology is the landscape ecosystems, i.e., inherently complex systems of landscapes and the ecological processes they support. Generally, landscape ecosystem studies insist a holistic analysis of the pattern, function and dynamics of landscapes and the agency of humans [13–15]. As complex human–earth systems, cultural landscapes are landscape ecosystems with specific patterns, functions, and dynamics at scales of interest, the evolution of which is influenced by the interaction between nature and humans [36]. In recent years, many scholars have introduced the notions of pattern, function, process, mechanism, and scale from landscape ecosystem theory into cultural landscape studies, analyzing the elements, functions, and characteristics of cultural landscapes, and exploring the evolutionary processes, patterns, and mechanisms of the spatio-temporal evolution of cultural landscapes [37,38]. Pattern means the spatio-temporal arrangement of landscape elements of different sizes and shapes. Function represents the interaction between landscape structures and ecological processes, or the interaction between landscape structural units. Process refers to the evolution of landscape structure and function over time. Mechanism explains the way different factors, such as the economic, political, social, cultural, and technological ones, drive the evolution of landscape elements, structure, and function. Scale refers to the size of the area of the studied ecosystem (spatial scale) or the time interval of its dynamic changes (temporal scale). Many scholars have pointed out that a solid understanding of the evolutionary process and mechanism of cultural landscapes requires an in-depth examination of the interaction of human and ecological components across the landscape evolution process [39,40]. In that regard, landscape ecosystem theory offers a comprehensive, multidimensional, and multi-scale analytical framework which integrates human and natural elements such as economic, social, geographic, and ecological factors into the analysis of the complex evolution processes of cultural landscapes. However, the driving factors of cultural landscape evolution do not act independently; rather, they interact with one another, jointly influencing the evolution of cultural landscapes [41]. The existing studies on the evolution of cultural landscapes focus on the process and pattern of the spatio-temporal evolution, noticing that the functions of the landscapes coevolve through the process [42], and that the process is driven by different influencing factors [43,44]. Nevertheless, the deep mechanism of how different factors individually and jointly drive the evolution of cultural landscapes, and how the functions and the patterns of the landscape ecosystems coevolve through the process, is less addressed. In this aspect, the studies of the cultural landscape evolution are probably still in need of complementary theories in addition to the landscape ecosystem theory to develop a more systematic theoretical framework.

Regarding that, path dependence theory, which is commonly used to examine causal processes, and path creation theory, which emphasizes the role of human actors, may provide supplementary inspirations on the evolutionary process and mechanism of cultural landscapes in addition to the landscape ecosystem theory. Path dependence and path creation are two different perspectives in evolutionary economic geography for analyzing the evolution of regional economic landscapes. In the 1980s, path dependence theory was first introduced into the field of economics from biological research by American economic historian Paul A. David (1985) [45], and gradually became an important theoretical tool for studying the laws of economic and social evolution [20]. Path dependence theory emphasizes the importance of time and history, focusing on the dependence of system development on its own historical path and the resources, institutions, knowledge, technology, etc., inherited from the path [19]. The common result of path dependence is lock-in, i.e., once a system enters a certain development path (whether a good or bad one), it will continuously reinforce and gradually lock-in to that particular path under the effect of inertia until an "external shock" occurs to unlock it [46]. However, the path dependence theory does not involve the discussion of whether and how the locked path can be transformed, nor does it explain whether and how a new alternative path can be generated while the economic landscape is locked to a development path [47]. In this aspect, path creation theory provides an effective supplementary angle. The path creation theory originated from economist Schumpeter's theoretical research on creative destruction, that is, the entrepreneur's willpower plays a decisive role in the creation of new development paths. Garud and Karne (2001) explicitly introduced the notion of "path creation", highlighting economic actors' behavior of "conscious deviation" from the existing path [48]. Generally, path creation theory focuses on the dynamic role of the actors and their actions, which are embedded in economic, social, and cultural contexts across scales, in the reconstruction of economic landscapes [49]. It rejects the historical deterministic view of landscape evolution and regards path locking as a controllable and temporary conditional equilibrium state [50]. With the joint impacts of actors' conscious deviation behavior and the shock brought by the change of external economic, social, and cultural contexts, economic landscapes may break out the state of lock-in to make their development paths deviate from the existing ones and open new rounds of evolution [51]. Nevertheless, new paths are not created out of a vacuum [48]. There must be a certain basis on which path creation may be developed. Path dependence and path creation are relations of the unity of opposites in the evolution of economic landscapes. Thus, the two theories are complementary to one another in explaining the evolutionary process and mechanism of cultural landscapes.

Cultural landscapes are inherently complex landscape ecosystems with heterogeneous historical legacies [52]. The evolution of cultural landscapes is not only a cultural phenomenon, but also a socio-economic process. It is embedded in multi-scale economics, social and cultural contexts, and driven by the behaviors and interactions of actors from the local, regional, national, and even international spheres [53]. For the evolution of cultural landscapes, it is not sufficient to examine only their geographical rootedness and temporal historicity, and neither the dynamic behaviors of associated actors [54,55]. In these regards, path dependence theory and path creation theory are complementary to each other in explaining the evolutionary process of cultural landscape heritage and its inner driving mechanisms. The "historicity" of path dependence theory and the "subjectivity" of path creation theory jointly offer an integrated approach to systematically expound the interplay of endogenous and exogenous forces in the spatio-temporal evolution of cultural landscapes, uncovering the laws and mechanisms behind the evolution process [56]. In recent years, some scholars have noted the phenomenon of path dependence in cultural landscape evolution [57–59], while few scholars have explored the phenomenon of path creation in the field of cultural landscapes. Therefore, how path dependence theory and path creation theory could jointly provide a comprehensive analytical framework for the analysis of the evolution of cultural landscapes remains in need of thorough academic examination.

Under the above theoretical background, this paper constructs a triple-level analytical framework of cultural landscape evolution on the basis of landscape ecosystem theory and with reference to theoretical propositions of path dependence theory and path creation theory (Figure 1). Following the research philosophy of critical realism, the framework highlights a stratified ontology of cultural landscape evolution that expounds the evolution from three layers, i.e., the layer of landscape, the layer of mechanism, and the layer of dynamics. First, the layer of landscape refers to the surface evolution of landscape patterns, elements, and functions that happened but were not necessarily experienced or observed by humans. Second, the layer of mechanism explores the generative mechanisms that produce the empirical changes at the landscape layer. In this paper, the generative mechanisms include the path-dependent mechanism and the path-creating mechanisms which influence and co-evolve with each other and have different effects on the landscape evolution. Third, the layer of dynamics examines the driving forces of the behaviors of actors and the changes of contextuality that enable the path-dependent mechanism and the path-creating mechanisms to take effects such that the surface evolution of cultural landscapes happens. On the one hand, the active behavior of actors is the primary driving force for the continuous evolution of cultural landscapes. The impacts of actors' behavior on the evolution of cultural landscapes may be positive or negative. The behavior that aligns with the historical development path will strengthen such a path, while the behavior that deviates from the existing evolutionary path makes the transformation of the path possible. On the other hand, the evolution of cultural landscapes and the behavior of associated actors are embedded in particular economic, social, and cultural contexts at local, regional, national, and even international scales. The change of contextuality across scales produces a supportive or restrictive environment for the evolution of cultural landscapes and the behavior of associated actors, which drives the cultural landscapes to evolve according to the existing path or in a new direction.

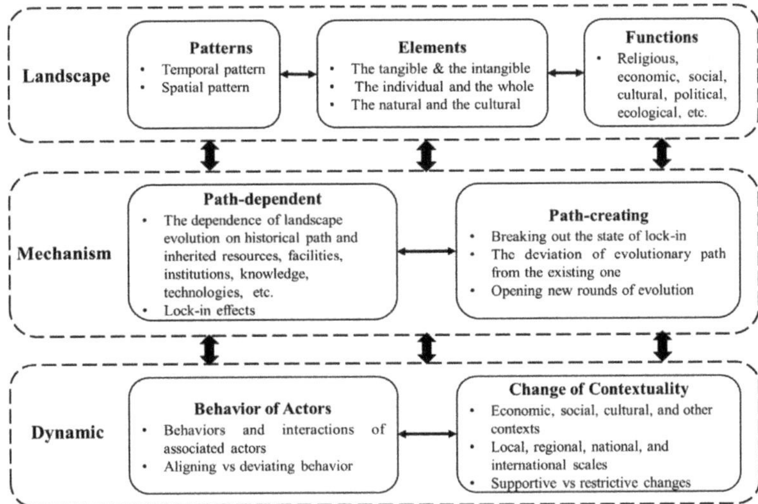

Figure 1. The Analytical Framework of Cultural Landscape Evolution from an Integrated Perspective of Landscape Ecology and Evolutionary Economic Geography (Source: The Authors).

3. Methods

3.1. Case Selection

This study takes Mount Lushan of China as an example. Mount Lushan is located in Jiujiang City, Jiangxi Province, China. It is the first World Cultural Landscape Heritage of China, listed by the United Nations Educational, Scientific and Cultural Organization in 1996. Bounded to the south by Poyang Lake and to the north by the Yangtze River, Mount Lushan displays an integrated panorama of hills, lakes, and rivers (Figure 2). It is a well-known beautiful scenic area with sheer precipices and peaks, changeable fogs and clouds, flying waterfalls and silver springs, as well as secluded forests and deep valleys (Figure 3). The beauty of this scene has been attracting religious, artistic, and intellectual figures for over two millennia. Mount Lushan is home to over two hundred antique buildings, the majority of which are prayer hall complexes that have been renovated and expanded throughout the centuries to serve as a dynamic hub for learning and worship. Among these are the complex of Buddhist East Grove Temple and West Grove Pagoda built in late 4th century, the Taoist Temple of Simplicity and Tranquility constructed in AD 461, and the Confucius Academy of White Deer Cave founded in AD 940. Up to the 19th century, a great number of libraries, temples, and study halls were added to this extensive complex, which continued to be destructed, restored, reopened, and extended many times. The stone single-span Guan Ying Bridge of AD 1015 and over 900 inscriptions on stone tablets and cliffs further make this area significant. In addition, around 600 villas were constructed in the area by Chinese visitors and international migrants between 1896 and 1935 during which Mount Lushan was developed into a famous resort and served as the Summer Capital of the Republic of China for a period. The villas, built in a variety of architectural styles, were set out in the landscape using then-conventional Western planning ideas.

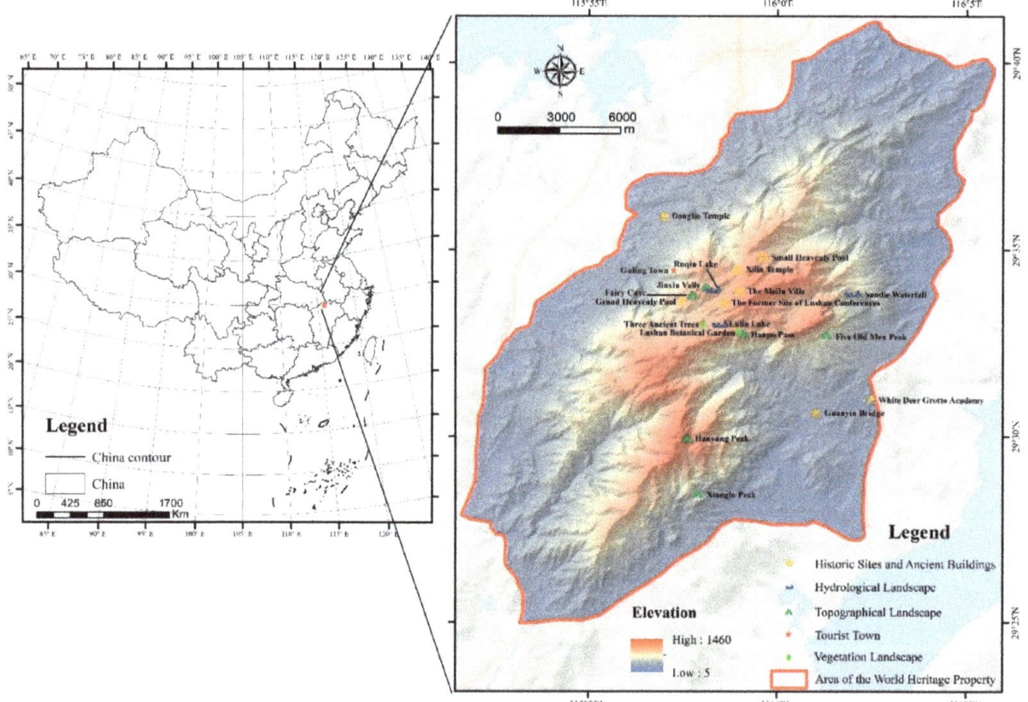

Figure 2. Geographical Location and Main Attractions of Mount Lushan (Source: The Authors).

Overall, through a long history of evolution, Mount Lushan has become a leading example of Chinese landscape culture, a unique model of Chinese academy-based education, and an outstanding representative of the successful fusion of Chinese and Western cultural traditions. Mount Lushan's natural beauty and historic architecture and features complement each other beautifully, resulting in a cultural landscape that is truly one-of-a-kind, and whose exceptional aesthetic worth is strongly identified with the spiritual and cultural life of China. Throughout its evolutionary history, Mount Lushan experienced the transformation of path several times, making it a good example for examining the evolutionary process and mechanism of cultural landscapes.

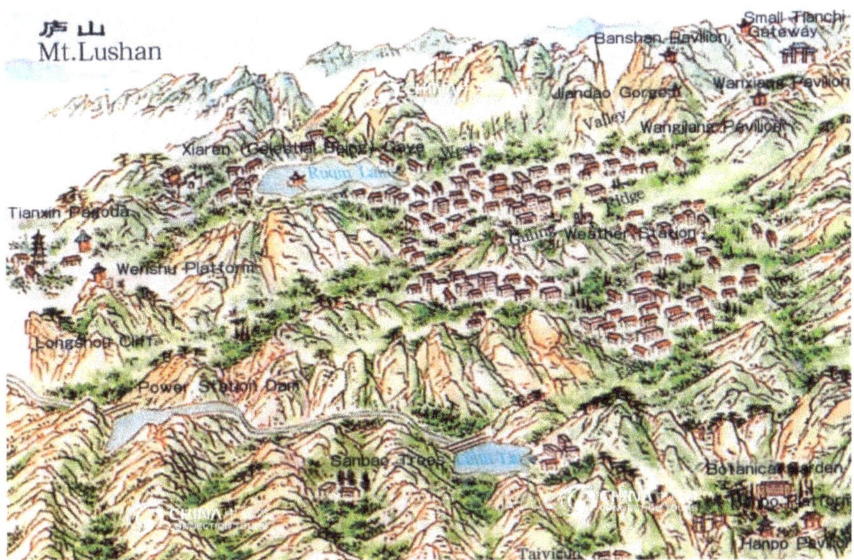

Figure 3. Main Natural Sceneries of Mount Lushan (Source: http://www.china-tour.cn/jiujiang/lushan-mountain-map.htm (accessed on 24 October 2022)).

3.2. Data Collection and Analysis

Researchers of this study are all highly familiar with Mount Lushan with the first author having conducted research surrounding the regional history of Mount Lushan for nearly 15 years, the second author being a local of Jiujiang city where Mount Lushan locates, and other authors having visited Mount Lushan a number of times. Moreover, in order to grasp the evolution history of Mount Lushan accurately and fully, this study employed the mixed method of field observation and historical data collection. First, researchers conducted three field trips to Mount Lushan between 2020 and 2022, with a total stay of seven days. During the field trips, researchers visited, observed, and photographed key sceneries of Mount Lushan, and drew basic information such as the locations, structures, construction years, and associated historical figures and events of the sceneries from the interpretation system of Mount Lushan. Second, researchers collected rich historical data regarding the evolution history of the cultural landscape of Mount Lushan from different secondary sources, for example, the Chronicle of Mount Lushan, the contemporary journal articles and monographs that focus on the history of Mount Lushan, and the literature relevant to Mount Lushan that were produced in different historical periods. Through field observation and collection of historical data, a repository of notes, transcripts, photos, and videos with respect to the functions, tangible and intangible elements, spatial and temporal patterns, as well as associated figures and events of the cultural landscape of Mount Lushan and its evolution was built.

Afterwards, the combination of thematic analysis and chronological organization was adopted to analyze the data generated from field observation and historical sources. A thematic analysis was carried out in order to discover, investigate, and report on recurring themes and patterns found within the data. At the horizontal level, thematic analysis makes it possible to describe and organize the data collected [60]. Chronological organization means that the earliest occurrences are listed first, while the most recent events are included last. It is the process of organizing and rearranging previously gathered information into a coherent, meaningful, and clear-eyed storyline [61]. Specifically, in this study, five steps of data analysis were performed. First, all of the materials that had been gathered were carefully examined, and each individual piece of data was initially coded and assigned a theme and a time mark. Second, the connections between the themes that were produced in the previous step were examined and then grouped into sub-categories and categories. Third, additional research was carried out to determine how the themes, sub-categories, and categories were related to one another. This was achieved through a mix of deductive and inductive reasoning, and the results were compiled into a narrative that discussed the elements and dimensions of the evolution of the cultural landscapes of Mount Lushan. Fourth, the codes, themes, categories, and sub-categories were reorganized chronologically; periodization analysis of the evolutionary process of the cultural landscapes of Mount Lushan was performed. Fifth, by incorporating the narrative that was generated in the third step, an integrated historical storyline regarding the evolution of the cultural landscapes of Mount Lushan was produced. To ensure the rigorousness of the data analysis process and the trustworthiness of the research findings, the triangulation of investigators as suggested by Denzin and Lincoln (2011) [62] was employed, with two different researchers performing data analysis individually and then comparing with each other their report of findings.

4. The Cultural Landscape Evolution of Mount Lushan

As with common cases, the cultural landscape of Mount Lushan is a combined work of men and nature. Its evolution is based on and shaped by its unique multi-genetic complex natural landmarks. Mount Lushan features typical quaternary glaciations, horst fault-blocking, metamorphic core complex structures, and water erosion landforms (Figure 4). First, during the Jurassic to Early Tertiary Period, when strong tectonic movements were observed, Lushan had formed a fault block landscape after the late Himalayan movement 20 million years ago. In response to the movement of the Pacific plate, Mount Lushan was uplifted in the Miocene due to crustal compression deformation. Consequently, cliffs (e.g., the Longshou Cliff), peaks (e.g., Wulaofeng Peak), and canyons (e.g., the Jinxiu Valley) were formed along the fault weak lines. Lakes (e.g., the Ruqin Lake and the Lulin lake), rivers, slopes and rock formations were created by elevation and subsidence in the strenuous fault-block movement. Second, the glacial motions during the Quaternary Period enriched the natural landscape of Mount Lushan with rich glacial remnants such as U-shaped valleys, ice tables, glacier horns, and ridges that record the process of glacial evolution and palaeoclimatic change. Third, with a large amount of rainfall, different kinds of hydraulic actions including abrasion and chemical weathering caused large amounts of erosion, forming the landforms of waterfalls (e.g., Sandie Waterfall) and natural caves (e.g., Fairy Cave). The many grotesque rocks, towering peaks and cascading waterfalls constitute a spectacular natural landscape, which provide a good environmental base for and shape the process of the cultural landscape of Mount Lushan.

Figure 4. Examples of natural landforms in Mount Lushan; an example of glacial remnants (Lushan Flyover) in the left, an example of tectonic landforms (Longshou Cliff) in the middle, and an example of water erosion landforms (Sandie Waterfall) in the right (Source: https://travel.qunar.com/p-cs298166-lushanshi-jingdian (accessed on 24 October 2022)).

Jointly shaped by the actions of humans who came to live in the region and the pre-formed unique natural landscape, Mount Lushan gradually experienced a dynamic complex process of cultural landscape evolution, which can be divided into three stages (Figure 5).

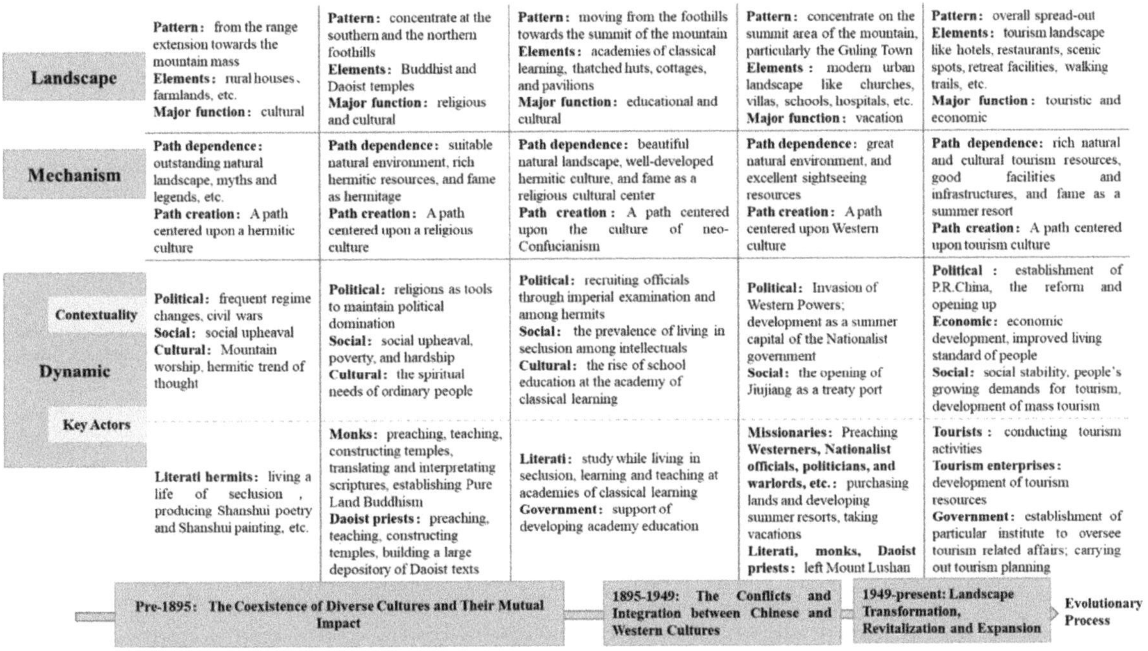

	Pre-1895: The Coexistence of Diverse Cultures and Their Mutual Impact			1895-1949: The Conflicts and Integration between Chinese and Western Cultures	1949-present: Landscape Transformation, Revitalization and Expansion
Landscape	**Pattern:** from the range extension towards the mountain mass **Elements:** rural houses, farmlands, etc. **Major function:** cultural	**Pattern:** concentrate at the southern and the northern foothills **Elements:** Buddhist and Daoist temples **Major function:** religious and cultural	**Pattern:** moving from the foothills towards the summit of the mountain **Elements:** academies of classical learning, thatched huts, cottages, and pavilions **Major function:** educational and cultural	**Pattern:** concentrate on the summit area of the mountain, particularly the Guling Town **Elements:** modern urban landscape like churches, villas, schools, hospitals, etc. **Major function:** vacation	**Pattern:** overall spread-out **Elements:** tourism landscape like hotels, restaurants, scenic spots, retreat facilities, walking trails, etc. **Major function:** touristic and economic
Mechanism	**Path dependence:** outstanding natural landscape, myths and legends, etc. **Path creation:** A path centered upon a hermitic culture	**Path dependence:** suitable natural environment, rich hermitic resources, and fame as hermitage **Path creation:** A path centered upon a religious culture	**Path dependence:** beautiful natural landscape, well-developed hermitic culture, and fame as a religious cultural center **Path creation:** A path centered upon the culture of neo-Confucianism	**Path dependence:** great natural environment, and excellent sightseeing resources **Path creation:** A path centered upon Western culture	**Path dependence:** rich natural and cultural tourism resources, good facilities and infrastructures, and fame as a summer resort **Path creation:** A path centered upon tourism culture
Dynamic – Contextuality	**Political:** frequent regime changes, civil wars **Social:** social upheaval **Cultural:** Mountain worship, hermitic trend of thought	**Political:** religions as tools to maintain political domination **Social:** social upheaval, poverty, and hardship **Cultural:** the spiritual needs of ordinary people	**Political:** recruiting officials through imperial examination and among hermits **Social:** the prevalence of living in seclusion among intellectuals **Cultural:** the rise of school education at the academy of classical learning	**Political:** Invasion of Western Powers; development as a summer capital of the Nationalist government **Social:** the opening of Jiujiang as a treaty port	**Political:** establishment of P.R.China, the reform and opening up **Economic:** economic development, improved living standard of people **Social:** social stability, people's growing demands for tourism, development of mass tourism
Dynamic – Key Actors	**Literati hermits:** living a life of seclusion, producing Shanshui poetry and Shanshui painting, etc.	**Monks:** preaching, teaching, constructing temples, translating and interpretating scriptures, establishing Pure Land Buddhism **Daoist priests:** preaching, teaching, constructing temples, building a large depository of Daoist texts	**Literati:** study while living in seclusion, learning and teaching at academies of classical learning **Government:** support of developing academy education	**Missionaries:** Preaching **Westerners, Nationalist officials, politicians, and warlords, etc.:** purchasing lands and developing summer resorts, taking vacations **Literati, monks, Daoist priests:** left Mount Lushan	**Tourists:** conducting tourism activities **Tourism enterprises:** development of tourism resources **Government:** establishment of particular institute to oversee tourism related affairs; carrying out tourism planning

Evolutionary Process

Figure 5. The Evolutionary Process of the Cultural Landscape of Mount Lushan (Source: The Authors).

4.1. From the Pre-Qin and Han Dynasties to the Ming and Qing Dynasties (Pre–1895): The Coexistence of Diverse Cultures and Their Mutual Impact

4.1.1. A Landscape Centered upon a Hermitic Culture

The arrival of human activities are the key factor that drove the preliminary changes of the original landscape spaces at Mount Lushan. From the perspective of landscape function, Mount Lushan gradually evolved from a natural landscape to a cultural one. Its natural material base and rich cultural heritage laid the possibility and the conditions for the development of a hermitic culture at Mount Lushan.

As a fault-block mountain, namely, one that stands by itself and does not form part of a mountain range, Mount Lushan is located in the Middle-Lower Yangtze Plain, next to Poyang Lake. Apart from its excellent geographic location, Mount Lushan has significant vertical differentiation in terms of its geology, landforms, hydrology, and climate, endowing it outstanding natural landscape. In the early days, the majestic and lofty Mount Lushan received deep reverence and sincere worship from nearby people who were in short of scientific knowledge and production capacity. They produced many myths and legends that were associated with Mount Lushan and paved the way for the development of culture and cultural landscape of the region. During the turbulent times in Wei and Jin dynasties, while regimes changed frequently and government officials competed fiercely among themselves for power and profit, the literati not only found it difficult to exercise their talents to make a difference in politics, but also constantly feared for their life and safety. Consequently, many from the learned class became disappointed with the political situation at the time and decided to abandon city life to live in seclusion, trying to seek a spiritual refuge in the vague and unreal realm of the immortals and to ease their frustration through living a minimalist life, drinking alcohol, feigning insanity, and so on.

Under the influences of that hermitic culture, Mount Lushan, with its advantage in terms of being away from cities but with good accessibility as well as having vast lands, high mountains, and dense forest, unsurprisingly became an ideal place where the literati could find refuge and indulge themselves in nature. As a result, a large number of learned men visited Mount Lushan and lived there in seclusion, thus adding a humanistic touch to the landscape evolution of Mount Lushan. Among them, Tao Yuanming, in particular, has had the greatest influence on later generations. At the beginning of the Yi Xi Era during the Jin dynasty (405 AD), Tao Yuanming started his life in seclusion in his hometown near Mount Yujing towards the southern foothills of Mount Lushan, due to his "reluctance to bow to nasty persons in the village merely for the sake of earning five buckets of rice". Focusing on the landscape of Mount Lushan, Tao Yuanming produced lots of poems which have been recognized as the first of its kind in Chinese landscape poems. Those poems, to a great extent, eulogized and promoted Mount Lushan. In the later era, numerous poets and intellectuals such as Li Bai, Zhou Dunyi, Zhu Xi, and many others chose to spend time in seclusion at Mount Lushan. Inspired by the mountain's beautiful natural sceneries, they produced countless literary works, which promoted Mount Lushan to become the birthplace of Shanshui poetry and Shanshui painting. In that regard, Mount Lushan became increasingly famous among those who wished to seek a life of seclusion, with the hermitic culture being further enriched and developed in the region.

During this stage, the contextuality of social upheaval and the behaviors of the literati hermits jointly provoked the birth and development of hermitic culture at Mount Lushan. The emergence of this culture, in turn, also contributed, as a decisive factor, to the formation of Mount Lushan's first cultural landscapes. Subsequently, the interplay of the literati's behavior and the dynamic contexts at the period redefined the mountain's landscape function. The landscape spaces demonstrated a dynamic pattern that moved in general from the range extension towards the mountain mass, with the landscape elements largely distributed in clusters at the southeastern foothills of the mountain. Being a major part of the spiritual culture of Mount Lushan, the hermitic culture prompted the formation of the evolutionary path of the cultural landscape of Mount Lushan. Additionally, as the cradle and bedrock of the cultural landscape of Mount Lushan, the hermitic culture was a culture of rootedness, and laid a strong foundation for the subsequent cultural landscape evolution of the mountain.

4.1.2. A Landscape Centered upon a Religious Culture

The settlement and prosperity of Buddhism and Daoism caused the cultural landscape of Mount Lushan to evolve from one centered on hermitic culture to one focused on religious culture. Due to the effects of path dependence, the natural resources, the landscape features, and the cultural foundation developed at Mount Lushan during the previous stage, offered a solid material and spiritual basis for the formation and development of the evolutionary path of a landscape that centered upon a religious culture.

During the Wei, Jin, and Southern-Northern dynasties, the social upheavals and the poverty and hardship people experienced, to a great extent, became a hotbed for the development of a religious culture. Indeed, the Buddhist and the Daoist cultures largely satisfied the spiritual needs of ordinary people, as the former promoted the idea that "those chanting the Buddha will go to the Western Pure Land of Ultimate Bliss", while the latter emphasized the possibility of "becoming immortal through cultivation". Moreover, during the Qin and Han dynasties, the emperors' offerings of official sacrifices and rites to pay homage to mountains, rivers, heaven and earth, processes that were themselves characterized by strong religious and political overtones, also accelerated the popularization of a religious culture among ordinary people, turning religion into a dominant part of social life.

Notably, both Buddhism and Daoism tend to emphasize the importance of seeking a pure and unadulterated natural environment, for the purpose of cultivation. In this regard, with great height and steepness, vast lands and rich resources, Mount Lushan, the "cottage of the immortals", offered an ideal environment where monks and Daoist priests, namely, the key driving actors of Cultural Landscape Evolution in Mount Lushan at this stage, could preach, teach, and work on the translation and interpretation of scriptures. During the Eastern Jin dynasty, Huiyuan, a senior monk, arrived at Mount Lushan to build the Donglin Temple and to establish Pure Land Buddhism by combining and fusing foreign Buddhist thought with the dominant doctrines of Confucianism and Daoism. Through Huiyuan's work, Mount Lushan became the center and icon of Southern China Buddhism. Then, in the Southern dynasty, Lu Jingxiu, a Daoist priest, arrived at Mount Lushan to practice cultivation. He also built the Jianji Temple. Then, he systemized the Daoist scriptures by summarizing and categorizing all the available scriptures he had found and built the largest depository of Daoist texts at that time, thus turning the Jianji Temple on Mount Lushan into a research center for studying Daoism. During the Tang and Song dynasties, Mount Lushan's religious culture further consolidated, based on its previous path development. This period saw a significant increase in numbers of both Buddhist and Daoist temples. Additionally, the frequent communication between the literati and the Buddhist monks and Daoist priests elevated the status and fame of Mount Lushan as a religious center. Subsequently, Mount Lushan attracted the attention and even won the support of the ruling class. Prompted by the historical contextuality at the period and favored by those in power, Mount Lushan's religious culture thus entered a stage of prosperity. Regarding the development of the Buddhist culture, this period saw the appearance of "three hundred guesthouses to accommodate the monks visiting Mount Lushan" and "three hundred and sixty temples in the Song dynasty". Regarding Daoist culture, Guangfu Temple, Taiping Palace, Baihe Temple, and Qizhen Temple were successively built, contributing to "gorgeous palaces and Daoist temples with magnificent views".

In summary, it was the support of political forces, the spiritual needs of the ordinary people, and the promotion work done by both the Buddhist and Daoist religionists that jointly drove the cultural landscape evolution of Mount Lushan to deviate from its original development path of that centered upon a hermitic culture and follow a new path that centered upon a religious culture. During this stage, with the appearance of new landscape elements such as Buddhist and Daoist temples, the cultural landscape of Mount Lushan acquired multiple functions, in contrast with its single function during the previous stage. Overall, the landscape elements tended to concentrate at the southern and the northern foothills of the mountain. As the material carrier of the religious landscape, the Buddhist and Daoist temples, along with their distribution and evolutionary process, were thus the spatial expression of the evolution of the religious cultural landscape at Mount Lushan. The dynamic evolution of the Buddhist temples should also be understood as a localized construct in the process of the localization and Sinicization of Buddhism in China. Consequently, Mount Lushan's increasing fame as a religious center and the intermingling and flourishing of Confucianism, Buddhism and Daoism, each being a self-contained system, contributed to the prosperity of the cultural landscape of Mount Lushan.

4.1.3. A Landscape Centered upon the Culture of Neo-Confucianism

As time passes and circumstances change, during the process of its development and inheritance, the core of a given culture may give rise to new core values, which can, in turn, cause changes in the evolutionary pathways of a cultural landscape. Thus, with rich culture and beautiful natural environment, as well as prompted by the prevalence of the imperial examination system, the support of governmental policies, and the intellectuals' efforts to seek official posts by first spending some time in seclusion, Mount Lushan gradually stepped into a path that centered upon the culture of neo-Confucianism.

The rise of the imperial examination system to a great extent facilitated the formation of Mount Lushan's neo-Confucian culture. Since the Sui dynasty, imperial examination had become an important method for identifying talent in feudal China. During the Song dynasty, the state's emphasis on culture and education further prompted the intellectuals to view the imperial examination as the key means for achieving personal goals. However, there were numerous intellectuals that failed the imperial examination again and again, despite their talent and capabilities to run the country well and bring peace and security to its people. With the ruling class tending to favor those intellectuals living in seclusion and treat them with respect and esteem, many learned men thus chose to study while living a secluded life. By doing this, they could, on the one hand, indulge themselves in nature and ease their frustration, while on the other hand, they could also create an elegant image of themselves and, in this way, be better prepared for government recruitment. Consequently, seeking official posts by first spending time in seclusion had thus become a shortcut for many intellectuals to achieve their ambitions.

Against this backdrop, Mount Lushan, with its beautiful natural landscape, its well-developed hermitic culture, and its fame as a religious cultural center, subsequently attracted many intellectuals to study in seclusion, where they could seek knowledge and search for the right path. For example, Li Bo, a Tang Dynasty scholar, studied at Mount Lushan for a long period and was appointed as a government official afterwards. Li discovered and managed the White Deer Grotto, turning this location into a secluded study place. Later, the place gradually became a famous scenic spot among intellectuals. During Southern Tang dynasty, the imperial court established a national academy of Chinese studies at the White Deer Grotto. The academy was a prestigious institute of higher education at that time. The establishment of the academy then paved the way for the later foundation of the White Deer Grotto Academy. After the Song dynasty, the academy of classical learning gradually assumed the responsibility of training and fostering future officials. At the same time, to varying degrees, it also acquired the functions of government-owned schools. During the Northern Song dynasty, under the encouragement of Emperor Taizu of Song, the White Deer Grotto Academy was finally established, soon attracting thousands of students and becoming one of the four most renowned Chinese academies of classical learning, with the other three being the Yuelu Academy, the Songyang Academy, and the Shigu Academy. Zhou Dunyi also created the Lianxi Academy at Lotus Peak of Mount Lushan, where he completed his Diagram of the Supreme Ultimate Explained and Penetrating the Classic of Change, two works that laid the foundation of neo-Confucianism. In addition, he also taught two disciples, Cheng Hao and Cheng Yi, who later both became leading figures in neo-Confucianism. Following Zhou Dunyi and his Lianxi Academy, the spread of neo-Confucian philosophy was accelerated by the rise of school education at the academy of classical learning, which served as a solid base for neo-Confucianism thought to flourish at Mount Lushan. At the beginning of the Southern Song dynasty, Zhu Xi, the preeminent Neo-Confucian master, which is generally ranked as second only to Confucius, was appointed as the governor of Nankang, an upper administrative district of Lushan. Zhu revitalized the academy of classical learning and created new academic rules. Based on the proposals made by Cheng Hao and Cheng Yi, Zhu further contributed to the development of neo-Confucianism and turned it into a mature theoretical system. Thanks to his efforts, the reputation of the White Deer Grotto Academy soared; subsequently, it not only attracted the most famous scholars in Chinese history, such as Lu Jiuyuan, Wang Yangming, and Li Mengyang, to come to teach at the Academy, but also successfully trained tens of thousands of outstanding intellectuals.

During this stage, the interaction between the imperial examination system and the rise of school education at the academy of classical learning, as well as intellectuals' wish to seek official posts by first living a secluded life, brought changes to Mount Lushan's cultural landscape. At the same time, these changes introduced new landscape elements such as academies of classical learning, thatched huts, cottages, and pavilions. The distribution of the landscape spaces was also altered, exhibiting a dynamic pattern with the spaces moving from the foothills towards the summit of the mountain. Compared to the previous stage, at this stage of its cultural landscape evolution, Mount Lushan had developed a richer cultural landscape where the connections between the different cultures seemed to be tighter and stronger. In addition, the agglomeration of academies of classical learning, including the White Deer Grotto Academy, the Lianxi Academy, and others, eventually turned Mount Lushan into the cradle of Chinese neo-Confucian philosophy.

4.2. From the End of the Qing Dynasty to the Republic of China (1895–1949): The Conflicts and Integration between Chinese and Western Cultures

For a long time, Chinese traditional culture remained the core of Mount Lushan's cultural landscape, changing little over time and giving rise to a "locked in" path. However, despite its inheritance of location advantages, great natural environment, and excellent sightseeing resources, at this stage, the heavy blow dealt by Western powers eventually forced Lushan's cultural landscape to integrate Western cultural elements along its path of evolution.

Towards the end of the Qing dynasty, China suffered from political corruption, a backward defense system, and both internal and external crises. Western powers took advantage of this situation and attempted to exert influences over China. After the Second Opium War, the Qing government entered the Treaty of Tientsin with Western forces, according to which various cities along the Yangtze River, such as Jiu Jiang and Han Kou, were to open to foreign powers as treaty ports. As a result, the Yangtze River Basin was forced open by external powers, subsequently allowing many Westerners to enter port cities along the Yangtze River. The middle and lower reaches of the Yangtze River are located in a subtropical monsoon climate zone, and cities such as Wu Han and Nan Jing are well-known "furnaces". In summertime, many Westerners living there thus suffered heavily from the hot weather and diseases caused by the weather. It was precisely against this backdrop that the search for a summer resort with beautiful surroundings and a rich culture became a real need for foreign residents.

Driven by all these factors, Mount Lushan, unsurprisingly, attracted the attention of certain foreign envoys, merchants, politicians, and in particular, foreign missionaries, given its location between two large port cities, Han Kou and Nanjing, beautiful scenery, cool climate, and easy access. Towards the end of 1895, the British missionary Li Deli (Edward Selby Little) first obtained permission to build a summer house at Guling Town, Mount Lushan. Following Li, foreign residents from eighteen Western countries, including the United Kingdom, the United States, France, Germany, and Russia, arrived at Mount Lushan successively. After the Northern Expedition, the Nationalist government made Mount Lushan a summer capital. Subsequently, politicians, government officials, warlords, and those from the wealthy class purchased land and built villas at Mount Lushan. Over time, an architectural landscape with buildings of different styles from all over the world and with varied functions including housing, education, and religion, for examples, villas, schools, churches, and hospitals, were built at Mount Lushan, turning the place into a summer resort whose fame not only spread throughout China, but also reached beyond its borders. Due to the invasion of Western cultures, Mount Lushan was no longer suitable for those seeking a life of seclusion, and the literati, monks, Daoist priests, and many others left. Following these people's evacuation from the area, cultural landscape elements such as the academies of classical learning, religious sites, and thatched huts either gradually decayed or were completely destroyed.

The foreign invasion of China and Westerners' attempts to purchase lands and build summer resorts subsequently prompted Mount Lushan's cultural landscape to gradually turn to an evolutionary path that was dominated by Western cultures. During this period, Mount Lushan saw the appearance of many landscape elements that were typical of Western cultures, such as Catholic churches and foreigners' villas, which tended to concentrate on the summit area of the mountain. With the conflicts, struggles, and integration between Chinese culture and Western culture, Mount Lushan had, eventually, evolved from a renowned center of ancient Chinese culture, where Confucianism, Buddhism, and Daoism coexisted and prospered, to a "capital of various nations" with a strong "Western touch".

4.3. After the Establishment of the New China (1949–Present): Landscape Transformation, Revitalization, and Expansion

Whether being a coincidence or not, the cultural landscape evolution of Mount Lushan that took place at a previous stage would allow for or create opportunities for new evolution possibilities at subsequent stages. In this regard, the evolutionary path of Mount Lushan as a "summer resort" that had taken shape previously obviously facilitated the transition of the mountain's cultural landscape towards one that would highlight the mountain's status as a "holiday resort", due to the strong connections, in terms of facilities, resources, and technology, between Mount Lushan's previous evolutionary path and its current one, i.e., a focus on tourism.

The integration between Chinese and Western cultures and the opening of Guling Town stimulated the development of Mount Lushan's culture of leisure and recreation. Following the establishment of the New China, as the "summer capital" of the previous Nationalist government, as well as the site where three important conferences of the Central Committee of the Chinese Communist Party were held, Mount Lushan has been widely recognized, both domestically and internationally, as a political center and a great summer resort. In particular, following experts' findings in 1956 that confirmed the status of Mount Lushan as the best natural sanatorium in an alpine climate in China, the proposal to turn Mount Lushan, which had already developed a healing function, into a large "nursing home" finally took shape. On the basis of this proposal, Mount Lushan soon went through careful planning and development stages, subsequently paving the way for the further evolution of the mountain's cultural landscape. Since the reform and opening up, social stability and economic development have not only improved people's living standard, but also changed people's day-to-day lifestyle in China. People's growing demands for traveling and visiting different places consequently triggered the development of mass tourism. Under such circumstances, the function of Mount Lushan as a site for healing and official retreats was thus undermined over time.

In 1982, Mount Lushan was among the first Key National Parks of China promoted by the State Council of the People's Republic of China. Later, in 1996, it was also included in World Heritage List as a "cultural landscape". In this way, Mount Lushan's beautiful natural landscape and rich culture thus gained wide recognition. At the beginning of 1980, the screening of the Chinese movie "Romance on Mount Lushan" further increased the mountain's fame. Against this backdrop, many tourists have visited Mount Lushan to sightsee and participate in excursions and other activities. The local government has set up a special department to facilitate the development and reinforce the management of Mount Lushan. Additionally, tourism businesses have also flocked to Mount Lushan, encouraged by its burgeoning tourism industry. With the influx of tourists and tourism businesses and support from the local government, Mount Lushan has gradually developed a tourism-based industrial structure, which has, in turn, modified the mountain's previous landscape spaces. The new changes mainly include the appearance of tourism landscape elements such as tourist attractions, retreat facilities, restaurants, and accommodations to address tourists' varying needs. In this way, Mount Lushan has gradually become a holiday resort where tourists' demand for leisure and recreation can be met and satisfied.

During this stage, the development of tourism was the dominant factor that drove the evolution of Mount Lushan's cultural landscape. The activities conducted by tourists, tourism businesses, and the local government also injected new vitality into the cultural landscape evolution of Mount Lushan. As a result, this stage has included the appearance of many landscape elements, such as walking trails and hotels, whose main purpose is to address the needs of tourists, and the distribution of such elements tends to demonstrate an overall spread-out pattern. Thus, in the context of the New Era, the co-evolution of the elements, functions, and patterns of Mount Lushan's cultural landscape has accelerated the development of a modern tourism industry in the region and brought the cultural landscape evolution of Mount Lushan into a new stage.

5. Discussion

A systematic examination of the process, mechanism, and influencing factors of the spatiotemporal evolution of cultural landscapes is significant for enhancing the ontological understanding of cultural landscapes and further providing scientific references for the effective conservation and sustainable development of cultural landscapes. Based on the traditional perspective of landscape ecology in cultural landscape research and synthesizing the mainstream perspectives of economic geography on regional landscape evolution, i.e., path dependence and path creation, this paper constructs a triple-layered integrated analytical framework of cultural landscape evolution. The framework is empirically applied to examine the evolutionary process of Mount Lushan at the layers of the landscape, mechanisms, and dynamics. The major findings of this study include:

First, as Arce-Nazario (2007) discussed in his analysis on the history of the reconstruction of Peruvian Amazon landscape [39] and Nickens (2022) implicated in his study of the evolution of cultural landscapes in North America [42], cultural landscapes are not fixed, static, and inert presentations, but non-linear, dynamic, and complex products of historical and cultural processes. Across the evolutionary process, the elements, functions, and patterns of cultural landscapes are constantly constructed and reconstructed, giving them rich, diversified, and profound historical and cultural values. The cultural landscape of Mount Lushan has gone through three stages of evolution: the coexistence of hermitic, religious, and Confucian culture, the fusion of Chinese and Western culture, and the transformation, revival, and development of landscape under the context of modern tourism development. In this process, Mount Lushan evolves from a simple natural landscape to a mountain that possesses rich cultural elements including temples, study halls, and villas, and further to a scenic area that was developed on the basis of its splendid natural and cultural landscape and equipped with excellent modern tourism facilities. Across the process, the key areas of its landscape changed from the southern foothills to the southern and northern foothills to the mountain tops and to all major areas of the mountain. In addition, its landscape functions became richer and richer with its core value changed from stage to stage.

Second, the evolution of cultural landscapes is the result of the synergistic effect of path dependence and path creation. On the one hand, the cultural landscape evolution of Mount Lushan highlights the effect of the path-dependent mechanism on constraining the evolutionary process. Above all, the geographical location, topography, ecological environment, and other natural factors provide the environmental basis for the cultural landscape evolution of Mount Lushan and play a vital role across the entire evolutionary process. Furthermore, although the evolutionary path of the cultural landscape of Mount Lushan has changed several times, the hermitic culture, religious culture, Confucian culture, and Western culture developed during the process has not been completely abandoned in subsequent evolutionary stages. Many of their associated tangible and intangible elements were preserved and reused such that subsequent and even the current evolutionary process of the cultural landscape of Mount Lushan has been influenced. For example, the temples, study halls, and villas built in the first two stages have become the resource basis for the development of modern tourism in Mount Lushan in the third stage. This is consistent with the argument of Nickens (2022) on the "layering" of past cultural landscapes, which is believed to have considerable impacts on the modern-day heritage management [42]. On the other hand, the cultural landscape evolution of Mount Lushan also highlights the role of the path-creating mechanism in generating a holistic, dynamic, and non-linear evolutionary process, which is reflected in the multiple changes of the evolutionary path and the significant differences among evolutionary stages in terms of landscape elements, functions, and patterns.

Third, existing studies have acknowledged the importance of the behavior of associated actors [54] and the change of contextuality [43] in shaping cultural landscape evolution. Analogously, this article found that the interplay of the behavior of associated actors and the change of contextuality is the fundamental dynamic that drives the evolution of cultural landscapes. In different stages of the cultural landscape evolution of Mount Lushan, litterateurs, monks, Taoists, Christian missionaries and other Westerners, tourists, tourism enterprises, governments, and other actors dynamically adapted to the changes of contextuality such as social unrest, the invasion of foreign powers, as well as modern economic and social development. Their actions, be they in terms of pursuing hermitic life or constructing temples, study halls, and villas, created new possibilities for further evolution and thus promoted changes of the evolutionary path. After a new evolutionary path was created, the continued aligning and even enhancing behaviors of the associated actors and the continuation of supportive changes of contextuality enabled further development and growth of the path.

6. Conclusions

This paper is only a preliminary attempt to apply the theories of path dependence and path creation into the examination of the evolution of cultural landscapes. It has not explored in depth the boundaries and points for attention while applying the theories in the context of cultural landscapes. Compared with the major applied research target of the theories of path dependence and path creation, i.e., the evolution of regional economic landscapes which generally lasts a few decades, the evolution of cultural landscapes often lasts hundreds or even thousands of years. The evolutionary process of cultural landscapes is therefore rightly more complex, and the factors, elements, and actors involved in the process are also more diverse. In this regard, further empirical studies on other cases of cultural landscapes are called to deepen and improve the application of path dependence and path creation theories in the field of cultural landscape evolution. In addition, due to the difficulties existing regarding the collection of full detailed historical data, the paper may not be specific enough in depicting the evolutionary process of the cultural landscape of Mount Lushan. The findings of this paper may offer some strategic references for heritage sites in monitoring the evolutionary process of cultural landscapes, responding to internal and external contextual changes, managing the behavior of different actors, and introducing proactive development strategies.

Author Contributions: Conceptualization, Z.G. and W.W.; methodology and formal analysis, Z.Z., J.Z. (Jianqin Zhou) and J.Z. (Jiami Zhou); investigation and resources, Z.G., Z.Z., J.Z. (Jianqin Zhou), J.Z. (Jiami Zhou) and W.W.; writing—original draft preparation, Z.Z.; writing—review and editing, Z.G., Z.Z. and W.W.; supervision, project administration, and funding acquisition, Z.G. and W.W. All authors have read and agreed to the published version of the manuscript.

Funding: This research was funded by the National Social Science Foundation of China (17BGL115), the Social Science Foundation of Jiangxi Province (21YJ27), the Project of Key Research Institutes of Humanities and Social Sciences at the Universities of Jiangxi Province (JD21013), and the Postgraduate Innovation Special Foundation of Jiangxi Province (YC2021-S063).

Data Availability Statement: Not applicable.

Acknowledgments: We acknowledge all people who contributed to the data collection and processing, as well as the constructive and insightful comments by the editor and **anonymous** reviewers.

Conflicts of Interest: The authors declare no conflict of interest.

References

1. Gfeller, A.E. Negotiating the meaning of global heritage: 'Cultural landscapes' in the UNESCO World Heritage Convention, 1972–1992. *J. Glob. Hist.* **2013**, *8*, 483–503. [CrossRef]
2. Graeme, A. world heritage cultural landscapes. *Int. J. Herit. Stud.* **2007**, *13*, 427–446.
3. Zhou, J.; Wang, W.; Zhou, J.; Zhang, Z.; Lu, Z.; Gong, Z. Management effectiveness evaluation of world cultural landscape heritage: A case from China. *Herit. Sci.* **2022**, *10*, 22. [CrossRef]
4. Maxim, C.; Chasovschi, C.E. Cultural landscape changes in the built environment at World Heritage Sites: Lessons from Bukovina, Romania. *J. Hosp. Mark. Manag.* **2021**, *20*, 100583. [CrossRef]
5. Hung, K.; Yang, X.; Wassler, P.; Wang, D.; Lin, P.; Liu, Z. Contesting the commercialization and sanctity of religious tourism in the Shaolin Monastery, China. *Int. J. Tour. Res.* **2017**, *19*, 145–159. [CrossRef]
6. Loulanski, T. Revising the concept for cultural heritage: The argument for a functional approach. *Int. J. Cult. Prop.* **2006**, *13*, 207–233. [CrossRef]
7. Wood, J.; Walton, J.K. *The Making of a Cultural Landscape: The English Lake District as Tourist Destination*; Routledge: New York, NY, USA, 2016.
8. Ferro-Vázquez, C.; Lang, C.; Kaal, J.; Stump, D. When is a terrace not a terrace? The importance of understanding landscape evolution in studies of terraced agriculture. *J. Environ. Manag.* **2017**, *202*, 500–513. [CrossRef]

9. Vakhitova, T.V. Rethinking conservation: Managing cultural heritage as an inhabited cultural landscape. *Built. Environ. Proj. Asset Manag.* **2015**, *5*, 217–228. [CrossRef]
10. Manningtyas, R.D.T.; Furuya, K. Traditional Ecological Knowledge versus Ecological Wisdom: Are They Dissimilar in Cultural Landscape Research? *Land* **2022**, *11*, 1123. [CrossRef]
11. Qu, M.; Liu, D.P. A Study on the Preservation of Cultural Landscape Heritage Based on Landscape Ecology as Exemplified in the Trunk Line of Chinese Eastern Railway. *J. Archit.* **2017**, *587*, 100–104.
12. Wu, J. Landscape of culture and culture of landscape: Does landscape ecology need culture? *Landsc. Ecol.* **2010**, *25*, 1147–1150. [CrossRef]
13. Turner, M.G. Landscape ecology in North America: Past, present, and future. *Ecology* **2005**, *86*, 1967–1974. [CrossRef]
14. Naveh, Z.; Lieberman, A.S. *Landscape Ecology: Theory and Application*; Springer Science & Business Media: New York, NY, USA, 2013.
15. Gergel, S.E.; Turner, M.G. *Learning Landscape Ecology: A Practical Guide to Concepts and Techniques*; Springer: New York, NY, USA, 2017.
16. Fang, Y.; Liu, J. Cultural landscape evolution of traditional agricultural villages in North China—Case of Qianzhai Village in Shandong Province. *Chin. Geogr. Sci.* **2008**, *18*, 308–315. [CrossRef]
17. Antrop, M.; Van Eetvelde, V. Mechanisms in recent landscape transformation. *WIT Trans. Built Environ.* **2008**, *100*, 183–192.
18. MacKinnon, D. Evolution, path dependence and economic geography. *Geogr. Compass* **2008**, *2*, 1449–1463. [CrossRef]
19. Martin, R.; Sunley, P. Path dependence and regional economic evolution. *J. Econ. Geogr.* **2006**, *6*, 395–437. [CrossRef]
20. Martin, R.; Sunley, P. The place of path dependence in an evolutionary perspective on the economic landscape. In *The Handbook of Evolutionary Economic Geography*; Boschma, R., Martin, R., Eds.; Edward Elgar: Northampton, MA, USA, 2010; pp. 62–92.
21. MacKinnon, D.; Dawley, S.; Pike, A.; Cumbers, A. Rethinking path creation: A geographical political economy approach. *Econ. Geogr.* **2019**, *95*, 113–135. [CrossRef]
22. Garud, R.; Karnoe, P. *Path Dependence and Creation*; Psychology Press: New York, NY, USA, 2001.
23. Garud, R.; Kumaraswamy, A.; Karnøe, P. Path dependence or path creation? *J. Manag. Stud.* **2010**, *47*, 760–774. [CrossRef]
24. Fredin, S. Regional path dependence and path creation: A conceptual way forward. In *Geography, Open Innovation and Entrepreneurship*; Gråsjö, U., Karlsson, C., Bernhard, I., Eds.; Edward Elgar: Northampton, MA, USA, 2018; pp. 308–327.
25. Singh, R.; Mathiassen, L.; Mishra, A. Organizational Path Constitution in Technological Innovation. *MIS Q.* **2015**, *39*, 643–666. [CrossRef]
26. Jones, M. The concept of cultural landscape: Discourse and narratives. In *Landscape Interfaces*; Palang, H., Fry, G., Eds.; Springer: Dordrecht, The Netherlands, 2003; pp. 21–51.
27. De Sousa, C. The greening of urban post-industrial landscapes: Past practices and emerging trends. *Local Environ.* **2014**, *19*, 1049–1067. [CrossRef]
28. Mitchell, N.J.; Barrett, B. Heritage values and agricultural landscapes: Towards a new synthesis. *Landsc. Res.* **2015**, *40*, 701–716. [CrossRef]
29. Song, B.; Robinson, G.M.; Bardsley, D.K. Measuring multifunctional agricultural landscapes. *Land* **2020**, *9*, 260. [CrossRef]
30. Nyseth, T.; Sognnæs, J. Preservation of old towns in Norway: Heritage discourses, community processes and the new cultural economy. *Cities* **2013**, *31*, 69–75. [CrossRef]
31. Alberts, H.C.; Hazen, H.D. Maintaining authenticity and integrity at cultural world heritage sites. *Geogr. Rev.* **2010**, *100*, 56–73. [CrossRef]
32. Pardo Abad, C.J. The post-industrial landscapes of Riotinto and Almadén, Spain: Scenic value, heritage and sustainable tourism. *J. Herit. Tour.* **2017**, *12*, 331–346. [CrossRef]
33. Haileslassie, A.; Mekuria, W.; Schmitter, P.; Uhlenbrook, S.; Ludi, E. Changing agricultural landscapes in Ethiopia: Examining application of adaptive management approach. *Sustainability* **2020**, *12*, 8939. [CrossRef]
34. Rössler, M. World Heritage cultural landscapes: A UNESCO flagship programme 1992–2006. *Landsc. Res.* **2006**, *31*, 333–353. [CrossRef]
35. Wu, J. Key concepts and research topics in landscape ecology revisited: 30 years after the Allerton Park workshop. *Landsc. Ecol* **2013**, *28*, 1–11. [CrossRef]
36. Plieninger, T.; Van der Horst, D.; Schleyer, C.; Bieling, C. Sustaining ecosystem services in cultural landscapes. *Ecol. Soc.* **2014**, *19*, 53–58. [CrossRef]
37. Bogaert, J.; Vranken, I.; André, M. Anthropogenic effects in landscapes: Historical context and spatial pattern. In *Biocultural Landscapes*; Hong, S.K., Bogaert, J., Min, Q., Eds.; Springer: Dordrecht, The Netherlands, 2014; pp. 89–112.
38. Tappeiner, U.; Leitinger, G.; Zariņa, A.; Bürgi, M. How to consider history in landscape ecology: Patterns, processes, and pathways. *Landsc. Ecol.* **2021**, *36*, 2317–2328. [CrossRef]
39. Arce-Nazario, J.A. Human landscapes have complex trajectories: Reconstructing Peruvian Amazon landscape history from 1948 to 2005. *Landsc. Ecol.* **2007**, *22*, 89–101. [CrossRef]
40. Bürgi, M.; Salzmann, D.; Gimmi, U. 264 years of change and persistence in an agrarian landscape: A case study from the Swiss lowlands. *Landsc. Ecol.* **2015**, *30*, 1321–1333. [CrossRef]
41. Dzik, A.J. Kangerlussuaq: Evolution and maturation of a cultural landscape in Greenland. *Bull. Geogr. Socio-Econ. Ser.* **2014**, *24*, 57–69. [CrossRef]
42. Nickens, P.R. Imagining the Multilayered Cultural Landscape: A Template from the Columbia Plateau of North America. *Land* **2022**, *11*, 1613. [CrossRef]

43. Bal, W.; Czalczynska-Podolska, M. The stages of the cultural landscape transformation of seaside resorts in Poland against the background of the evolving nature of tourism. *Land* **2020**, *9*, 55. [CrossRef]
44. Labbaf-Khanii, M. The Evolution of the Cultural Landscape of Maymand Based on Historical Studies and Archaeological Findings. *J. Archaeol. Stud.* **2017**, *8*, 111–130.
45. David, P.A. Clio and the Economics of QWERTY. *Am. Econ. Rev.* **1985**, *75*, 332–337.
46. Cecere, G.; Corrocher, N.; Gossart, C.; Ozman, M. Lock-in and path dependence: An evolutionary approach to eco-innovations. *J. Evol. Econ.* **2014**, *24*, 1037–1065. [CrossRef]
47. Vergne, J.P.; Durand, R. The missing link between the theory and empirics of path dependence: Conceptual clarification, testability issue, and methodological implications. *J. Manag. Stud.* **2010**, *47*, 736–759. [CrossRef]
48. Garud, R.; Karnøe, P. Path creation as a process of mindful deviation. In *Path Dependence and Creation*; Garud, R., Karnoe, P., Eds.; Psychology Pressa: New York, NY, USA, 2001; pp. 1–38.
49. Steen, M. Reconsidering path creation in economic geography: Aspects of agency, temporality and methods. *Eur. Plan. Stud.* **2016**, *24*, 1605–1622. [CrossRef]
50. Schienstock, G. From path dependency to path creation: Finland on its way to the knowledge-based economy. *Curr. Sociol.* **2007**, *55*, 92–109. [CrossRef]
51. Hassink, R.; Isaksen, A.; Trippl, M. Towards a comprehensive understanding of new regional industrial path development. *Reg. Stud.* **2019**, *53*, 1636–1645. [CrossRef]
52. Farina, A. The cultural landscape as a model for the integration of ecology and economics. *BioScience* **2000**, *50*, 313–320. [CrossRef]
53. Wang, W.; Qian, J. An Evolutionary-relational Analysis of the Mechanism of Destination Evolution: The Case of Cheung Chau Island, Hong Kong, China. *Tour. Trib.* 2022, *online first*.
54. Mercuri, A.M. Genesis and evolution of the cultural landscape in central Mediterranean: The 'where, when and how' through the palynological approach. *Landsc. Ecol.* **2014**, *29*, 1799–1810. [CrossRef]
55. Bürgi, M.; Bieling, C.; von Hackwitz, K.; Kizos, T.; Lieskovský, J.; Martín, M.G.; McCarthy, S.; Müller, M.; Palang, H.; Plieninger, T.; et al. Processes and driving forces in changing cultural landscapes across Europe. *Landsc. Ecol.* **2017**, *32*, 2097–2112. [CrossRef]
56. Sydow, J.; Windeler, A.; Müller-Seitz, G.; Lange, K. Path constitution analysis: A methodology for understanding path dependence and path creation. *Bus. Res.* **2012**, *5*, 155–176. [CrossRef]
57. Röhring, A.; Gailing, L. *Linking path dependency and resilience for the analysis of landscape development. Resilience and the Cultural Landscape: Understanding and Managing Change in Humanshaped Environments*; Cambridge University Press: New York, NY, USA, 2012; pp. 146–163.
58. Zariņa, A. Path dependence and landscape: Initial conditions, contingency and sequences of events in Latgale, Latvia. Geografiska Annaler: Series, B. *Geogr. Ann. B. Hum. Geogr.* **2013**, *95*, 355–373. [CrossRef]
59. Palang, H.; Zariņa, A.; Printsmann, A. Making sense of breaks in landscape change. *Landsc. Ecol.* 2022, *online first*. [CrossRef]
60. Boyatzis, R.E. *Transforming Qualitative Information: Thematic Analysis and Code Development*; Sage: London, UK, 1998.
61. Danto, E.A. *Historical Research*; Oxford University Press: New York, NY, USA, 2008.
62. Denzin, N.K.; Lincoln, Y.S. *The Sage Handbook of Qualitative Research*, 4th ed.; Sage: London, UK, 2011.

 land

Article

Walking in China's Historical and Cultural Streets: The Factors Affecting Pedestrian Walking Behavior and Walking Experience

Mimi Tian [1], Zhixing Li [1], Qinan Xia [1], Yu Peng [1], Tianlong Cao [1], Tianmei Du [1] and Zeyu Xing [2,*]

[1] School of Design and Architecture, Zhejiang University of Technology, Hangzhou 310023, China
[2] School of Management, Zhejiang University of Technology, Hangzhou 310023, China
* Correspondence: xingzeyusmile@zjut.edu.cn; Tel.: +86-19103522318

Abstract: The urban street has evolved into an important indicator reflecting citizens' living standard today, and pedestrian walking activity in the streets has been proved to be a major facilitator of public health. Uncertainties, however, exist in the factors affecting pedestrian walking behavior and walking experience in streets. Especially, the factors affecting pedestrian walking behavior and walking experience in the historical and cultural streets. For the study of their main influencing factors, Hefang Street business block and Gongchen Bridge life block in Hangzhou are selected here as the study objects. Both non-participatory and participatory research methods are adopted to collect pedestrian information and observe pedestrians' ambiguous behavior, specific behavior, and stopping behavior. According to the study result, walking preference, walking time, environmental characteristics, and land-use mix (LUM) significantly impact pedestrian walking motivation. The type differences between Gongchen Bridge life block and Hefang Street business block leads to the difference in pedestrians' behaviors and their stopping time in business. Meanwhile, gender differences bring pedestrians' significant differences in walking motivation. Pedestrian walking preference and walking time are positively correlated with walking motivation in both streets. Environmental characteristics and LUM have also been proved to be important influencing factors of pedestrians' walking motivation. In this article, design and planning strategies are proposed for streets of different types in an attempt to provide reference for the revitalization and utilization of cultural heritage streets.

Keywords: historical and cultural streets; walking experience; walking behavior; public health

Citation: Tian, M.; Li, Z.; Xia, Q.; Peng, Y.; Cao, T.; Du, T.; Xing, Z. Walking in China's Historical and Cultural Streets: The Factors Affecting Pedestrian Walking Behavior and Walking Experience. *Land* **2022**, *11*, 1491. https://doi.org/10.3390/land11091491

Academic Editors: Bellotti Piero and Alessia Pica

Received: 10 August 2022
Accepted: 2 September 2022
Published: 5 September 2022

1. Introduction

The urban street is a necessary part of the city. Since ancient times, urban streets have reflected citizens' living standards, and related behavioral activities have been conducted by street pedestrians in the street space. This serves as a big promoter of public health. Today, as material civilization becomes increasingly mature, the streets morph into the dividing line between administrative areas, commercial areas, living areas and traffic areas, as well as an important carrier for functions of leisure, commerce, and entertainment. Therefore, the planning, construction and management of urban streets have become subjects worthy of key consideration. Studies on urban streets have been initiated at an early time, with a wide range of areas involved. As early as 1961, Jane Jacobs [1] in "The Death and Life of Great American Cities" conducted studies and made recommendations from the perspective of maintaining urban diversity and vitality. It's proposed that streets and sidewalks should be the main public areas in cities, and an in-depth analysis of the street's vitality and safety was conducted. Moreover, researchers extensively analyze the street space and its impact on pedestrian walking behavior and walking experience within the space from two dimensions: pedestrian's own conditions and environments. The impact of spatial environment on pedestrian satisfaction is explained via assessment and analysis [2,3]. Other studies have been carried out based on the relationship between people and streets.

Ways of creating spatial environments are provided via the study on pedestrian needs, the evaluation on the quality of pedestrian spaces, as well as the development and optimization of spatial design strategies [4–7].

In the study, Hangzhou of Zhejiang Province, a city with rich history and culture, is taken as the research background. Here, Hefang Street business block, and Gongchen Bridge life block are selected in Hangzhou for comparative analysis. These two streets are chosen for the case study because these two, in the same administrative district, exhibit different spatial and economic patterns of streets, thus the author can conduct an in-depth study on the pedestrian behavior, walking experience and their influencing factors in different blocks. An in-depth analysis is made on the street's history, passenger flow, spatial form, business distribution, and public facilities to lay the foundation for a study on analysis methods of these two blocks' differences. Literature from home and abroad is analyzed from four perspectives: pedestrian walking behavior, walking experience, pedestrian cities, and pedestrian streets, as well as the relationship between walking and public health, thus knowing the current situation and shortcomings of existing studies. Meanwhile, the pedestrian information is collected and analyzed through questionnaires. These questionnaires are distributed on weekdays, and followed up on weekends, with a view to obtaining as many samples as possible. Finally, 100 valid samples are collected from the streets. The basic information of pedestrians in these two streets is analyzed. In addition, SPSS data software is used to analyze the relationship between age and pedestrian ambiguous behavior, specific behavior and business stopping behavior. Besides, in the article, the impact of pedestrians' age, education level, walking preference, walking motivation, environmental characteristics and LUM on pedestrian walking time, is discussed, and the key factors impacting pedestrians' walking time are analyzed.

From the study perspective, the environmental assessment of the historic district is conducted by the observation of pedestrian walking behavior. In terms of study methodology, both participatory and non-participatory research and analyses are proposed to provide a basis for street researches of the same type. As for study application, life block and business block in historical streets are selected for comparison. The case selection containing cultural attribute, provides reference for the revitalization and utilization of cultural heritage streets from home and abroad.

2. Literature Review

2.1. Pedestrian Walking Behavior and Experience

While studying pedestrian walking behavior—a hot topic in recent years, some researchers have discussed it from different perspectives and multifaceted viewpoints based on the division of different pedestrian populations, largely in an attempt to figure out factors affecting walking, and study walking safety. For example, Ross [8], via the documented observations of relationship between children's walking behavior and the factors associated with it, discussed the impact of children's gender and walking time on their walking behavior. In the study of the built environment and walking behavior, Mirzaei [9] argued that most previous studies on walking behavior have focused on utilitarian or recreational walking behavior. Considering the differences in walking purposes, she explored the different effects of the built environment on walking behavior, thus confirming the necessity of study on walking motivation. Marisamynathan [10] discussed the influence of personal information, income, and road facilities on walking behavior of pedestrians crossing the road by studying the factors that affect walking behavior of pedestrians crossing the road, providing a method by which walking behavior of pedestrians crossing the road and its safety levels can be predicted. While conducting a comprehensive study of factors affecting walking behavior of pedestrians crossing the road, Aghabayk [11] discussed the differences between the signalized crossing and the unsignalized one, and examined overall the effects of gender, age, crossing awareness, technical equipment, and carried items on pedestrian crossing behaviors at the signalized crosswalk and the unsignalized one. In urban life, there are many road accidents involving pedestrian behavior that plays a central part in these accidents. Jay Mathilde [12] studied the walking behavior of French and Japanese populations based on their differences, with a view to analyzing accidental risks caused by pedestrian behavior and providing selections for safe road crossing behavior. Mukherjee [13], from the survey data about video images from the signalized crossings, extracted individual pedestrian acts in violation of regulations and individual pedestrian road crossing behaviors, thereby predicting possible accidents.

The perspective of study on walking experience now is generally the combination of environmental psychology, walking experience and modern technology. Simulation and analysis and more are generally adopted as study methods [14,15]. The study content is divided into two parts: study and definition on the types of pedestrians [16], and study from perspectives including the intervention of the physical environment, the impact on pedestrians' psychological experience, and the humanitarian for special populations. For example, Bornioli, Anna [17], in the article, explored whether the physical environment affects the pedestrian walking experience and pedestrian psychology based on theories related to environmental psychology, concluding that the physical environment affects the pedestrian sensory experience- a key element of the walking experience. Then strategies can be established to create a good sensory experience through the construction of the physical environment. Stevenson and others [18] investigated the pedestrian walking experience via the interviews, as well as conceptualized and captured the pedestrian walking experience, in a bid to help pedestrians further deepen sense of leisure experience in a dynamic space, connect their physical environment, and bodily and mental environment to others, as well as strengthen their experience of the space through a wide range of connections. JIYOUNG [19] studied the negative pedestrian walking experience caused by subway stations by collecting materials, as well as analyzing and examining the function of subway stations. Here, the implication of walking as an experience was redefined, with the study purpose of turning a subway station into a positive space and creating a good pedestrian experience through music, sensors, and interesting facilities. Wong, Jeremy D [20] proposed that the preferred gait during walking can be adjusted by the nervous system to facilitate the reduction of the physical fatigue that pedestrians experience while walking, and the enhancement of the pedestrian's walking status from a neurological perspective. Cambra, Paulo [21] studied the interventions and effects of the built environment on adult walking behavior by modifying the physical environment, and also conducted a postmortem analysis. The study suggested that environmental interventions serve as an important factor influencing walking behavior, and possibly small-scale interventions in the walking environment can more effectively improve the walking experience. Tz-Yang Chao [22] improved the walking experience by adopting mixed reality technology. Pedestrians were guided to walk and interact with virtual characters within a prescribed range of mixed virtual reality technology, providing a new paradigm for the improvement of the walking experience. Ohjisuck [23] provided a realistic walking experience for the visually impaired by creating a new virtual walking experience environment, and setting up walking paths and braille devices, in the study on the walking experience of special populations from a sociological perspective. Jiyoung, Kwahk [24] developed pedestrian experience guidelines by analytical research and interviews with special populations with impairments in hearing, vision, speech, and physical mobility. These guidelines were used as a reference standard for the design and evaluation of pedestrian friendliness in pedestrian environments.

2.2. Cities and Pedestrian Streets

Based on urban pedestrian planning, the study is conducted from aspects of transportation, economy, and policy in the literature of this category. In these studies, the effective use of public transportation systems and information and communication technologies are advocated; innovations are made in implementation methods; and valuable proposals for urban economy, transportation, and environmental sustainability as well as the improvement of living spaces for urban residents, are offered based on the pedestrian city concept. For instance, Varma [25] suggested that the effective use of public transport systems and pedestrian cities should be becoming a priority for urban development. Varma also explored the impact of new trends in urban mobility and information and communication technologies on the cities' future development. With Seoul as an example, Young, Kim Sun, and others [26] analyzed the gait features of pedestrians by photographing walking streets in order to find the relationship between the rational selection of pedestrians and the extracted walking environment. Rebecchi [27] proposed a framework for assessing the walkability of cities to study the strengths and weaknesses of the urban environments and to improve healthy living spaces. Yassin [28] proposed an innovative practice specially used for pedestrians, in a bid to re-attract people to the downtown and pedestrian environment. Sustainable urban development is realized by the restoration of urban walkability. Fallahranjbar [29] proposed that putting people first should be a key development principle for all cities that aim to improve urban quality; and a healthy environment should be fundamentally related to a pedestrian community. Fancello [30] explored how citizens' preferences and values vary spatially, as well as designed walking policies that improve citizens' quality, providing new recommendations for mapping out walkability-oriented urban policies. Hui He [31] pointed out that population density was negatively correlated with sport frequency and total sport time, and proposed intervention strategies for an aging-appropriate urban environment. The research focus was the relationship between vehicle conditions, visual signals such as monitors and environmental perception signals, and walking decisions. Fanny Malin [32] evaluated the short- and long-term impact of speed display signs in pedestrian street on the speed of moving vehicles in a low-speed urban environment. If the speed displays are installed at pedestrian crossings, the speed of moving vehicles will be reduced, and pedestrian safety will be guaranteed. With historical street of Shapowei, Xiamen, as an example, Lemin Zhang [33] constructed a street vitality evaluation system based on spatio-temporal data of pedestrians, as well as systematically examined the impact of the built environment on street vitality in historical streets using multi-modal analysis techniques.

2.3. Impact of Walking on Public Health

Considering the inevitable link between walking and public health, there are increasing studies on how urban transportation planning meets the need of public health. Most literatures focus on the benefits of a walking lifestyle for public health, including physical and mental benefits. On this basis, the discussion is made based on specific influencing factors in these literatures. Also, some researchers have studied the adverse effects of walking on public health by attributing them to the external environments on basis of a comprehensive consideration. The positive impact of walking on health has been studied mainly from internal factors. For instance, D. Merom [34] argued that less dependence on cars facilitates public health, and that active behavioral activities provide a solution to sedentariness and lack of physical activities. James F Sallis [35] concluded from his analysis that most environmental attributes are positively correlated with physical activities. The study results are similar in each and every city. James pointed out, the design of urban environments contributes a lot to physical activities, and the global health burden brought by physical inactivity can be reduced through the participation of all departments. Other researchers discussed on basis of specific walking-based influencing factors of the health benefits. Mohammad Javad Koohsari [36] noted that, public open space brings many health benefits physically and mentally. In the study, multilevel logistic regression models are adopted to examine the correlation between measured values of public open space and walking and depression, and the importance of the potential impact of these assessment criteria on health is emphasized. By conducting experiments, Ming-YiHsu [37] proved that brisk walking, as an effective physical activity promoting mental health in adolescents, can reduce depression and anxiety as well as improve self-assessment. However, the walking lifestyle is a double-edged sword for health. The negative impacts are mainly studied from external factors, such as air quality. Giorgos Giallouros [38] argued that, commuting by walking, compared with vehicles, may increase the intake of fine particles for pedestrians, and cause negative health effects. The arguments were started from an integrated relative risk perspective. Caihua Zhu [39] et al. thought, PM2.5 in the air impact physical health when pedestrians are fully exposed to the outside environment. They studied the specific impacts as a way to evaluate the risk features of walking and provide recommendations for the improvement of residents' health and decision making about walking trips.

It's found from the analysis of the literature that. Two main features of the existing studies are as follows:

The study is conducted on external factors of pedestrians. Pedestrians are impacted by many factors amid walking. Especially, the physical environment can directly impact pedestrians' physical and psychological experiences. Thus it can be said that pedestrian walking experience is related to the physical environments (roads, signals, facilities, etc.). It is confirmed that, amid the street construction, it is necessary to provide an excellent physical and sensory experience for pedestrians via the construction of the physical environment on the one hand. The impact of the physical environment on different groups, as well as their applicability in the environment, need to be taken into account, thereby meeting the walking experience of people of different types on the other hand. On this basis, the factors influencing the pedestrian walking experience are further investigated via the interventions and technological means that change the physical environment.

The study on the pedestrian ontology. On the one hand, it's confirmed physiologically and psychologically that a walking lifestyle can improve the emotional state, thus promoting public health. Amid construction, physical and mental impact on the public should be considered, and the negative impact from external environment should be minimized. An all-round street space environment, which facilitates daily traveling, and promotes public health, should be constructed. On the other hand, the study is conducted on the division of pedestrian types, and pedestrian behavior differences; the construction of a sustainable city is explored from the urban walkability; policies based on walking city are advocated; importance is attached to the planning of walking environment; citizens' walking behavior is guided. However, in most existing studies, non-participatory pedestrian behavior is referred, and attention is more likely to be paid to pedestrians' behavior and experience in a single physical environment. While the subjective intention of pedestrians is neglected. The policies about street design and pedestrian urban planning are proposed without support of enough studies on pedestrian ontology. Additionally, there is still much room for improvement in comparative studies on walking behaviors in different places. Therefore, in the study, both non-participatory and participatory approaches are adopted, with emphasis on the subjective evaluation of pedestrians. Pedestrian behavior states on streets of different types are observed and recorded, and a multi-dimensional study of pedestrian walking behavior is conducted, so as to put forward constructive suggestions in a targeted manner.

3. Methods and Measures

Literature from home and abroad is analyzed to learn the current situation and shortcomings of existing studies. In the study, two blocks-Hefang Street business block and Gongchen Bridge life block are selected. The mainly investigated roads in Hefang Street include the main road of Hefang Street, Qinghefang Community, Gaoyin Street, Huimin Road, Dingan Road, and Huaguang Road. The mainly investigated roads in Gongchen Bridge involve Xiaohelu Community, Lishui Road, and Scenic Street Area. Here major factors impacting pedestrian behaviors are explored in the block.

The investigation is divided into two parts: field investigation and user investigation. The former consists of two parts: site investigation and behavior observation. Interview, photo-taking and other forms are adopted to obtain the distribution information of street businesses, including the distribution locations of businesses of different types and the proportion of the business of a certain type in the business as a whole. In terms of building scale and street paving and decoration, the building height, road width and paving material are investigated and analyzed. While behavior observation activity is performed, pedestrian behavior is observed and recorded in tracking investigation, that is, pedestrians are tracked in real time in the street and their status and staying time in shops are recorded. Different business types include essential necessities, commercial consumption facilities, leisure and entertainment facilities and public service facilities. In the user survey, pedestrians, staff of surrounding shops, recreational residents in the community and other street users are invited to fill out questionnaires for information collection. While some of pedestrians with poor educational background or the aged ones can't read the questionnaires, so they are inquired orally and recorded. The questionnaire contents include pedestrian behaviors, socio-demographic characteristics, total walking time, walking preference, walking motivation and environmental characteristics. Information including the participants' demographic data, their environment satisfaction and daily travelling behavior preferences are obtained in the questionnaire of the study (Figure 1).

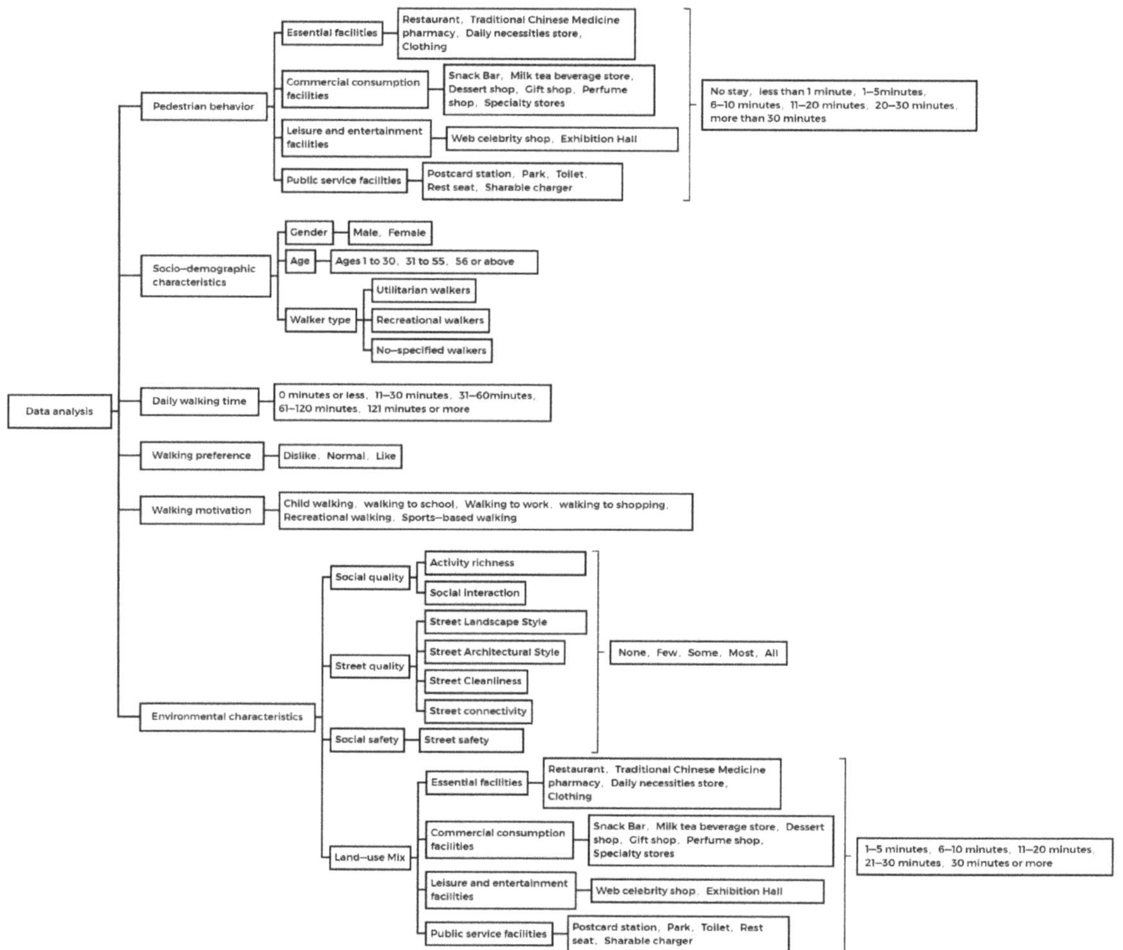

Figure 1. Data analysis framework.

"Socio-demographic characteristics" include gender, age, education, profession, and income. There are two categories of gender: male and female. The ages are divided into three categories: 1–30, 31–55, 55 or above. Street walkers are divided into utilitarian walkers, recreational walkers, and no-specified walkers. Utilitarian walkers mainly walk for commuting between home and school or workplace. Recreational walkers mainly walk and jog. The no-specified walkers mainly walk for daily affairs, such as going shopping, going to hospital, visiting friends, and doing some commute-related things.

The specific businesses where pedestrians stay are: (1) essential necessities such as restaurants, traditional Chinese medicine pharmacies, daily necessities stores, and clothing stores; (2) commercial consumption facilities such as snack bars, milk tea beverage stores, dessert shops, gift shops, perfume shops, and specialty shops; (3) leisure and entertainment facilities such as web celebrity shops, and exhibition halls; (4) public service facilities: postcard stations, parks, toilets, rest seats, and sharable chargers.

"Total walking time" is assessed by individual items, with a 5-point system (1 = 10 min or below; 2 = 11–30 min; 3 = 31–60 min; 4 = 61–120 min; 5 = 121 min or above) adopted to measure respondents' "daily walking time".

"Walking preference" is assessed by an individual item, with a 3-point system (1 = "dislike"; 2 = "normal"; 3 = "Like") adopted to answer the question "How much do you enjoy walking on a daily basis?".

"Walking motivation" is assessed by six items, with a 5-point system adopted to measure (1 = none; 2 = few; 3 = some; 4 = most; 5 = all), and six categories include "walking with children", "walking to study", "walking to work", "walking to shop", "taking a walk to play", and "walking exercise".

"Environmental characteristics" refer to four dimensions, of which social quality, street quality and safety issues, are scored based on a 5-point system (1 = none; 2 = minority; 3 = some; 4 = majority; 5 = all). For the LUM, the 5-point system (1 = 1–5 min; 2 = 6–10 min; 3 = 11–20 min; 4 = 21–30 min; 5 = over 30 min) is adopted to measure the walking time of respondents using facilities of 4 types. Here, social quality refers to two dimensions: namely, "activity richness" and "social interaction degree"; street quality includes four dimensions: namely, street landscape style, street architectural style, street cleanliness and street connectivity; safety issue is street safety; LUM involves four dimensions: i.e., essential facilities, commercial consumption facilities, leisure and entertainment facilities, and public service facilities.

The LUM consists of 17 items used to assess land-use mix, "LUM (Land-use Mix)"and a 5-point system (1 = 1–5 min; 2 = 6–10 min; 3 = 11–20 min; 4 = 21–30 min; 5 = over 30 min) is adopted to measure respondents' walking time in 17 amenities, with lower scores indicating higher levels of land-use mix. The 19 items of LUM are grouped into four dimensions: (1) essential facilities; (2) commercial consumption facilities; (3) leisure and entertainment facilities; and (4) public service facilities.

4. Results

4.1. Historical Evolution

4.1.1. The Historical Evolution of Hefang Street and Gongchen Bridge Life Block

In this study, the Hefang Street business block and the Gongchen Bridge life block, both of which are located in the main city zone of Hangzhou, are the study subjects. These two differ in street forms and economic patterns.

As shown in Figure 2, Hefang Street located at the foot of Wushan, is part of Qinghefang and belongs to the old city of Hangzhou. Located in the northern part of Hangzhou city, the whole Gongchen Bridge is built above the ancient Beijing-Hangzhou Grand Canal, mainly including two parts: Gongchen Bridge block and the historical and cultural block in the west of the bridge.

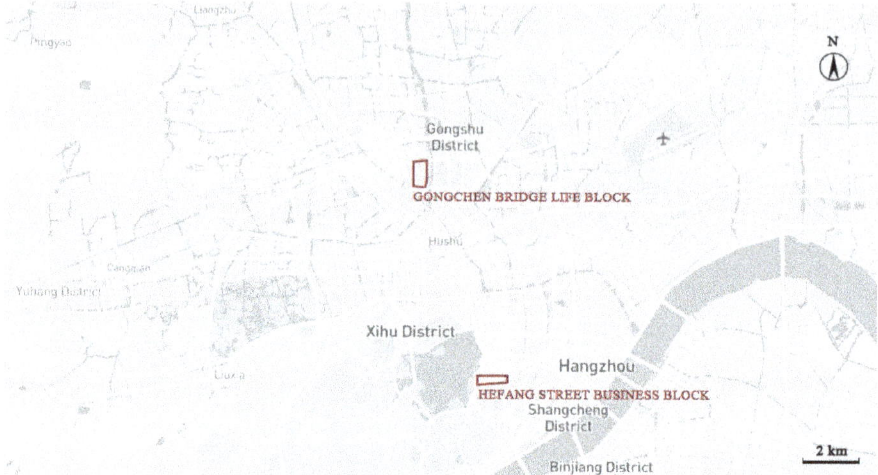

Figure 2. Location map of Hefang Street business block and Gongchen Bridge life block (according to Baidu map).

Once upon a time, Hefang Street was the "root of the imperial city" of Hangzhou, the capital of ancient times, and also the cultural center and economic and trade center of the Southern Song Dynasty. As shown in Figure 3, as the only old street in Hangzhou that maintains the historical appearance of the old city, Hefang Street, where time-honored brands (stores) stood in great numbers, integrated the most representative historical culture, commercial culture, marketplace culture and architectural culture in Hangzhou, so that it was saved from the fate of total demolition during the transformation of the old city of Hangzhou.

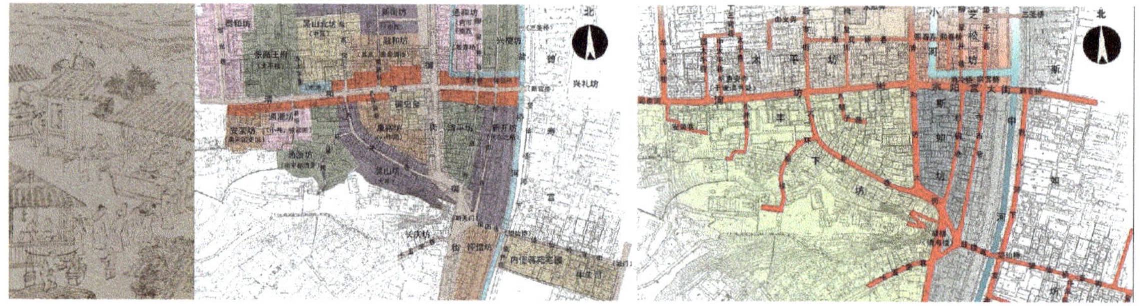

Figure 3. paintings in the past (**left**), Street pattern map of southern Song Dynasty (**middle**) and Street pattern map of Ming and Qing Dynasties (**right**). (online website https://jz.docin.com/p-2811341437.html accessed on 22 November 2021).

In 1999, the Hangzhou Municipal Government decided to redevelop the 13.6-hectare block adjacent to the Wushan (Town God's Temple) square into an antique pedestrian street of commerce and tourism. The renovated Hefang Street was opened in October 2002, as shown in Figure 4. Nowadays, Hefang Street reveals the style of the late Qing Dynasty and early Ming Dynasty. The cultural value is highlighted here to create a marketplace culture of commerce, pharmaceutical industry, and architecture. Its historical authenticity, cultural continuity and the overall appearance of the landscape are maintained. On both sides of Hefang Street, there are mainly local products, antiques, paintings and tourist souvenirs, and a craft pavilion is set up in the center of the street to display folk handicraft performances.

Figure 4. Photos of Hefang Street in 1966 (**left**) and the current (**right**). (online website http://www.360doc.com/content/20/0812/10/17132703_929812656.shtml accessed on 25 November 2021).

Now, there are many problems in Hefang Street. First of all, the public design at the monotonous node of the entrance space of Hefang Street is relatively simple. Most of them are seats combined with tree pools and simple humanoid sculptures. They are not attractive to walkers. The whole street is similar in format and has a high repetition rate. The feeling is bland and cannot produce great fluctuations in the sensory psychology of walkers. The viewing speed of respondents has not changed significantly. Secondly, in the design process of Hefang Street, there is not enough leisure space at the door of the shop which is easy to arouse pedestrians' interest, and the influence of natural factors such as season and climate is not considered, which is easy to cause local space congestion due to natural factors such as sunshine and rain, which affects pedestrians' walking experience. Finally, due to the high degree of commercialization of blocks in the old city bring vitality, the excessive commercialization and destruction of the traditional charm of space in Hefang Street reflects the traditional culture. Except for the architectural appearance, most of them show the traditional life scene by sculpture. Such facilities cannot interact with pedestrians, cannot render the traditional cultural atmosphere of the street, cannot leave a deep impression on pedestrians, and lack of interactive interest.

The relatively desolate original site of Gongchen bridge featured by a dotted scattering of the street layout was later evolved into the terminal dock of the Grand Canal in ancient times(Figure 5). The street was developed linearly along the river and extended radially to the interior. The Qing government, after the Sino-Japanese War, signed the Treaty of Shimonoseki with the Japanese imperialists, and Gongchen bridge was set up as a Japanese concession. Later, as the renovation project of the old city of Gongchen bridge area advanced, the faceted blocks and grouped factories took shape. In modern times, Gongchen bridge becomes a cultural preservation unit of Zhejiang Province, and also a part of the downtown. Unlike Hefang Streat with a mature business block mode, Gongchen Bridge, as a living pedestrian street, mainly serves pedestrians and surrounding residents.

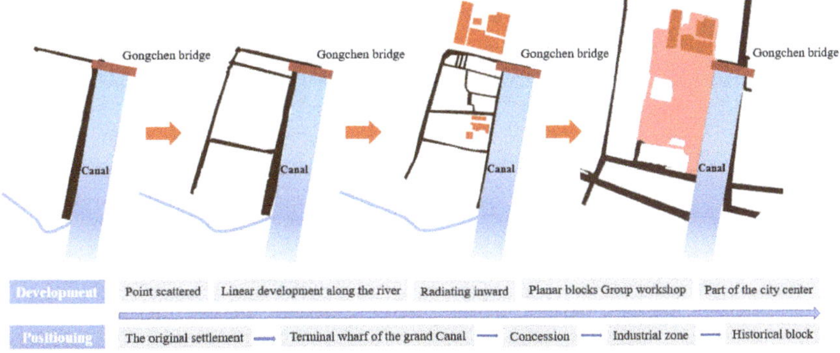

Figure 5. Morphological evolution of Gongchen Bridge Block. (Self-painted according to the historical evolution of Gongchen Bridge).

At present, the main problems of Gongchen Bridge block are that some houses were built earlier, and the street landscape environment quality in some roadways was not good. Because of the lack of protection awareness of historical and cultural blocks, the color volume of newly built buildings could not be well integrated with the style of historical and cultural blocks, which seemed to be somewhat out of place. Secondly, due to the high population density and mostly elderly people, the problem of aging in streets is prominent, and the blocks lack vitality and provide leisure and entertainment facilities are also in short supply.

4.1.2. Block Spatial Scale

In the form of street space, two streets have different forms of site distribution (Figure 6).The ratio of distance (D) and height (H) of building along Hefang Street-D/H = 13/6 = 2.167 > 2. The less closed and more open space as a whole is spacious and low, with an overall trend of horizontal extension generating a sense of openness and flow, but less affinity. Therefore, store seats in Hefang Street have been set up in the middle section of the road, increasing pedestrians' sense of safety for this place and creating spacious human touch in the block.

Figure 6. Maps of Hefang Street business block and Gongchen Bridge life block (according to Baidu map).

The streets in historical Gongchen Bridge block are mainly divided into two types-Qiaoxi Straight Street—a commercial block and Jixiang Temple Alley-a living alleyway. The ratio of distance (D) and height (H) of buildings along the commercial block—Qiaoxi Straight Street is D/H = 6/9 = 0.667 < 1, and the ratio of distance (D) and height (H) of buildings along the living alleyway Jixiang Temple Alley is D/H = 2.5/8 = 0.3125 < 1. It can be seen that the overall space of Gongchen Bridge block brings a strong sense of closure, and the long and narrow street creates a sense of profundity in the space.

4.1.3. Block Business Distribution

In Gongchen Bridge life block, there are totally 9 essential facilities accounting for 11%, 14 commercial consumption facilities taking up 20%, 10 leisure and entertainment facilities accounting for 13%, and 41 public service facilities taking up 55%. In Hefang Street business block, there are totally 63 essential facilities accounting for 20.3%, 109 commercial consumption facilities taking up 35%, 16 leisure and entertainment facilities accounting for 5.1%, and 123 public service facilities taking up 39.5%. In terms of essential facilities, restaurants play the biggest part in both Gongchen Bridge and Hefang Street, while there is only one daily necessities store in Gongchen Bridge. There is the same number of Traditional Chinese Medicine pharmacies in both of them. In terms of commercial consumption facilities, gift shops (5 in Gongchen Bridge and 65 in Hefang Street) play the biggest part in both of them. There is only one snack bar, one perfume shop and one specialty store in Gongchen Bridge, while there are 13 snack bars, 19 specialty stores, and only 1 dessert shop in Hefang Street. In terms of leisure and entertainment facilities, there is almost the same number of web celebrity shops in these two. There are 11 exhibition halls in Hefang Street and there is 6 in Gongchen

Bridge. In terms of public service facilities, there are 15 rest seats, and 22 sharable chargers in Gongchen Bridge, and 49 rest seats and 70 sharable chargers in Hefang Street. The number of postcard stations, parks and toilets is the same in these two.

According to the statistics of the number and proportion of businesses in the two blocks (Table 1), Hefang Street business block, relatively, boasts rich businesses, and the proportion of facilities of different types is balanced. Only leisure and entertainment facilities play a very small part there. Essential necessities and commercial consumption facilities in Hefang Street business block, outnumber that in Gongchen Bridge life block. Public service facilities have the highest proportion in Gongchen Bridge life block, while essential facilities account for the least there, so that the daily needs of the surrounding residents can't be met. In terms of the proportion of leisure and entertainment facilities and public service facilities, Gongchen Bridge outperforms Hefang Street.

Table 1. Number formats of Hefang Street business block and Gongchen Bridge life block.

Type	Formats	Gongchen Bridge Life Block (n)	Gongchen Bridge Life Block (%)	Hefang Street Business Block (n)	Hefang Street Business Block (%)
Essential facilities	Restaurant	4	5%	50	16%
	Traditional Chinese Medicine pharmacy	3	4%	3	1%
	Daily necessities store	1	1%	8	2.5%
	Clothing	1	1%	2	0.6%
Total		9	11%	63	20.3%
Commercial consumption facilities	Snack Bar	1	1%	13	4.2%
	Milk tea beverage store	3	4%	9	2.9%
	Dessert shop	3	4%	1	1%
	Gift shop	5	10%	65	21%
	Perfume shop	1	1%	2	0.6%
	Specialty stores	1	1%	19	6%
Total		14	20%	109	35%
Leisure and entertainment facilities	Web celebrity shop	4	5%	5	1.6%
	Exhibition Hall	6	8%	11	3.5%
Total		10	13%	16	5.1%
Public service facilities	Post card station	1	1%	1	1%
	Park	1	1%	1	1%
	Toilet	2	3%	2	0.6%
	Rest seat	15	20%	49	15.8%
	Sharable charger	22	30%	70	23%
Total		41	55%	123	39.5%

4.2. Basic Pedestrian Information

Descriptive statistics of pedestrians in both blocks are shown in Table 2. The number of male respondents is almost equal to that of female respondents in Hefang Street, with females accounting for 52% and males taking up 48%. Pedestrians aged 1–30, the main part of pedestrians, account for 54.7% of all, followed by those aged 31–55 accounting for 31.3%; those aged 56 and above account for only 14%. Pedestrians in Hefang Street are mainly recreational walkers, accounting for 43.3% of the total, while utilitarian walkers play a small part, taking up only 16% of the total. Among respondents in Gongchen Bridge, females account for 43.3% and males 56.7%. Those aged 31–55, the main part of respondents, account for 56.7% of all, followed by those aged 1–30 taking up 30%; those aged 56 and above, as the small part of the whole, only account for 13.3%. Pedestrians in Gongchen Bridge are mainly recreational walkers accounting for 55.3% of the total, followed by utilitarian walkers taking up 25.3%. No-specified walkers account for only 19.3% of the total.

Table 2. Pedestrian information in Hefang Street business block and Gongchen Bridge life block.

Pedestrian Information	Describe	GCQ		HFJ	
		n	%	n	%
All	All	150	100%	150	100%
Gender	Male	85	56.7%	72	48%
	Female	65	43.3%	78	52%
Age	1–30	45	30%	82	54.7%
	31–55	85	56.7%	47	31.3%
	56 or above	20	13.3%	21	14.0%
	Utilitarian walkers	38	25.3%	24	16%
	Recreational walkers	83	55.3%	65	43.3%
Pedestrian type	No-specified walkers	29	19.3%	61	40.7%

4.3. Pedestrian Behavior Analysis

1. The data about different genders and business staying time show that (Table 3):

Table 3. Mean and standard deviation of dwell time for pedestrian street furniture of different genders.

Gender	The Overall				Male				Female			
	HFJ		GCQ		HFJ		GCQ		HFJ		GCQ	
	N = 150		N = 150		N = 72		N = 85		N = 78		N = 65	
	Mean	SD	Mean	SD	Mean	SD	Mean	SD	Mean	SD	Mean	SD
Pedestrian behavior (stay)												
Essential facilities												
Restaurant	0.880	1.634	2.200	2.288	1.014	1.756	2.318	2.341	0.756	1.513	2.046	2.225
Traditional Chinese Medicine pharmacy	0.860	1.810	1.693	2.397	0.861	1.739	1.788	2.508	0.859	1.884	1.569	2.257
Daily necessities store	0.520	0.981	0.947	1.310	0.597	1.057	0.765	1.151	0.449	0.907	1.185	1.467
Clothing	0.927	1.457	1.540	1.709	1.083	1.545	1.141	1.465	0.782	1.364	2.062	1.870
Commercial consumption facilities												
Snack Bar	1.267	1.294	1.453	1.256	1.403	1.411	1.447	1.305	1.141	1.170	1.462	1.200
Milk tea beverage store	0.627	1.127	1.093	1.318	0.667	1.163	1.071	1.289	0.590	1.098	1.123	1.364
Dessert shop	0.887	1.156	0.973	1.080	0.806	1.182	1.106	1.124	0.962	1.133	0.800	1.003
Gift shop	1.507	1.646	1.767	1.826	1.583	1.642	1.729	1.880	1.436	1.656	1.815	1.767
Perfume shop	0.507	0.910	0.747	1.112	0.583	1.071	0.494	0.840	0.436	0.731	1.077	1.327
Specialty stores	0.967	1.234	0.887	1.308	1.056	1.299	0.882	1.331	0.885	1.173	0.892	1.288
Leisure and entertainment facilities												
Web celebrity shop	0.993	1.445	1.593	1.655	1.125	1.501	1.612	1.655	0.872	1.390	1.569	1.667
Exhibition Hall	1.787	2.298	2.827	2.716	1.819	2.222	3.106	2.699	1.756	2.381	2.462	2.716
Public service facilities												
Post card station	0.867	1.544	1.653	1.791	1.000	1.653	1.835	1.883	0.744	1.436	1.415	1.648
Park	0.993	1.679	1.887	1.859	1.042	1.736	1.941	1.911	0.949	1.635	1.815	1.802
Toilet	0.480	0.857	0.880	0.989	0.556	0.933	0.706	0.828	0.410	0.780	1.108	1.134
Rest seat	1.653	1.843	2.207	2.162	1.653	1.602	2.271	2.055	1.654	2.050	2.123	2.308
Sharable charger	0.253	0.657	0.380	0.720	0.333	0.751	0.400	0.727	0.179	0.552	0.354	0.717

SD—standard deviation. Pedestrians (staying): "0" means "no staying"; "1" means "within 1 min"; "2" represents "1–5 min"; "3" means "6–10 min"; "4" represents "11–20 min"; "5" means "20–30 min"; "6" means "over 30 min".

Generally, the staying time in all business types is for pedestrians in Hefeng Street is shorter than that in Gongchen Bridge. Pedestrians in Hefang Street have the longest stay in commercial consumption facilities (5.762), while those in Gongchen Bridge have the longest stay in public service facilities (7.007). In terms of pedestrian's staying time in essential facilities (Hefang Street = 3.187, Gongchen Bridge = 6.380), the difference in both blocks is the most obvious. Pedestrians in both blocks have the shortest staying time in recreational facilities (Hefang Street = 2.780, Gongchen Bridge = 4.420).

Male pedestrians in Hefang Street spend the longest time in commercial consumption facilities (6.098), while the male pedestrians in Gongchen Bridge spend the longest time in public service facilities (7.153). In addition, there is an obvious difference between male pedestrians' staying time in public service facilities (Hefang Street = 4.584, Gongchen Bridge = 7.153) in the two. For example, the park staying time in Hefang Street (1.042) is less than that in Gongchen Bridge (1.941).

The female pedestrians in both blocks spend the longest time in commercial consumption facilities (Hefang Street = 5.450, Gongchen Bridge = 7.169), while there is the most obvious difference between female pedestrians in both blocks in terms of the time spent on essential necessities (Hefang Street = 2.846, Gongchen Bridge = 6.862). For example, pedestrians' staying time in restaurants in Hefang Street (0.756) is shorter than that in Gongchen Bridge (2.046).

2. The data of age and business staying time (Table 4):

Table 4. Mean and standard deviation of dwell time for pedestrian street facilities of different ages.

Age	**1–30**				**31–55**				**56 or Above**			
	Hefang Street		**Gongchenqiao Street**		**Hefang Street**		**Gongchenqiao Street**		**Hefang Street**		**Gongchenqiao Street**	
	N = 82		N = 45		N = 47		N = 85		N = 21		N = 20	
	Mean	**SD**	**Mean**	**SD**	**Mean**	**SD**	**Mean**	**SD**	**Mean**	**SD**	**Mean**	**SD**
Pedestrian behavior (stay)												
Essential facilities												
Restaurant	0.610	1.368	2.022	2.426	1.000	1.694	2.235	2.213	1.667	2.176	2.450	2.373
Traditional Chinese Medicine pharmacy	0.476	1.269	1.311	2.087	0.809	1.715	1.576	2.362	2.476	2.786	3.050	2.819
Daily necessities store	0.366	0.854	0.689	1.145	0.809	1.191	1.012	1.314	0.476	0.814	1.250	1.585
Clothing	0.793	1.438	1.667	1.895	1.085	1.501	1.541	1.666	1.095	1.446	1.250	1.482
Commercial consumption facilities												
Snack Bar	1.378	1.330	1.644	1.317	1.213	1.250	1.365	1.184	0.952	1.244	1.400	1.429
Milk tea beverage store	0.476	0.933	1.089	1.240	0.766	1.088	1.000	1.185	0.905	1.729	1.500	1.906
Dessert shop	0.976	1.165	1.156	1.127	0.766	1.237	0.929	1.067	0.810	0.928	0.750	1.020
Gift shop	1.220	1.457	1.933	1.900	1.745	1.811	1.741	1.774	2.095	1.786	1.500	1.933
Perfume shop	0.585	0.942	1.111	1.210	0.511	0.997	0.635	1.056	0.190	0.402	0.400	0.940
Specialty stores	0.854	1.101	0.822	1.336	1.106	1.463	0.859	1.329	1.095	1.179	1.150	1.182
Leisure and entertainment facilities												
Web celebrity shop	1.061	1.651	2.156	1.953	0.830	1.129	1.400	1.474	1.095	1.221	1.150	1.387
Exhibition Hall	1.500	2.218	3.000	2.820	1.617	2.202	2.800	2.689	3.286	2.348	2.550	2.704
Public service facilities												
Post card station	0.927	1.661	2.289	2.030	0.681	1.337	1.376	1.647	1.048	1.532	1.400	1.501
Park	0.744	1.464	1.489	1.727	1.196	1.928	2.035	1.930	1.524	1.778	2.150	1.785
Toilet	0.427	0.847	0.822	0.936	0.489	0.804	0.906	0.971	0.667	1.017	0.900	1.210
Rest seat	1.390	1.762	2.289	2.052	1.809	1.789	1.953	2.098	2.333	2.129	3.100	2.511
Sharable charger	0.317	0.752	0.489	0.589	0.106	0.312	0.294	0.651	0.333	0.796	0.500	1.147

SD—standard deviation. Pedestrians (staying): "0" means "no staying"; "1" represents "within 1 min"; "2" means "1–5 min"; "3" represents "6–10 min"; "4" means "11–20 min"; "5" represents "20–30 min"; "6" means "over 30 min".

Pedestrians aged 1–30 in both blocks spend the longest time in commercial consumption facilities (Hefang Street = 5.489, Gongchen Bridge = 7.755). Pedestrians aged 1–30 in Hefang Street spend the shortest time in essential facilities (2.245), while pedestrians aged 1–30 in Gongchen Bridge spend the shortest time in the leisure and entertainment facilities (5.156). In addition, there is the most obvious difference in the staying time of pedestrians aged 1–30 in public service facilities (Hengfang Street = 3.805, Gongchen Bridge = 7.738) between these two. For example, pedestrians in Gongchen Bridge stay longer in the postcard station (2.289) than that in Hefang Street (0.927).

Pedestrians aged 31–55 in both blocks have the shortest staying time in leisure and entertainment facilities (Hefang Street = 2.447, Gongchen Bridge = 4.200). Pedestrians aged 31–55 in Hefang Street spend the longest time in commercial consumption facilities (6.107), while pedestrians aged 31–55 in Gongchen Bridge spend the longest time in public service facilities (6.564). Moreover, the most obvious difference can be seen in the staying time of essential facilities for pedestrians aged 31–55 in these two blocks (Hefang Street = 3.703, Gongchen Bridge = 6.364). For example, pedestrians stay longer in restaurants in Gongchen Bridge (2.235) than those in Hefang Street (1.000).

Pedestrians aged 56 and above in both blocks have the shortest stay time in leisure and entertainment facilities (Hefang Street = 4.381, Gongchen Bridge = 3.700). Pedestrians aged 56 and above in Hefang Street spend the longest time in commercial consumption facilities (6.047), while pedestrians aged 56 and above in Gongchen Bridge spend the longest time in public service facilities (8.050). Besides, there is the most obvious difference in staying time in essential facilities for pedestrians aged 56 and above in both blocks. (Hefang Street = 5.714, Gongchen Bridge = 8.000). For example, pedestrians stay longer in daily necessities store in Gongchen Bridge (1.250) than those in Hefang Street (0.476).

3. The business staying time data about walkers of different types show (Table 5):

Table 5. Mean and standard deviation of dwell time in street furniture for different pedestrian types.

	Utilitarian Walkers				**Recreational Walkers**				**No-Specified Walkers**			
	Hefang Street		**Gongchenqiao Street**		**Hefang Street**		**Gongchenqiao Street**		**Hefang Street**		**Gongchenqiao Street**	
	N = 24		N = 38		N = 65		N = 83		N = 61		N = 29	
	Mean	**SD**	**Mean**	**SD**	**Mean**	**SD**	**Mean**	**SD**	**Mean**	**SD**	**Mean**	**SD**
Pedestrian behavior (stay)												
Essential facilities												
Restaurant	0.667	1.373	2.895	2.252	0.954	1.605	2.193	2.361	0.885	1.771	1.310	1.834
Traditional Chinese Medicine pharmacy	0.208	1.021	2.447	2.882	0.815	1.776	1.759	2.366	1.164	2.026	0.517	0.986
Daily necessities store	0.333	0.702	0.974	1.325	0.569	1.060	1.036	1.392	0.541	0.993	0.655	1.010
Clothing	0.750	1.511	1.053	1.723	1.138	1.609	1.807	1.797	0.770	1.244	1.414	1.268
Commercial consumption facilities												
Snack Bar	1.042	1.160	1.395	1.480	1.492	1.301	1.506	1.173	1.115	1.318	1.379	1.208
Milk tea beverage store	0.500	0.885	0.737	1.131	0.569	1.104	1.277	1.417	0.738	1.237	1.034	1.180
Dessert shop	0.833	1.274	0.895	1.110	0.800	1.121	0.940	1.016	1.000	1.155	1.172	1.227
Gift shop	1.000	1.351	1.395	1.636	1.938	1.767	1.904	1.992	1.246	1.524	1.862	1.529
Perfume shop	0.375	1.013	0.368	0.883	0.431	0.770	0.940	1.253	0.639	1.001	0.690	0.806
Specialty stores	0.417	0.830	0.737	1.131	1.185	1.345	0.831	1.305	0.951	1.189	1.241	1.504
Leisure and entertainment facilities												
Web celebrity shop	0.417	1.018	1.447	1.589	0.985	1.474	1.614	1.738	1.230	1.510	1.724	1.533
Exhibition Hall	0.958	2.010	2.158	2.707	2.385	2.415	3.277	2.760	1.475	2.142	2.414	2.428
Public service facilities												
Post card station	0.792	1.414	1.184	1.449	0.738	1.492	1.795	1.924	1.033	1.653	1.862	1.747
Park	0.625	1.610	1.868	1.891	1.219	1.906	2.036	1.916	0.902	1.422	1.483	1.639
Toilet	0.292	0.624	0.711	0.898	0.508	0.937	0.940	1.052	0.525	0.849	0.931	0.923
Rest seat	0.917	1.472	1.737	2.140	1.985	1.663	2.398	2.252	1.590	2.077	2.276	1.888
Sharable charger	0.042	0.204	0.237	0.431	0.292	0.824	0.422	0.798	0.295	0.558	0.448	0.783

SD—standard deviation. Pedestrians (staying): "0" means "no staying"; "1" represents "within 1 min"; "2" means "1–5 min"; "3" represents "6–10 min"; "4" means "11–20 min"; "5" represents "20–30 min"; "6" means "over 30 min".

Utilitarian walkers (4.167) and recreational walkers (6.415) in Hefang Street stay in commercial consumption facilities for the longest time, while utilitarian walkers in Gongchen Bridge stay in essential necessities (7.369) for the longest time and recreational walkers there stay in public service facilities (7.519) for the longest time. No-specified walkers stay in commercial consumption facilities (Hefang Street = 5.689, Gongchen Bridge = 7.378) for the longest time. The is the most obvious difference in no-specified walkers' staying time in public service facilities (Hefang Street = 4.345, Gongchen Bridge = 7) in both blocks. There is the most obvious difference between utilitarian walkers (HFJ = 3.476, GCQ = 6.795) and recreational walkers (Hefang Street = 3.476, Gongchen Bridge = 6.795) in terms of the staying time in essential facilities.

4.4. Pedestrian Walking Experience

(1) The data of pedestrian walking experience of both genders demonstrate that (Table 6):

Table 6. Correlation analysis of daily walking time and multiple factors of pedestrians of different genders on two streets.

Gender	The Overall				Male				Female			
	HFJ	t	GCQ	t	HFJ	t	GCQ	t	HFJ	t	GCQ	t
Walking preference	0.352	0.000 ***	0.274	0.001 ***	0.227	0.137	0.294	0.031 **	0.340	0.008 ***	0.173	0.114
Walking motivation												
Walking with children	−0.034	0.646	0.000	0.997	−0.156	0.166	−0.028	0.832	0.072	0.496	0.032	0.784
Walking to school	0.078	0.301	0.090	0.309	−0.106	0.376	−0.139	0.340	0.213	0.056 *	0.246	0.032 **
Walking to work	−0.079	0.262	−0.067	0.431	−0.032	0.764	−0.341	0.018 **	−0.016	0.880	0.099	0.366
Walking to shop	0.099	0.247	0.141	0.084 *	0.040	0.766	0.224	0.084 *	0.210	0.066 *	0.116	0.296
Recreational walking	0.240	0.008 ***	0.047	0.579	0.250	0.092 *	−0.210	0.111	0.193	0.123	0.338	0.004 ***
Sports-based walking	−0.043	0.601	−0.008	0.920	0.081	0.572	0.008	0.952	−0.063	0.568	−0.163	0.153
Environmental characteristics												
Social quality	−0.029	0.746	0.014	0.864	0.026	0.850	0.003	0.984	0.011	0.937	0.031	0.792
Street quality	0.106	0.318	−0.048	0.552	0.387	0.020 **	0.056	0.666	−0.181	0.237	0.004	0.971
Social safety	−0.034	0.646	−0.024	0.769	−0.156	0.166	−0.157	0.287	0.072	0.496	−0.009	0.932
LUM												
Essential facilities	0.057	0.405	0.135	0.100	0.038	0.735	0.310	0.022 **	0.020	0.842	−0.038	0.725
Commercial consumption facilities	−0.110	0.126	−0.213	0.063 *	−0.162	0.181	−0.164	0.358	−0.140	0.189	−0.279	0.074 *
Leisure and entertainment facilities	0.102	0.194	0.165	0.079 *	0.052	0.643	0.219	0.132	−0.086	0.461	0.168	0.190
Public service facilities	−0.064	0.444	−0.115	0.303	0.047	0.714	−0.011	0.953	−0.216	0.075 *	−0.120	0.431

*** $p < 0.01$, ** $p < 0.05$, * $p < 0.1$.

Generally, walking preferences (Hefang Street: r = 0.352, $p < 0.01$, Gongchen Bridge: R = r = 0.274, $p < 0.01$) significantly impact pedestrian walking experience in a positive manner in both blocks. In terms of walking motivation, recreational walking (r = 0.240, $p < 0.01$) impacts pedestrian walking experience in a positive manner in Hefang Street. Walking to shop (r = 0.141, $p < 0.1$) significantly impacts pedestrian walking experience in a positive manner in Gongchen Bridge. In terms of environmental characteristics, the distribution of commercial consumption facilities (r = −0.213, $p < 0.1$) significantly impacts the pedestrian walking experience in a negative manner in Gongchen Bridge. While the distribution of leisure and entertainment facilities (r = 0.165, $p < 0.1$) is negatively correlated with pedestrian walking experience.

Walking preferences (r = 0.294, $p < 0.05$) significantly impact the walking experience of male pedestrians in a positive manner in Gongchen Bridge. In terms of walking motivation, recreational walking (r = 0.250, $p < 0.1$) greatly impacts the walking experience of male pedestrians in a positive manner in Hefang Street. Walking to work (r = −0.341, $p < 0.05$)

and walking to shop (r = 0.224, $p < 0.1$) greatly influence the walking experience of male pedestrians in Gongchen Bridge. The former is negatively correlated with the walking experience, while the latter is positively correlated with it. In terms of environmental characteristics, street quality (r = 0.387, $p < 0.05$) significantly impacts the walking experience of male pedestrians in a positive manner in Hefang Street. The distribution of essential facilities (r = 0.310, $p < 0.05$) has a significant impact on the walking experience of male pedestrians in a positive manner in Gongchen Bridge.

Walking preference (r = 0.340, $p < 0.01$) has a significant impact on the walking experience of female pedestrians in a positive manner in Hefang Street. In terms of walking motivation, walking to school (r = 0.213, $p < 0.1$) and walking to shop (r = 0.210, $p < 0.1$) have significant influence on the walking experience of female pedestrians in a positive manner in Hefang Street. Walking to school (r = 0.246, $p < 0.05$) and recreational walking (r = 0.338, $p < 0.01$) have significant influence on the walking experience of female pedestrians in a positive manner in Gongchen Bridge. In terms of environmental characteristics, the distribution of public service facilities (r = −0.216, $p < 0.1$) significantly impacts the walking experience of female pedestrians in a negative manner in Hefang Street. The distribution of commercial consumption facilities (r = −0.279, $p < 0.1$) significantly impacts the walking experience of female pedestrians in a negative manner in Gongchen Bridge.

(2) The data of pedestrian walking experience of different age groups show that (Table 7):

Table 7. Correlation analysis of daily walking time and multiple factors of pedestrians of different ages on two streets.

Age	**1–30**				**31–55**				**56 or Above**			
	HFJ	t	**GCQ**	t	**HFJ**	t	**GCQ**	t	**HFJ**	t	**GCQ**	t
Walking preference	0.389	0.003 ***	0.355	0.065 *	0.218	0.256	0.196	0.096 *	−0.283	0.443	0.300	0.290
Walking motivation												
Walking with children	−0.035	0.718	−0.033	0.902	0.017	0.899	−0.089	0.462	0.212	0.386	−0.063	0.928
Walking to school	0.144	0.222	0.066	0.799	−0.092	0.551	0.067	0.562	−0.154	0.665	0.611	0.248
Walking to work	0.012	0.913	0.006	0.980	−0.445	0.006 ***	−0.054	0.637	0.082	0.804	−0.050	0.888
Walking to shop	0.206	0.097 *	0.100	0.616	0.264	0.214	**0.186**	0.128	**0.346**	0.292	0.516	0.201
Recreational walking	0.221	0.115	0.084	0.646	0.405	0.015 **	−0.090	0.464	0.007	0.986	0.111	0.841
Sports-based walking	0.060	0.605	0.136	0.479	−0.066	0.694	−0.123	0.319	−0.602	0.118	**0.785**	0.222
Environmental characteristics												
Social quality	−0.217	0.132	**−0.268**	0.212	−0.063	0.659	**0.158**	0.227	**1.042**	0.093 *	0.687	0.345
Street quality	**0.234**	0.130	−0.167	0.402	**0.181**	0.043 **	0.015	0096 *	0.212	0.186	**0.823**	0.094 *
Social safety	−0.035	0.718	−0.167	0.335	0.017	0.899	0.044	0.708	−0.154	0.665	−0.261	0.503
LUM												
Essential facilities	0.064	0.504	−0.012	0.942	−0.053	0.722	0.140	0.241	−0.110	0.663	0.387	0.264
Commercial consumption facilities	−0.084	0.382	**−0.203**	0.415	−0.089	0.647	**−0.158**	0.352	0.310	0.204	0.068	0.927
Leisure and entertainment facilities	−0.044	0.662	−0.055	0.791	**0.227**	0.126	0.072	0.574	0.113	0.787	−0.009	0.990
Public service facilities	**0.100**	0.423	−0.043	0.875	−0.115	0.530	−0.022	0.884	**−0.528**	0.033 **	**−0.786**	0.250

*** $p < 0.01$, ** $p < 0.05$, * $p < 0.1$.

Pedestrian preference of pedestrians aged 1–30 in Hefang Street and Gongchen Bridge (Hefang Street: r = 0.389, $p < 0.01$, Gongchen Bridge: r = 0.355, $p < 0.1$) is positively correlated with the total walking time. The correlation between pedestrian preference and total walking time in Hefang Street is stronger than that in Gongchen Bridge. It indicates that the walking experience of pedestrians aged 1–30 in business block is more likely to be influenced by walking preference. The walking motivation of pedestrians aged 1–30 in Hefang Street (r = 0.206, $p < 0.1$) impacts their walking experience the most, while that of pedestrians aged 1–30 in Gongchen Bridge has no significant influence on their walking experience. The environmental characteristics have no significant impact on pedestrians aged 1–30 in both blocks.

In addition, walking preference has no significant impact on pedestrians aged 31–55 in both blocks. The walking motivation of walking to work for pedestrians aged 31–55 (r = −0.445, $p < 0.01$) in Hefang Street is negatively correlated with their walking experience, while the walking motivation of recreational walking (r = 0.405, $p < 0.05$) for those aged 31–55 in Hefang Street is positively correlated with their walking experience. The walking motivation has no obvious impact on walking experience for those aged 31–55 in Gongchen Bridge. Social quality in the environmental characteristics (Hefang Street: r = 0.181, $p < 0.05$, Gongchen Bridge: r = 0.015, $p < 0.1$) significantly impacts the walking experience of pedestrians aged 31–55 in a positive manner in both blocks.

Besides, walking preference and walking motivation have no significant impact on pedestrians aged 56 and above in both blocks. The social quality in the environmental characteristics (r = 0.823, $p < 0.1$) has a significant impact on the walking experience of pedestrians aged 56 and above in a positive manner in Gongchen Bridge. LUM of public service facilities (r = −0.528, $p < 0.05$) is negatively correlated with the walking experience of pedestrians aged 56 and above in Gongchen Bridge.

(3) The data of walking experience of walkers of different types show that (Table 8):

Table 8. Correlation analysis of daily walking time and multiple factors of different walker types on two streets.

	Utilitarian Walkers				Recreational Walkers				No-Specified Walkers			
	HFJ	t	GCQ	t	HFJ	t	GCQ	t	HFJ	t	GCQ	t
Walking preference	0.772	0.129	0.104	0.502	0.443	0.001 ***	0.290	0.013 **	0.264	0.152	0.636	0.036 **
Walking motivation												
Walking with children	−0.248	0.462	0.225	0.190	−0.108	0.347	−0.120	0.314	−0.189	0.230	0.068	0.808
Walking to school	−0.383	0.215	0.394	0.039 **	−0.002	0.984	0.122	0.341	0.157	0.274	0.136	0.673
Walking to work	−0.907	0.158	−0.045	0.776	0.001	0.990	−0.092	0.437	−0.126	0.332	−0.180	0.483
Walking to shop	0.862	0.139	−0.192	0.227	−0.026	0.827	0.271	0.026 **	0.122	0.496	0.188	0.470
Recreational walking	0.197	0.647	0.296	0.172	−0.011	0.930	−0.013	0.913	0.345	0.057 *	0.003	0.991
Sports-based walking	0.396	0.321	−0.329	0.121	−0.207	0.060 *	0.009	0.942	−0.058	0.749	0.207	0.480
Environmental characteristics												
Social quality	−0.362	0.292	0.162	0.398	0.007	0.955	−0.063	0.594	0.071	0.678	−0.130	0.703
Street quality	0.502	0.277	−0.235	0.181	0.121	0.537	0.039	0.731	−0.005	0.977	0.075	0.785
Social safety	−0.248	0.462	0.025	0.871	−0.108	0.347	−0.095	0.422	−0.189	0.230	0.385	0.237
LUM												
Essential facilities	0.093	0.835	0.657	0.001 ***	0.060	0.581	−0.036	0.745	0.076	0.530	0.211	0.436
Commercial consumption facilities	0.285	0.461	−0.114	0.560	−0.021	0.847	−0.199	0.171	−0.184	0.240	−0.349	0.495
Leisure and entertainment facilities	0.174	0.640	0.454	0.016 **	−0.015	0.895	0.083	0.540	0.010	0.949	0.177	0.563
Public service facilities	0.936	0.129	−0.412	0.128	−0.353	0.008 ***	−0.176	0.193	−0.062	0.690	−0.077	0.867

*** $p < 0.01$, ** $p < 0.05$, * $p < 0.1$.

Walking preference (Hefang Street: r = 0.443, $p < 0.01$, Gongchen Bridge: r = 0.037, $p < 0.05$) significantly impacts the walking experience of recreational walkers in a positive manner.

Additionally, the walking motivation and environmental characteristics have no significant influence on the walking experience of utilitarian walkers in Hefang Street but have great influence on the walking experience of utilitarian walkers in Gongchen Bridge. Here, walking to school (r = 0.394, $p < 0.05$), street safety (r = 0.025, $p < 0.1$), distribution of essential necessities (r = 0.657, $p < 0.01$) and leisure and entertainment facilities (r = 0.454, $p < 0.05$) are positively correlated with the walking experience of utilitarian walkers.

In addition, walking motivation and environmental characteristics significantly impact the walking experience of recreational walkers in both blocks. The street quality (Hefang Street: r = 0.121, $p < 0.1$, Gongchen Bridge: r = 0.037, $p < 0.1$) is positively correlated with the walking experience of recreational walkers in both blocks. Sports-based walking (r = −0.207, $p < 0.1$) and the distribution of public service facilities (r = −0.353, $p < 0.01$) are negatively correlated with the walking experience of recreational walkers in Hefang Street. Walking to shop (r = 0.271, $p < 0.05$) and street quality (r = 0.039, $p < 0.05$) are positively correlated with the walking experience of recreational walkers in Gongchen Bridge.

Moreover, environmental characteristics have no significant impact on the walking experience of no-specified walkers in Hefang Street, while the walking motivation has a significant impact on the walking experience of no-specified walkers in Hefang Street. Recreational walking (r = 0.345, $p < 0.1$) is positively correlated with the walking experience of no-specified pedestrians in Hefang Street. It is worth noting that walking motivation and environmental characteristics have no significant influence on the walking experience of utilitarian walkers in Gongchen Bridge.

5. Discussion

To find specific factors impacting pedestrian walking behavior, create a favorable environment in the historical and cultural block, and benefit the pedestrian walking experience in the street, an objective approach is adopted in this paper to analyze pedestrian behavior in the street and demonstrate the correlation between (pedestrians' own factors and the street environment) and (pedestrian behavior and pedestrian walking experience). As for study methods, questionnaires and Statistical Product and Service Solutions (SPSS) data analysis are used in the study to more objectively and effectively decompose the correlations between various factors. In terms of the study content, the attention was paid to the walking of dynamic level in recent study in recent years. While this paper focuses more on the walking behavior of pedestrians who are lingering, and a relevant study is therefore conducted. Meanwhile, in previous studies, more attention was paid to the environmental construction and traffic planning of physical environment characteristics, and less attention was paid to the perspectives of the street cultural atmosphere and historical appearance. From the perspective of pedestrian walking behavior and walking experience, this paper explores the comprehensive characteristics of pedestrian walking in cultural heritage streets qualitatively and quantitatively. Multi-angle research can more accurately grasp the factors affecting pedestrian walking, provide empirical evidence for better building street space environment, provide research data support for the renovation and renewal of heritage streets in the future, and promote the related research on the improvement of street walking environment to a certain extent.

Based on the cultural attributes of historical and cultural blocks, the historical evolution, spatial scale and business distribution of the two blocks in the article are studied. Then the differences of pedestrian walking behaviors and walking experiences of different age groups are specifically analyzed. It's confirmed that the walking preference is generally positively correlated with the total walking time, with the influencing factor of pedestrian walking preference in Hefang Street business block being 0.352, and that of Gongchen Bridge life block being 0.274. In addition, due to the difference of pedestrians' walking motivation and block environment characteristics, and the difference in pedestrian's walking time in these two blocks, there are different correlations between LUM and walking time. The influence coefficient of social quality dimension in Hefang Street is −0.029, while that in Gongchen Bridge is (+)0.014. Meanwhile the influence coefficient of street quality dimension in Hefang Street is 0.106, but that in Gongchen Bridge is −0.048. Above all, environmental characteristics have totally different influences on walking time in different blocks.

5.1. Differences of Pedestrian Walking Behaviors

It's found from the multiple linear regression analysis of both genders, daily walking time and multiple factors in Table 6, that for different genders of pedestrians, different pedestrian walking experiences will be produced.

(1) The correlation between gender and walking behaviors

Pedestrians' staying time in Gongchen Bridge life street is longer than that in Hefang Street business block. Pedestrians around the business block prefer staying in commercial consumption facilities, while pedestrians around the life block prefer staying in public service facilities. Male pedestrians prefer staying in commercial consumption facilities in business block, while they prefer staying in public service facilities in life block. Female pedestrians prefer staying in commercial consumption facilities in both blocks, which was possibly triggered by marginal effects. Previous studies have demonstrated that travelling time significantly impacts walking behaviors of both genders. Distance significantly impacts pedestrians' walking tendency. The marginal effect of distance is greater among women than among men in work trips, while the opposite is true in shopping trips [40];

(2) The relationship between different age groups and walking behaviors.

Commercial consumption facilities have a great influence on the walking behaviors of pedestrians in both blocks. It is worth noting that commercial consumption facilities have a great influence on the walking behaviors of pedestrians over 30 years old in business block, and public service facilities have a great influence on the walking behaviors of pedestrians over 30 years old in life block. It is shown in a study that young consumers show increasing spending power, and their spending rate is much higher than that of previous generations [41]. Compared with the elderly consumers, young people have the highest level of income, savings, and expenditure. Consumers over 65 years old show stronger emotional brand attachment than those aged 50 to 65 [42]. At present, the newly established business model is more likely to attract pedestrians aged 1–30;

(3) The relationship between walkers of different types and walking behaviors.

In both blocks, commercial consumption facilities have a strong influence on the pedestrian walking behaviors of different types. Existing studies have shown that blocks with various shopping options will attract pedestrians to stay for a long time, and shopping activities in blocks will positively impact pedestrians [43]. In addition, the essential facilities have a strong influence on the walking behaviors of utilitarian pedestrians in life block, while the public service facilities have an influence on the walking behaviors of recreational walkers and no-specified walkers.

5.2. Differences in Pedestrian Walking Experience

(1) The relationship between both genders and walking behaviors

Male pedestrians' walking experience in life block is influenced by their walking preferences. In business block, the walking motivation that has a great influence on male pedestrians' walking experience is recreational walking. In terms of environmental characteristics, street quality impacts their walking experience. For example, street landscape, architectural style, street cleanliness and connectivity are likely to influence their walking experience. Walking motivations that have great influence on male pedestrians' walking experience in life block involve walking to work and walking to shop. The influence of environmental characteristics on their walking experience can be reflected in the distribution of essential facilities. That's possibly because that most male pedestrians in Hefang Street are students who go for entertainment. The existing research shows that the student group has evolved into the main consumption force [44], and the walking environment impacts pedestrians' sense of security. While males who like walking boast a higher sense of security [45], and the marginal effect of males' shopping distance is greater than that of females' [40]. Female pedestrians' walking experience in business block is influenced by their walking preference. Walking motivations in business block that have great influence on female pedestrians' walking experience include walking to school and walking to shop. The influence of environmental characteristics on their walking experience is reflected in the distribution of public service facilities. Walking motivations that have great influence on female pedestrians' walking experience in life block are walking to school and recreational walking. The influence of environmental characteristics on their walking experience is reflected in the distribution of commercial consumption facilities. Previous studies have shown that women's extroversion and openness to experience give them a stronger desire to buy compared with men [46], while there is not much commercial development in Gongchen Bridge life block, and most female walkers are college students, mainly travelling with friends. Close social distance promotes pedestrians' consumption behaviors [47]. While in Hefang Street business block with concentrated businesses, the consumption purpose is very clear. While the scattered businesses of Gongchen Bridge make pedestrians more likely to wander;

(2) Relationship between different age groups and walking experience

Previous studies have shown that younger teenagers tend to be more physically active and prefer walking [48]. The data of this study show that the walking preference of pedestrians aged 1–30 has a greater impact on their total walking time. Most pedestrians aged 1–30 in Hefang Street and Gongchen Bridge, attend elementary school or junior high school. They hang out in the pedestrian block together with their family members. In the above research, social quality is positively correlated with the overall walking time of pedestrians aged 31–55 in Hefang Street and pedestrians aged 31–55, 56 and above in Gongchen Bridge. Previous studies have shown that the relationship between built environment and elderly people's travelling behaviors can be explained by peer effect or collective socialization [49]. The block activity richness and social interaction have a strong influence on middle-aged and elderly people. Neighborhood is an outdoor space, a transportation place, and an important area of the elderly friendly community. Studies have shown that there is a positive correlation between walking activities and social interaction [50]. This view is verified in our research. For the elderly, the harm caused by their lack of spiritual comfort is even more serious than that caused by physical diseases. With the acceleration of population aging, the geographical relationship between neighbors and friends in public space is particularly important for the mental health of the elderly;

(3) Relationship between walkers of different types and walking experience.

The walking experience of recreational walkers will be strongly influenced by their walking preferences. Additionally, utilitarian walkers' walking experience is strongly influenced by street safety, the distribution of essential facilities: restaurants, Traditional Chinese Medicine pharmacies, daily necessities store and clothing stores, as well as the number of leisure and entertainment facilities: web celebrity shops and exhibition halls. Studies have shown that utilitarian walkers, as the name suggests, are utilitarian [51]. They walk mainly for commuting but pay little attention to social quality and street quality. Some studies have also shown that there is a causal relationship between walking experience and street safety values [52]. In addition, the walking experience of recreational walkers is strongly influenced by street quality referring to street landscape style, street architectural style, street cleanliness and street connectivity. The research shows that the key factor impacting the evaluation of recreational walking comfort is street quality [53]. The construction of infrastructures such as green road has a positive impact on the increase of pedestrians' weekly walking time [54]. The walking experience of recreational walkers in business block will be impacted by the distribution and quantity of public service facilities, such as postcard stations, parks, toilets, rest seats and sharable chargers. Recreational walking is a factor that significantly influences the walking experience of non-specified pedestrians in business block. Previous studies have shown that recreational walking is more likely to produce positive and healthy emotions than utilitarian walking [55];

(4) Comparison of existing studies

The study result here is similar to that from researchers who study on pedestrians' walking behavior and walking experience. The block business distribution tends to have a greater impact on street dynamics than the block cultural characteristics. Compared with other environment characteristics, the dimension of functional feature nurtures street dynamics. The business reflects pedestrians' immediate needs. It's found the study that higher LUM facilitates commuting and leisure walking, instead of walking of practical nature [56]. In this study, a variability exists in the effect of LUM on pedestrian walking behavior for streets of different types. In Hefang Street block of greater walkability, the utility of walking to work and walking to study is not significantly impacted by LUM, and the leisure walking coefficient of walking to shop sees a significant rise. While a different result is demonstrated in the life-oriented Gongchen Bridge block, where the utility of walking to school is significantly impacted by LUM.

6. Conclusions and Implications

6.1. Conclusions

The main conclusions of this study can be summarized as follows:

(1) Pedestrian walking behavior

The walking behavior of those aged 1–30 is greatly influenced by commercial consumption facilities. The walking behavior of pedestrians aged over 30 in the business block is greatly influenced by the commercial consumption facilities, and the walking behavior of those aged over 30 in the life block is greatly influenced by the public service facilities. In both blocks, recreational walkers, utilitarian walkers, and no-specified walkers will stay in commercial consumption facilities for a long time. Essential facilities will have an impact on the walking behavior of pedestrians in life block, and public service facilities will have an impact on the walking behavior of pedestrians in business block;

(2) Pedestrian experience

Men prefer walking in life block, while women prefer walking in business block. The street quality of business block is positively correlated with men's walking time. The street landscape, architectural style, cleanliness and connectivity have a significant impact on men's walking experience in business block. Men's walking experience in life block is greatly influenced by the distribution of essential facilities, while women's walking experience is greatly influenced by the distribution of public service facilities in business block. Walking preference has a strong influence on the walking experience of pedestrians aged 1–30, but a weak influence on the walking experience of pedestrians over 30. Street environmental characteristics, especially social quality, are positively correlated with the total walking time of pedestrians aged 56 and above. Activity richness and social interaction have a significant impact on the walking experience of pedestrians aged 56 and above. Walking preference will impact the walking experience of recreational walkers, and street quality will influence the walking experience of recreational walkers. Here, street landscape style, street architectural style, street cleanliness and street connectivity are important factors (in terms of street quality).

As the closest way to daily life and the healthiest way to environmental protection, it is necessary and meaningful to study the correlation between environment and pedestrian behavior, update and improve walking environment, stimulate pedestrian walking behavior, and enhance pedestrian walking experience. First, it can improve the health level of individuals and promote more pedestrians to participate in the cultural heritage blocks by improving the walking environment of cultural heritage streets, so that their body and mind can develop together. Second, the construction of neighborhood atmosphere and the improvement of quality of life create more opportunities for pedestrians to stay, build street space with humanistic care and social atmosphere, enhance the overall vitality of cultural heritage streets, and improve the quality of life of pedestrians, which is also of great significance to public health.

6.2. Implications

The main recommendations of the study can be summarized as followed:

(1) For pedestrian walking behaviors

As the commercial consumption facilities have a universal influence on the walking behaviors of pedestrians of all types, attention should be paid to the role of commercial consumption facilities in block. The consumption preferences of pedestrians of all age groups should be considered in the commercial consumption facilities of business block. But in both types of blocks, the setting of business types can be mainly aimed at pedestrians aged 1–30. The commercial consumption facilities inside the block should be considered. In addition, amid the construction of the block, the connection with the surrounding commercial complexes must be strengthened.

There is a strong correlation between the walking behavior of pedestrians over 30 years old and the block types. When the life block is designed, meeting the basic living needs of the surrounding residents should be considered, essential facilities should be diversified, so that residents can enjoy greater convenience. In addition, public service facilities should be fully improved in life block. When the block environment is renewed and constructed, the design that is friendly to the elderly can be prioritized. The full play is given to the historical and cultural functions of streets, so that the historical and cultural functions of street spaces can impact the physical and mental health of pedestrians, promote walking exercise, and improve public health;

(2) For pedestrian walking experience

Pedestrians of different genders have different preferences for block types. When various types of blocks are planned and designed, attention should be paid to pedestrians of the target gender and the preferences of pedestrians of the other gender should be catered in other aspects. For example, if life blocks mainly attract male pedestrians, amid construction, the walking experience of male pedestrians in essential facilities and street quality must be mainly considered. Besides, more attention should be paid to the walking experience of female pedestrians in commercial consumption facilities and public service facilities, so as to meet the walking needs of pedestrians of both genders.

As the pedestrian's walking preference over 30 years old has a weak influence on the walking experience, the influence of other factors on pedestrians' walking experience should be fully considered in the process of block construction. For example, the needs of the elderly for social quality, activity richness and social interaction should be considered. Specifically, the circulation and landscape of the block can be improved. A platform for communication can be provided to strengthen the block interaction.

6.3. Limitations and Future Research

There is a higher correlation between the walking experience of recreational walkers and various influencing factors, so the opinions of such walkers are more likely to deserve attention. In addition, before the block is designed, the structures of pedestrians of different types should be investigated. Corresponding design strategies must be made according to the pedestrian structure after the site properties and pedestrian opinions are fully considered.

There are some limitations in the study. For example, pedestrians are picked up selectively rather than randomly to complete the questionnaires. Therefore, there may be some bias in the whole process of data statistics. Moreover, some respondents are impatient or pay less attention while filling out questionnaires. Thus, the subsequent analysis is possibly hampered by the lack of authenticity and reliability of the information.

Due to the difference in street attributes, or the variability in the streets where respondents are, the study conclusion is impacted to some degree. The pedestrians in the article can't fully represent all populations due to the differences in personal circumstances, insufficient sample size, and uneven distribution of population of different information. Thereby, the validity and generalizability of the conclusions obtained in the article are possibly affected. In addition, direct observation is adopted, and the observation result is more likely to be influenced by subjective factors and the specific surrounding environments. The observer's attention and memory alone cannot make sure the complete record of pedestrians' complex behavioral activities. Also, the observation result is impacted by various factors.

It's found from the pedestrian information of the research samples, there is a small number of young and old respondents. There is not sufficient research on these two age groups, which then will be supplemented in subsequent studies. Besides, due to the investigation time, it is not possible to determine the impact of seasonal and climatic factors on pedestrian behaviors in the street space. It is also a key consideration for future research.

Author Contributions: In this paper, M.T. performed the conceptualization, Z.L. and Z.X. proposed the methodology, Q.X., Y.P. and T.C. performed the site investigation and data calculation, T.D. reviewed it and refined the diagram. All authors (Z.L., M.T., Q.X., Y.P., T.C., T.D. and Z.X.) organized the paper structure, wrote and edited it. All authors have read and agreed to the published version of the manuscript.

Funding: This research was funded by Humanities and Social Sciences Research Project of the Chinese Ministry of Education, grant number 18YJC760078; Zhejiang University of Technology Liberal Arts Laboratory, Database Construction Project "Zhejiang Historical and Cultural Villages and Towns Spatial Humanities Database", grant number GZ21681180011; Humanities and Social Sciences Fund of Zhejiang Education Department, grant number Y202146952.

Institutional Review Board Statement: Not applicable.

Informed Consent Statement: Not applicable.

Data Availability Statement: Not applicable.

Acknowledgments: The author gratefully acknowledges the editors and referees for their positive and constructive comments in the review process.

Conflicts of Interest: The authors declare that they have no known competing financial interests or personal relationships that could have appeared to influence the work reported in this paper.

References

1. Row, A.T. The death and life of great American cities. *Yale Law J.* **1962**, *71*, 1597–1602. [CrossRef]
2. Gehl, J. *Life between Buildings*; Van Nostrand Reinhold: New York, NY, USA, 1987.
3. Taverno Ross, S.E.; Clennin, M.N.; Dowda, M.; Colabianchi, N.; Pate, R.R. Stepping it up: Walking behaviors in children transitioning from 5th to 7th grade. *Int. J. Environ. Res. Public Health* **2018**, *15*, 262. [CrossRef] [PubMed]
4. Kang, B. Identifying street design elements associated with vehicle-to-pedestrian collision reduction at intersections in New York City. *Accid. Anal. Prev.* **2019**, *122*, 308–317. [CrossRef]
5. Liu, Y.; Yang, D.; Timmermans, H.J.; de Vries, B. Analysis of the impact of street-scale built environment design near metro stations on pedestrian and cyclist road segment choice: A stated choice experiment. *J. Transp. Geogr.* **2020**, *82*, 102570. [CrossRef]
6. Kang, Y.; Fukahori, K.; Kubota, Y. Evaluation of the influence of roadside non-walking spaces on the pedestrian environment of a Japanese urban street. *Sustain. Cities Soc.* **2018**, *43*, 21–31. [CrossRef]
7. Jung, H.; Lee, S.-Y.; Kim, H.S.; Lee, J.S. Does improving the physical street environment create satisfactory and active streets? Evidence from Seoul's Design Street Project. *Transp. Res. Part D Transp. Environ.* **2017**, *50*, 269–279. [CrossRef]
8. Torun, A.Ö.; Göçer, K.; Yeşiltepe, D.; Argın, G. Understanding the role of urban form in explaining transportation and recreational walking among children in a logistic GWR model: A spatial analysis in Istanbul, Turkey. *J. Transp. Geogr.* **2020**, *82*, 102617. [CrossRef]
9. Mirzaei, E.; Kheyroddin, R.; Behzadfar, M.; Mignot, D. Utilitarian and hedonic walking: Examining the impact of the built environment on walking behavior. *Eur. Transp. Res. Rev.* **2018**, *10*, 20. [CrossRef]
10. Perumal, V. Study on pedestrian crossing behavior at signalized intersections. *J. Traffic Transp. Eng.* **2014**, *1*, 103–110.
11. Aghabayk, K.; Esmailpour, J.; Jafari, A.; Shiwakoti, N. Observational-based study to explore pedestrian crossing behaviors at signalized and unsignalized crosswalks. *Accid. Anal. Prev.* **2021**, *151*, 105990. [CrossRef]
12. Jay, M.; Régnier, A.; Dasnon, A.; Brunet, K.; Pelé, M. The light is red: Uncertainty behaviours displayed by pedestrians during illegal road crossing. *Accid. Anal. Prev.* **2020**, *135*, 105369. [CrossRef] [PubMed]

13. Mukherjee, D.; Mitra, S. A comparative study of safe and unsafe signalized intersections from the view point of pedestrian behavior and perception. *Accid. Anal. Prev.* **2019**, *132*, 105218. [CrossRef] [PubMed]
14. Ma, Y.; Lee, E.W.; Hu, Z.; Shi, M.; Yuen, K.K.R. An Intelligence-Based Approach for Prediction of Microscopic Pedestrian Walking Behavior. *IEEE Trans. Intell. Transp. Syst.* **2019**, *20*, 3964–3980. [CrossRef]
15. Hussein, M.; Sayed, T. A Methodology for the Microscopic Calibration of Agent-Based Pedestrian Simulation Models. In Proceedings of the 2018 21st International Conference on Intelligent Transportation Systems (ITSC), Maui, HI, USA, 4–7 November 2018; pp. 3773–3778. [CrossRef]
16. Feng, Y.; Duives, D.C.; Hoogendoorn, S.P. Developing a VR tool for studying pedestrian movement and choice behavior. In Proceedings of the 2020 IEEE Conference on Virtual Reality and 3D User Interfaces Abstracts and Workshops (VRW), Atlanta, GA, USA, 22–26 March 2020; pp. 814–815.
17. Bornioli, A.; Parkhurst, G.; Morgan, P.L. Affective experiences of built environments and the promotion of urban walking. *Transp. Res. Part A Policy Pract.* **2019**, *123*, 200–215. [CrossRef]
18. Stevenson, N.; Farrell, H. Taking a hike: Exploring leisure walkers embodied experiences. *Soc. Cult. Geogr.* **2018**, *19*, 429–447. [CrossRef]
19. Hazzard, A.; Spence, J.; Greenhalgh, C.; McGrath, S. The Rough Mile: A Design Template for Locative Audio Experiences. *J. Audio Eng. Soc.* **2018**, *66*, 231–240. [CrossRef]
20. Wong, J.D.; Selinger, J.C.; Donelan, J.M. Is natural variability in gait sufficient to initiate spontaneous energy optimization in human walking? *J. Neurophysiol.* **2019**, *121*, 1848–1855. [CrossRef]
21. Cambra, P.; Moura, F. How does walkability change relate to walking behavior change? Effects of a street improvement in pedestrian volumes and walking experience. *J. Transp. Health* **2020**, *16*, 100797. [CrossRef]
22. Chao, T.Y.; Lin, S.F. A mixed reality system to improve walking experience. In Proceedings of the 2018 International Conference on System Science and Engineering (ICSSE), New Taipei, Taiwan, 28–30 June 2018; pp. 1–4.
23. Oh, J.; Bong, C.; Kim, J. Design of Immersive Walking Interaction Using Deep Learning for Virtual Reality Experience Environment of Visually Impaired People. *J. Korea Comput. Graph. Soc.* **2019**, *25*, 11–20. [CrossRef]
24. Kang, Y.W.; Cho, W.A.; Hong, S.A.; Lee, K.; Ko, H. A Study of 'Hear Me Later' VR Content Production to Improve the Perception of the Visually-Impaired. *J. Korea Comput. Graph. Soc.* **2020**, *26*, 99–109. [CrossRef]
25. Varma, G.R. A study on New Urbanism and Compact city and their influence on urban mobility. In Proceedings of the 2017 2nd IEEE International Conference on Intelligent Transportation Engineering (ICITE), Singapore, 1–3 September 2017; pp. 250–253.
26. Kim, S. A Study on the Relationship between the Pedestrian Environment and Pedestrian Behavior of Pedestrian Passages in Seoul. *Arch. Des. Res.* **2017**, *30*, 145–147. [CrossRef]
27. Rebecchi, A.; Buffoli, M.; Dettori, M.; Appolloni, L.; Azara, A.; Castiglia, P.; D'Alessandro, D.; Capolongo, S. Walkable Environments and Healthy Urban Moves: Urban Context Features Assessment Framework Experienced in Milan. *Sustainability* **2019**, *11*, 2778. [CrossRef]
28. Yassin, H.H. Livable city: An approach to pedestrianization through tactical urbanism. *Alex. Eng. J.* **2019**, *58*, 251–259. [CrossRef]
29. Fallahranjbar, N.; Dietrich, U.; Pohlan, J. Development of a Measuring Tool for Walkability in the Street Scale—The case study of Hamburg. *IOP Conf. Ser. Earth Environ. Sci.* **2019**, *297*, 012047. [CrossRef]
30. Fancello, G.; Congiu, T.; Tsoukiàs, A. Mapping walkability. A subjective value theory approach. *Socio-Econ. Plan. Sci.* **2020**, *72*, 100923. [CrossRef]
31. He, H.; Lin, X.; Yang, Y.; Lu, Y. Association of street greenery and physical activity in older adults: A novel study using pedestri-an-centered photographs. *Urban For. Urban Green.* **2020**, *55*, 126789. [CrossRef]
32. Malin, F.; Luoma, J. Effects of speed display signs on driving speed at pedestrian crossings on collector streets. *Transp. Res. Part F Traffic Psychol. Behav.* **2020**, *74*, 433–438. [CrossRef]
33. Zhang, L.; Zhang, R.; Yin, B. The impact of the built-up environment of streets on pedestrian activities in the historical area. *Alex. Eng. J.* **2021**, *60*, 285–300. [CrossRef]
34. Merom, D.; Humphries, J.; Ding, D.; Corpuz, G.; Bellew, W.; Bauman, A. From 'car-dependency'to 'desirable walking'–15 years trend in policy relevant public health indicators derived from Household Travel Surveys. *J. Transp. Health* **2018**, *9*, 56–63. [CrossRef]
35. Sallis, J.F.; Cerin, E.; Conway, T.L.; Adams, M.A.; Frank, L.D.; Pratt, M.; Salvo, D.; Schipperijn, J.; Smith, G.; Cain, K.L.; et al. Physical activity in relation to urban environments in 14 cities worldwide: A cross-sectional study. *Lancet* **2016**, *387*, 2207–2217. [CrossRef]
36. Koohsari, M.J.; Badland, H.; Mavoa, S.; Villanueva, K.; Francis, J.; Hooper, P.; Owen, N.; Giles-Cortice, B. Are public open space attributes associated with walking and depression? *Cities* **2018**, *74*, 119–125. [CrossRef]
37. Hsu, M.Y.; Lee, S.H.; Yang, H.J.; Chao, H.J. Is Brisk Walking an Effective Physical Activity for promoting Taiwanese Adolescents' Mental Health? *J. Pediatric Nurs.* **2021**, *60*, e60–e67. [CrossRef] [PubMed]
38. Giallouros, G.; Kouis, P.; Papatheodorou, S.I.; Woodcock, J.; Tainio, M. The long-term impact of restricting cycling and walking during high air pol-lution days on all-cause mortality: Health impact Assessment study. *Environ. Int.* **2020**, *140*, 105679. [CrossRef] [PubMed]
39. Zhu, C.; Fu, Z.; Liu, L.; Shi, X.; Li, Y. Health risk assessment of PM2.5 on walking trips. *Sci. Rep.* **2021**, *11*, 1–11. [CrossRef]

40. Hatamzadeh, Y.; Habibian, M.; Khodaii, A. Walking mode choice across genders for purposes of work and shopping: A case study of an Iranian city. *Int. J. Sustain. Transp.* **2020**, *14*, 389–402. [CrossRef]
41. Francis, A.; Sarangi, G.K. Sustainable consumer behaviour of Indian millennials: Some evidence. *Curr. Res. Environ. Sustain.* **2022**, *4*, 100109. [CrossRef]
42. Gordon-Wilson, S.; Modi, P. Personality and older consumers' green behaviour in the UK. *Futures* **2015**, *71*, 1–10. [CrossRef]
43. Hahm, Y.; Yoon, H.; Choi, Y. The effect of built environments on the walking and shopping behaviors of pedestrians; A study with GPS experiment in Sinchon retail district in Seoul, South Korea. *Cities* **2019**, *89*, 1–13. [CrossRef]
44. Zhang, Y.; Zhao, P.; Lin, J.-J. Exploring shopping travel behavior of millennials in Beijing: Impacts of built environment, life stages, and subjective preferences. *Transp. Res. Part A Policy Pract.* **2021**, *147*, 49–60. [CrossRef]
45. Basu, N.; Haque, M.; King, M.; Kamruzzaman, M.; Oviedo-Trespalacios, O. The unequal gender effects of the suburban built environment on perceptions of security. *J. Transp. Health* **2021**, *23*, 101243. [CrossRef]
46. Tarka, P.; Kukar-Kinney, M.; Harnish, R.J. Consumers' personality and compulsive buying behavior: The role of hedonistic shopping experiences and gender in mediating-moderating relationships. *J. Retail. Consum. Serv.* **2022**, *64*, 102802. [CrossRef]
47. Chen, X.; Kassas, B.; Gao, Z. Impulsive purchasing in grocery shopping: Do the shopping companions matter? *J. Retail. Consum. Serv.* **2021**, *60*, 102495. [CrossRef]
48. Kamargianni, M.; Dubey, S.; Polydoropoulou, A.; Bhat, C. Investigating the subjective and objective factors influencing teenagers' school travel mode choice—An integrated choice and latent variable model. *Transp. Res. Part A Policy Pract.* **2015**, *78*, 473–488. [CrossRef]
49. Cheng, L.; De Vos, J.; Zhao, P.; Yang, M.; Witlox, F. Examining non-linear built environment effects on elderly's walking: A random forest approach. *Transp. Res. Part D Transp. Environ.* **2020**, *88*, 102552. [CrossRef]
50. Wang, Z.; Zhang, H.; Yang, X.; Li, G. Neighborhood streets as places of older adults' active travel and social interaction—A study in Daokou ancient town. *J. Transp. Health* **2022**, *24*, 101309. [CrossRef]
51. Kang, B.; Moudon, A.V.; Hurvitz, P.M.; Saelens, B.E. Differences in behavior, time, location, and built environment between objectively measured utilitarian and recreational walking. *Transp. Res. Part D Transp. Environ.* **2017**, *57*, 185–194. [CrossRef]
52. Kento, Y.; Khaimook, S.; Kenjie, D.; Takahiro, Y. Study on influence of walking experience on traffic safety attitudes and values among foreign residents in Japan. *IATSS Res.* **2022**, *46*, 161–170.
53. Ma, X.; Chau, C.K.; Lai, J.H.K. Critical factors influencing the comfort evaluation for recreational walking in urban street envi-ronments. *Cities* **2021**, *116*, 103286. [CrossRef]
54. He, D.; Lu, Y.; Xie, B.; Helbich, M. Large-scale greenway intervention promotes walking behaviors: A natural experiment in China. *Transp. Res. Part D Transp. Environ.* **2021**, *101*, 103095. [CrossRef]
55. Mondal, A.; Bhat, C.R.; Costey, M.C.; Bhat, A.C.; Webb, T.; Magassy, T.B.; Pendyala, R.M.; Lam, W.H.K. How do people feel while walking? A multivariate analysis of emotional well-being for utilitarian and recreational walking episodes. *Int. J. Sustain. Transp.* **2021**, *15*, 419–434. [CrossRef]
56. Bauman, Z. Wasted Lives. In *Modernity and Its Outcasts*; Polity Press: Cambridge, UK, 2004.

 land

Review

Landscape Modifications Ascribed to El Niño Events in Late Pre-Hispanic Coastal Peru

Marco Delle Rose

Institute of Atmospheric Sciences and Climate, National Research Council of Italy, 73100 Lecce, Italy; m.dellerose@le.isac.cnr.it

Abstract: Coastal Peru, one of the driest deserts in the world, is a key region to investigate the connection between climate processes and Earth surface responses. However, the trends in space and time of the landscape effects of El Niño events throughout the last millennium are hard to outline. A deeper understanding of geological and archaeological data in pre-Hispanic time can help to shed light on some critical questions regarding the relationship between such a coupled atmosphere–ocean phenomenon and landscape modifications. The bibliographic sources required for this purpose are scattered throughout various disciplines, ranging from physical to human sciences, and thus comprehensive databases were used to identify and screen relevant studies. The performed examination of these documents allowed us to assess strengths and weaknesses of literature hypotheses and motivate additional studies on targeted research objectives.

Keywords: desert landscape; coastal plain; paleoflood record; El Niño proxies; debris flow; slack-water deposit; braided streams; desert pavement; regolith denudation

Citation: Delle Rose, M. Landscape Modifications Ascribed to El Niño Events in Late Pre-Hispanic Coastal Peru. *Land* **2022**, *11*, 2207. https://doi.org/10.3390/land11122207

Academic Editors: Bellotti Piero and Alessia Pica

Received: 27 October 2022
Accepted: 1 December 2022
Published: 5 December 2022

1. Introduction

Due to the dynamic of the El Niño-Southern Oscillation (ENSO), the coastal desert of Peru is a key region to investigate the connections between climate processes and Earth surface responses [1–3]. El Niño precipitation events cause abrupt and rapid landscape modifications on the whole central Pacific coast of South America [4,5]. However, the trends in space and time of the landscape effects of El Niño events on the coastal Peru throughout the last millennia are hard to outline [6,7].

Coastal Peru has recently experienced dramatic landscape changes and asset destruction during precipitation events due to El Niño. Widespread landslides and extensive floods are the more relevant hydrogeomorphic signatures [8–12]. With reference to the late Holocene, geoarchaeological studies are crucial to explore El Niño proxies and paleoflood record [4,9,13]. As a matter of fact, several archaeological and geological studies have focused on the strong impact on settlements and irrigation systems of severe precipitation events named "Super" and "Mega" El Niños [14–19], sometimes believed to be the cause of the collapse of centuries-old cultures [20–23]. However, for some case studies, alternative interpretations involving gradual evolution of the landscape or questioning the occurrence itself of the catastrophic events have been provided in the literature [24–26].

A deeper understanding of the geological and archaeological data can help to shed light on some critical questions about the relationship between landscape modifications and El Niño events. This review aims to explore the period from the rise of "modern" periodicity of ENSO [13,27] and beginning of the Little Ice Age (roughly coincident with the arrival of the Spanish conquistadors) [28,29], herein called "late pre-Hispanic". The bibliographic sources required for this purpose are scattered throughout the literature of various disciplines ranging from physical to human sciences. Thus, the identification and screening of pertinent documents (Sections 3 and 4) have been performed by the citation databases of Web of Science (WoS) and Scopus, in accordance with the Preferred Reporting Items for

Systematic Reviews and Meta-Analyses statement (PRISMA) [30,31]. Episodes of landscape change ascribed to severe precipitations in the late pre-Hispanic time are discussed, and their consistencies and inconsistencies are exposed (Section 5). Some examples of variation in landscape response due to extensive human intervention are also reported. The careful examination of geoarchaeological data allowed strengths and weaknesses of literature hypotheses on landscape effects of El Niño events to be assessed and motivates additional studies on targeted research objectives (Sections 5 and 6).

2. Background Knowledge about the Topic

Due to the interdisciplinary nature of the subject matter, some basic concepts must be given before addressing the Materials and Methods (Section 3).

2.1. Physical Setting

The landscape of coastal Peru (Figure 1) is characterized by hill ridges—the western Andean offshoots named *lomas*—made up of pre-Tertiary carbonate, igneous, and metamorphic rocks, and flat areas crossed by rivers—the *pampas*—made up by Tertiary-Quaternary clastic rocks produced from the dismantling of the Andes during orogenesis. Both are devoid of soil and vegetation and partially covered by deposits produced by eolian processes (dunes, desert pavements, and reg soils). Typical landforms are the *quebradas*, which dry braided stream and ravine systems mainly located at the Andean footzone (over alluvial fans) or next to the coast.

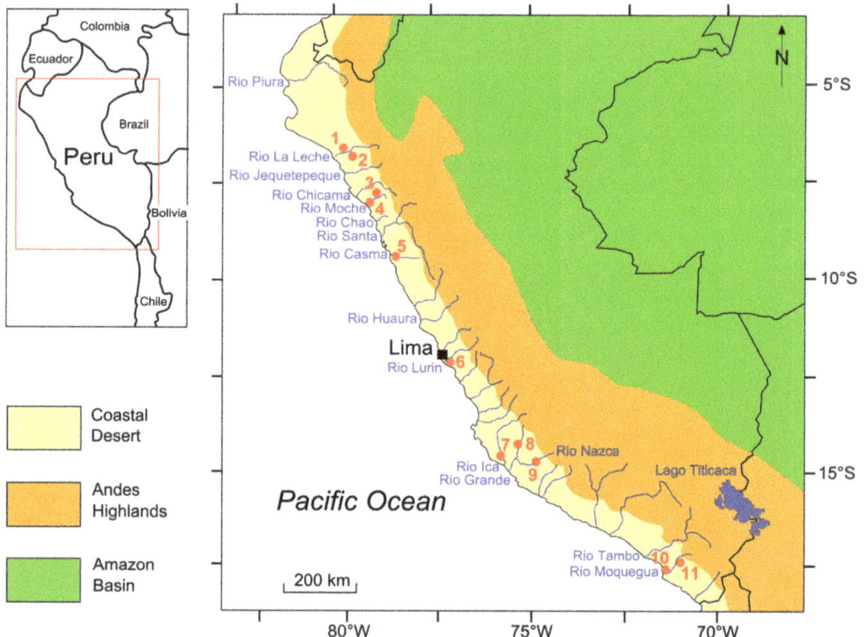

Figure 1. Main geographical regions of Peru. Numbers indicate the location of the cases discussed in the text; 1, Batan Grande; 2, Racarumi; 3, Caballo Muerto and Quebrada de los Chinos; 4, Huacas de Moche; 5, Casma and Cerro Sechin; 6, Pachacamac and Urpi Kocha; 7, Samaca; 8, Palpa; 9, Cahuachi; 10, Quebrada Miraflores; 11, Rio Muerto.

Coastal Peru (3°30′–18°30′ S, 81°30′–70°30′ W) is one of the driest regions in the world. Such a desert environment results from the SE Pacific anticyclone and the Humboldt Current coming from the Antarctic Sea, because they both prevent rain [32,33]. Annual precipitation values are rather uniform along the entire coast, averaging 10–25 mm per year.

The desert along the Pacific is narrow, only about 120 km wide before the land rises into the Andes Highlands, where precipitation increases with elevation. Tens of perennial or seasonal rivers, coming down from the Andes, pass through the desert on the way to the ocean (Figure 1). They allow both biological life and human settlement within and in the surrounding riparian oases. Once every 7–15 years, an El Niño brings warm sea surface temperatures and torrential rainfall on the coastal region, breaking the hyperarid state. Within the last century, severe El Niño events occurred during the years 1925/1926, 1940/1941, 1957/1958, 1972/1973, 1982/1983, 1997/1998, and 2015/2016, the latter with an extraordinary prolongation in 2017 [7,10,34–37].

2.2. El Niño Event and Landscape Response

The term El Niño was initially applied to a weak warm ocean current that runs southward along the Ecuador and Peru coastline around Christmastime. It is the warm phase of the El Niño–Southern Oscillation (ENSO), an interannual fluctuation in sea surface temperature and air pressure across the equatorial Pacific Ocean [38–41]. The cold phase is called La Niña, and the neutral phase is the intermediate one. Such coupled atmosphere–ocean phenomena dominate the interannual variability of planetary climate system and cause strong drought and flood around the world [42,43]. ENSO has dramatically changed in occurrence and magnitude during the Holocene, increasing in frequency to reach its current features about 3.0 Ka ago [27,44,45], roughly corresponding with the beginning of the Early Horizon (ca. 800–200 BCE).

Coastal Peru is responsible for a direct correspondence among El Niño severity, flooding, and landscape modifications. During El Niño, the South Pacific anticyclone weakens while the northern boundary of the Humboldt current migrates southward. In such a condition, significant precipitation is able to reach the coastal desert, where it is reinforced due to the rain-shadow effect of the Andes [17]. However, the flood magnitude does not affect the different desert areas equally [34,35,37]. Latitude 12° S roughly divides north-central coastal Peru, which is hardest hit by El Niño events, from southern coastal Peru, which is usually less affected [46]. Nevertheless, large events, such as that in 1925/1926, can extend much farther south [47].

Severe precipitation and flooding may be also produced by other atmospheric patterns. The eastern desert margin is involved in monsoonal fluctuations, and thus thunderstorms can occasionally reach it and cause convective precipitation, especially over southern Peru [48,49]. Again, enhanced precipitation across the Andes highlands during the La Niña event may produce flooding on main valleys [50,51]. From the perspective of paleoflood analysis, such processes make it more challenging to detect the triggering cause of hydrogeomorphic events preserved in geoarchaeological sequences [4,9].

The strength of ENSO events is measured by conventional physical indices. Regional Sea Surface Temperature (SST) indices and the surface atmospheric pressure-based Southern Oscillation index (SOI) are the most widely used indexes to classify ENSO events. However, there is no agreement among scholars about which index best defines ENSO strength, timing, and duration, and fits severe precipitations satisfactorily [38,42,52–54]. To address the change in frequency and intensity of events ascribed to El Niño events that hit the central Pacific coast of South America in pre-instrumental time, various approaches have been used, including geoarchaeological studies [10,55]. After the event of 1982/1983, scholars have focused on the power and occurrence of extremely strong events. Events with a recurrence interval up to 400 years were labeled as Super El Niño, while the ones that happen once every 1000 years as Mega El Niño [19,56].

The frequency and spatial variability of El Niño events is very difficult to establish and predict. Historical climatic and hydrological records may reflect specific atmospheric conditions of each valley and parts of them. To give some examples, the Lower Chicama Valley (northern Peru, Figure 1) saw extensive flooding as a result of the 1997/1998 El Niño event and none during the 2015/2016 El Niño event but again flooding during its anomalous prolongation that occurred during 2017 [57]. Instead, the upper Moquegua

Valley (southern Peru, Figure 1) experienced torrential rain and flood during the 1982/1983 El Niño event, but no significant surface process occurred 15 years later [7,22,47].

In arid regions, a landscape response to rainfall events is typical. In fact, severe precipitations *"are more likely to cause landform change than are floods of similar magnitude elsewhere"* [58]. In contrast, weathering and erosion rates are orders of magnitude lower than in less dry environments, and soil development processes differ markedly from those found on the vast majority of the Earth's surface [5,59,60]. As a result, in coastal Peru, landscape modifications are slow compared to non-desert regions, except during major alluvial and eolian events [61–63]. With a multi-century recurrence interval for precipitation events, exceptional debris flows are reported in dry landscapes [64,65]. Under favorable conditions, layers deposited as a result of high magnitude flood events can be used as horizon markers in ancient landscape reconstruction. Slack-water deposits are significant examples of such useful beds [66]. In the last century, an event that produced a notable landscape impact in the valleys of northern Peru was El Niño of 1925/1926. For what concerns Moche Valley (see Figure 1 for location), the river spilled over vast tracts of alluvial plain, destroying bridges and irrigation systems and threatening the stability of the *Huaca del Sol*, the great archaeological monument of the Early Intermediate Period (ca. 100–800 CE) [14].

3. Materials and Methods

The screening of the documents relevant to this review was conducted according to the PRISMA statement [30,31], which assists reviewers and meta-analysts in transparently reporting why the review was conducted, what the authors accomplished, and what they discovered. To identify and select the documents, Scopus and WoS databases were processed [67–69]. To increase the chances to find useful documents, the topics "Ecuador" and "Chile" were added in the search modes. Moreover, the tag "*" was used in the search phrases to cover as many keyword combination as possible (Table 1). The databases were last consulted on 30 August 2022.

Table 1. WoS/Scopus database search modes.

	Search Field	Search Phrase	Document Type
WoS	TS = Topic (Title, Abstract, Author Keywords, and Keyword Plus)	(((TS = ("El Nino*" OR Nino* OR ENSO OR "El Nino Southern Oscillation")) AND TS = (Peru OR Ecuador OR Chile)) AND TS = (landscape* OR settlement* OR site* OR archaeolog*))	Articles, Review Articles, Proceedings Papers, Early Access, Book Chapters
Scopus	TITLE-ABS-KEY (Article Title, Abstract, and Keywords)	(TITLE-ABS-KEY (El Nino* OR Nino* OR ENSO OR El Nino Southern Oscillation) AND TITLE-ABS-KEY (Peru OR Ecuador OR Chile)) AND TITLE-ABS-KEY (landscape* OR settlement* OR site* OR archaeolog*))	Article, Conference Paper, Conference Review, Review, Book Chapter

Despite their great advantages, Scopus and WoS platforms may have introduced biases that favor Natural Sciences, Engineering, and Biomedical Research at the expense of Social Sciences, Arts, and Humanities. In a similar manner, documents written in English predominate over those written in other languages [68]. In any case, to find further articles and other works pertinent to the review, a careful examination was made of the reference list of the documents whose full text was examined. For the full-text analysis, basic criteria to establish how consistently literature data support the occurrence of El Niño events had to be defined. Since the review deals with geological and archaeological *data*, epistemological features of the respective disciplines (see [70–72] for geology and [73–75] for archaeology), as well as of the cross-disciplinary geoarchaeology [76,77], have driven the choice of these criteria. The works cited above give particular attention to the conceptual meaning of the "data" in geology and archaeology and to the distinction between "data interpretation" and "data explanation" as a premise for the different perception of what counts as knowledge. Nevertheless, the peculiarity of the logical procedures inherent in the way of reasoning of geologists and archaeologists to interpret or explain the data is focused (see especially Frodeman [70] and Fogelin [74]). Accordingly, for the literature analysis, the following criteria were established and used: (a) geomorphological and stratigraphic features of the related deposits; (b) relationships between deposits and archaeological remains; (c) method of dating used to determine the age; (d) presence or absence of converging evidence; (e) consistency with data from other studies.

4. Screening Results

Among the documents extracted from the databases, only one duplicate was found, thus 458 works were identified in total (Table 2).

Table 2. Summary of identification, selection, and full-text examination stages.

	WoS		Scopus
Database mined documents	383		306
After duplicate check	383		305
Documents found in both databases		230	
Identified documents		458	
Excluded after abstract evaluation	337	402	261
Selected documents	46	56	44
From bibliography of selected works		14	
Reviewed documents		70	

Then, careful abstract evaluation allowed the selection of 56 documents. A total of 402 documents were disregarded because they dealt with topics in other subject areas (i.e., medicine, public health, sustainability, ethnology, anthropology, oceanography, tectonic, geochemistry, psychic atmosphere, meteorology, glaciology, dendroclimatology, palynology, biology, zoology, botany, ecology, astronomy, history, economy, and sociology) or were irrelevant to the review's objective in space and/or in time (i.e., study cases not on the central Pacific coast of South America or outside the considered time). Fourteen documents identified by examination of the bibliography of the selected works were added for the review analysis (Table 2); thus, a total of 70 documents were finally reviewed (Tables 3 and 4).

Table 3. Summary of reviewed documents. BSD = Bibliography of selected documents.

Author(s) and Year	Source	Sites/Study Areas
Nials et al., (1979 a, b) [14,15]	BSD	Huaca del Sol and 2 other sites (Moche Valley)
Samaniego et al. (1985) [78]	BSD	Cerro Sechin (Casma Valley)
Craig and Shimada (1986) [16]	Scopus	Batan Grande (La Leche Valley)
Sandweiss (1986) [79]	Scopus	Las Salinas (North coast of Santa mouth)
Rollins et al. (1986) [80]	Scopus	Las Salinas (North coast of Santa mouth)
DeVries (1987) [81]	BSD	*Review article*
Wells (1987) [82]	BSD	7 sites (Casma Valley)
Wells (1990) [17]	BSD	7 sites (Casma Valley)
Grodzicki (1992) [20]	BSD	Pampa de Nazca (Nazca Valley)
Moseley and Richardson (1992) [83]	BSD	Huaca del Sol (Moche Valley)
Moseley et al. (1992) [84]	BSD	Las Salinas (North coast of Santa mouth)
Ortlieb and Machare (1993) [85]	WoS, Scopus	*Review article*
Uceda and Canziani Amico (1993) [86]	BSD	Huaca de la Luna (Moche Valley)
Grodzicki (1994) [21]	BSD	Cahuachi and other 3 sites (Nazca Valley)
Keefer et al. (1998) [87]	BSD	Quebrada Tacahuay (Moquegua Valley)
Wells and Noller (1999) [88]	Scopus	*Review article*

Table 4. Summary of reviewed works. This table continues from Table 3.

Author(s) and Year	Source	Sites/Study Areas
Veit (2000) [89]	Scopus	*Review article*
Franco and Paredes (2000) [90]	BSD	Pachacamac (Lurin Valley)
Satterlee et al. (2000) [22]	BSD	Quebrada Miraflores (North coast of Moquegua mouth)
Magilligan and Goldstein (2001) [24]	WoS, Scopus	Rio Muerto (Moquegua Valley)
Sandweiss et al. (2001) [55]	WoS, Scopus	*Review article*
Van Buren (2001) [91]	WoS	*Review article*
Calderoni et al. (2002) [61]	Scopus	Mejia (Tambo Valley)
Huckleberry and Billman (2003) [92]	WoS, Scopus	Quebrada de los Chinos (Moche Valley)
Keefer et al. (2003) [93]	Wos, Scopus	*Review article*
Dillehay et al. (2004) [94]	WoS, Scopus	Los Mochicas del Norte Valleys
Federici and Rodolfi (2004) [95]	WoS, Scopus	Ensenada de Atacames, *Ecuador*
Keefer and Moseley (2004) [96]	WoS, Scopus	lower Moquegua Valley
Brooks et al. (2005) [97]	WoS, Scopus	Santa Rita (Chao Valley)
deFrance and Keefer (2005) [98]	WoS, Scopus	Quebrada Tacahuay (South coast of Moquegua mouth)
Eitel et al. (2005) [4]	WoS, Scopus	Quebrada Palpa (Grande Valley)
Zaro and Alvarez (2005) [99]	WoS, Scopus	Moquegua Valley and North coast of Moquegua mouth
Fabre et al. (2006) [100]	WoS, Scopus	Mollendo (North coast of Tambo mouth)
Machtle et al. (2006) [101]	WoS, Scopus	upper and middle Grande Valley
Manners et al. (2007) [50]	WoS	middle Moquegua Valley
Andrus et al. (2008) [102]	WoS, Scopus	*Review article*
Magilligan et al. (2008) [51]	WoS	3 sites (Moquegua Valley)
Beresford-Jones et al. (2009a) [103]	Wos, Scopus	Samaca (Ica Valley)
Beresford-Jones et al. (2009b) [104]	Scopus	Samaca (Ica Valley)
deFrance et al. (2009) [105]	WoS	North and south coast of Moquegua mouth
Eitel and Machtle (2009) [106]	WoS	upper and middle Grande Valley
Mettier et al. (2009) [1]	WoS, Scopus	middle Piura Valley
Reindel and Wagner (2009) [107]	WoS	upper Grande Valley
Abbuhl et al. (2010) [108]	WoS, Scopus	middle Piura Valley
Beresford-Jones (2011) [109]	Scopus	Samaca (Ica Valley)
Bernal et al. (2011) [110]	WoS	Pastaza Valley, *Ecuador-Peru area of Amazon Basin*
Goldstein and Magilligan (2011) [111]	WoS, Scopus	upper and middle Moquegua Valley
Gayo et al. (2012) [112]	WoS, Scopus	Pampa del Tamarugal, *Chile*
Huckleberry et al. (2012) [113]	WoS, Scopus	middle La Leche Valley
Sandweiss and Kelley (2012) [114]	WoS, Scopus	*Review article*
Sandweiss and Quilter (2012) [115]	WoS	*Review article*
Winsborough et al. (2012) [23]	WoS, Scopus	Urpi Kocha Lagoon (Rio Lurin)
Etayo-Cadavid et al. (2013) [116]	WoS, Scopus	North and south coast of Moche mouth
Hanzalova and Pavelka (2013) [117]	Scopus	Ciudad Perdida de Huayuri (upper Grande Valley)
Macthle and Eitel (2013) [118]	WoS, Scopus	upper Grance Valley and middle Nazca Valley
Kalicki et al. (2014) [119]	Scopus	Lomas de Lachay (South of middle Huaura Valley)
Nesbitt (2016) [120]	WoS	Caballo Muerto Archaeological Complex (middle Moche Valley)
Caramanica and Koons (2016) [121]	WoS, Scopus	Pampa de Mocan (Chicama Valley)
Pavelka et al. (2016) [122]	WoS	Cantalloc (upper Nazca Valley)
Christol et al. (2017) [123]	WoS, Scopus	West coast of La Leche mouth
Wang et al. (2017) [124]	WoS	Salar Grande, *Chile*
Kalicki et al. (2018) [125]	WoS, Scopus	Lomas de Lachay (South of middle Huaura Valley)
Delle Rose et al. (2019) [26]	WoS, Scopus	Cahuachi (middle Nazca Valley)
Caramanica et al. (2020) [126]	WoS, Scopus	Pampa de Mocan (Chicama Valley)
Kalicki and Kalicki (2020) [127]	WoS	Lomas de Lachay (South of middle Huaura Valley)
Sandweiss et al. (2020) [13]	WoS, Scopus	*Review article*
Uceda et al. (2021) [128]	WoS, Scopus	*Review article*
Sandweiss and Maasch (2022) [129]	WoS, Scopus	*Review article*
Rubinatto Serrano et al. (2022) [130]	WoS, Scopus	Rio Muerto and 3 other sites (Moquegua Valley)

Thirteen review articles are present among the documents whose full text was analyzed (Tables 3 and 4). None of these focused on landscape modifications due to El Niño events along the whole coastal Peru. Moreover, no useful data for this review were found in twenty reviewed works, including three articles on Ecuador and Chile case studies

(Table A1 in Appendix A). The major subject in archaeology studies centers on human response. The reviewed archaeological studies mainly aimed to understand the occupation history of sites or the function of issues related to monumental buildings, where the recovery of deposits ascribable to El Niño was apparently incidental. However, useful insights, for the purpose of this review, can be argued also from such literature. To expose and discuss the findings, data on landscape changes extracted from the reviewed documents were grouped and related according to main study areas (Section 5).

5. Main Study Areas for Landscape Changes

5.1. Los Mochicas del Norte Valleys

This area includes the valleys between Rio La Leche and Rio Jequetepeque and has an ethno-historical significance [131]. The earliest work identified by citation database queries is the one of Craig and Shimada [16] on the Batan Grande archaeological complex (see Figure 1 for location). Recent Quaternary stratigraphy, analysed along a modern hydraulic excavation, suggests to the authors that few deposits survive from the 1925/1926 El Niño event, except for slack-water beds roughly dated between 650 and 1000 CE by associated funerary pottery. Such deposits accumulate in areas of reduced velocity during flood flows and may be related to episodes of fluvial morphological adjustment and reshaping channel morphology [66]. Unluckily, the contained archaeological finds do not allow more precise dating, and thus the use of the above deposit as a horizon marker is prevented. Such a question may be addressed by absolute dating of several samples.

South of Batan Grande, along the Jequetepeque Valley (Figure 1), Dillehay et al. [94] documented several sediment release signatures of slack-water deposits containing Late Moche and Chimu ceramics and ^{14}C dated between 415 and 1420 CE. These authors also documented different debris flow deposits as well as erosional truncation of floors at several archaeological sites, all likely associated with El Niño events. The multidisciplinary character of this study and the numerous radiocarbon dating of organic material extracted from alluvial deposits make the findings reliable. A long-term landscape shaping due to severe ENSO-related floods characterized the northern coastal desert before the arrival of the Spanish conquistadors. The human responses to the destructive effects of El Niño events are evaluated as "*highly sophisticated*" by the authors. Large rebuilding activities on damaged hydraulic structures, inferred by stratigraphic analysis at different archaeological sites, seem in fact to have allowed communities to avoid socio-economic repercussions of the hydrogeomorphic calamities. Similar results are obtained by Huckleberry et al. [113] for the inter-valley canal system named Racamuri (upper La Leche Valley, Figure 1), a millennial construction that reached its maximum extension between 900 and 1470 CE. Despite ENSO-driven floods and droughts, such a hydraulic structure would continue to work for centuries, likely even after the beginning of the colonial occupation. As a whole, for the northernmost valleys of coastal Peru, the selected literature data do not suggest drastic landscape changes or dramatic human responses to strong hydrogeomorphic episodes. Only the ca. 775 CE episode of abandonment of the Pampa Grande site (45 km southeast of Batan Grande) may be associated with a strong El Niño with good confidence [129]. It must be highlighted that this date is located within the time interval of the event postulated in Ref. [16]. Thus, a detailed geoarchaeological study on this case is suggested.

5.2. Huacas de Moche (Lower Moche Valley)

The settlement of Huacas de Moche is an early capital of the Moche state. It is located in the lower Moche Valley and includes the well-known buildings named Huaca del Sol and Huaca de la Luna (Figure 2). In their pioneering works, Nials et al. (a) [14] report "*the discovery of an El Niño catastrophe of a magnitude far greater and more devastating than all other such natural disasters striking the coast since the conquistadors first arrived in 1532. Transpiring about 1100 A.D., this prehistoric Niño was of unprecedented magnitude, and the devastation it wrought taxes the imagination of geologists and archaeologists alike*". Since about 1000 CE, the Chimu culture had completely transformed the desert landscape, building a complicated system of reclamation channels over an area of tens of square kilometers. According to the authors (who have surveyed hundreds of kilometers of ancient waterways), all the channels experienced massive erosion and dropped from use before the beginning of the successive sub-cultural phase (a few tens of years later). This allowed an approximate dating of the hydrogeomorphic disaster, which was consequently called the Chimu Flood. Later, much of the reclaimed area would revert to desert in short time. However, the lack of absolute dating makes this reconstruction questionable.

Figure 2. Huacas de Moche; 2004 satellite image (8°07′25″–8°08′22″ S, 79°00′17″–78°58′35″ W). Huaca del Sol and Huaca de la Luna are marked by a red circle and blue circle, respectively. The depositional origin of late Holocene layers covering the plain between the monumental buildings should be ascertained, taking into account the landscape processes argued by the authors [14,83,128].

The water level of the Chimu Flood would be still recognizable on the western side of the adobe brick pyramid *Huaca del Sol* (see Figure 7 on page 12 of Nials et al. (a) [14]), marked by a notch about 10–15 m above the present Rio Moche level. However, according to the geomorphological reconstruction of the authors, the major landscape change happened at the middle Moche Valley, 5–10 km upstream Huaca del Sol, with the lowering of the alluvial plain due to erosion estimated in ca. 5 m [15] (see Section 5.3). Moreover, the authors stated that *"a very conservative estimate would be flood waters at least 2 to 4 times the size of the 1925[/1926] floods, the worst in the last 400 years"* [15].

Several works selected from Scopus and WoS databases by the method described in Section 3 cite the works of the research group of Nials. Craig and Shimada [16] hint at a possible regional correlation with the slack-water deposit they found along the Rio La Leche Valley (see Section 5.1), thus laying the groundwork for the conceptualization of an 11th century El Niño event. Wells [17] suggests the possible coincidence of the Chimu Flood with the so-called Naymlap Flood, an ethno-historically recorded hydrological disaster associated with the name of a cultural hero [7,132]. Again, Wells and Noller [88] use the Chimu Flood to explicate the recurrence interval of the "Mega" El Niños in northern coastal Peru. Finally, Van Buren [91], Nesbitt [120], Huckleberry et al. [113], and Caramanica et al. [126] cite Nials et al. [14,15], simply to describe the destructive effects of the major El Niño events on the ancient irrigation systems and the consequences on the human communities. None of these studies deals with the reliability of the reconstruction provided by Nials et al. [14,15].

By archaeological excavations on *Huaca de la Luna*, Uceda and Canziani Amico [86] argued for the occurrence of moderate El Niño events before the Chimu Flood. They interpreted three layers of sediment with remnants of painted washed-off walls, found on successive floors of the temple, as evidence of intense precipitation and runoff processes, the most recent dated around 600 CE. This latter should have produced significant flow within the braided streams of the Moche Valley without, however, causing either changes in the landscape or the abandonment of the settlement as asserted by Moseley and Richardson [83]. Instead, according to these authors, between 500 and 600 CE, *"flood water brought by El Niño struck the Moche capital, they leveled much of the city, stripping as much as 15 feet"* (about 4.6 m) *"off some areas. It is unclear if the magnitude of destruction reflects more than one El Niño events, perhaps exacerbated by an earthquake. The survivors rebuilt their city only to see it gradually inundated by sand dunes that swept inland after forming on the beach at the mouth of the Moche River"*. It is apparent how the interpretation of geoarchaeological data coming from the same site can even lead to conflicting hypotheses [128]. Thus, the impact of the El Niño event around 600 CE on settlement and landscape at Huacas de Moche must be further investigated, especially with the aim of filling possible gaps in knowledge and addressing the above question.

The study by Prieto et al. [133] on a mass-sacrifice event discovered at a site of the lower Moche Valley, in addition to reporting a possible landscape process due to an El Niño event dated between 1400 and 1450 CE, is a remarkable example of the indissoluble connection between geological and archaeological data in the late pre-Hispanic coastal Peru. According to these authors, *"stratigraphic evidences suggests that the sacrifice was made following a heavy rain/flood that deposited a layer of mud on top of the clean sand in which the children and camelids were buried"*. The conspicuous number of ^{14}C dated samples ensures a high reliability of the chronological attribution.

5.3. Caballo Muerto and Quebrada de los Chinos (Middle Moche Valley)

Caballo Muerto Archaeological Complex and Quebrada de los Chinos are located in the middle valley of Rio Moche (Figure 1). The first includes the building named Huaca Cortada, from which data on the effects of El Niño events on settlement and landscape have recently been collected by Nesbitt [120]. According to this author, the whole archaeological complex *"is situated within an environment susceptible to El Niño flooding"*. Due to its geomorphological setting, especially the alluvial fan located 10.5 km north-east of Huacas de Moche (Figure 3) may preserve significant geoarchaeological proxies.

Figure 3. Caballo Muerto Archaeological Complex; 2021 satellite image (8°03′59″–8°04′53″ S, 78°55′19″–78°53′44″ W). The alluvial fan on which lies the main settlement of the complex is marked by a red line. Its stratigraphy should contain a wealth of data regarding El Niño events (see text).

Archaeological excavations at Huaca Cortada documented the occurrence of four El Niño events throughout the second half of the second millennium BCE. The identified El Niño proxies are laminated muddy layers deposited from runoff water. They contain thin laminas of white paint, which formed as the rain washed off the painted, plastered surfaces of the temple walls. These layers are sandwiched between pre- and post-event structures and were exposed to weathering for a short time, as inferred from their sedimentological features. By the ^{14}C method, a date of 1600–1450 BCE is established for the earliest El Niño proxy of Huaca Cortada. The age of the subsequent events lies between 1100 and 900 BCE, as inferred from pottery finds. Finally, it must be noted that, as Caballo Muerto is surrounded by *quebradas*, changes in the shape of the braided streams for each event may be supposed, which would be significant modifications for a desert environment. Such a question should be addressed with geological studies. Nevertheless, according to Nesbitt [120], the communities of the Initial Period were able to rapidly reconstruct and enlarge buildings, and thus "*the social, religious and economic mechanisms that allowed for the mobilisation of labour [. . .] were not negatively impacted by El Niño*".

The geoarchaeological record of the middle Moche Valley preserves a frequent recurrence of El Niño also for the two millennia of the Common Era. Huckleberry and Billman [92] describe 13 ENSO-related flood and debris flow deposits beneath the present floodplain surface of Quebrada de los Chinos, which are younger than 2000 cal y BP. The findings of these authors seem to confirm the inference by Nials et al. [15] on apparent fluvial landscape changes due to severe precipitations (see Section 5.2), and reflect the late Holocene increases in El Niño activity [10,55]. With reference to methodological issues, the authors underline the need for a correct correlation between the geological characteristics of the deposits and the causative climatic events to make strong hypotheses [92].

5.4. Casma and Cerro Sechin (Lower Casma Valley)

With reference to radiocarbon dating, Wells [17,82] analysed in detail the alluvial terraces system, which characterizes the lower Rio Casma Valley (Figure 1). The system presents three floodplain surfaces whose ages of formation are between 3000 and 200 cal y BP (see Table 1 and Figure 4 of Reference [17], pp. 1135–1136), thus suggesting significant geomorphological changes during the late pre-Hispanic time driven by river sediment transport and tectonic uplift. These results are consistent with data on the beach ridge accretion of the Pampas Las Salinas [79,80], about 50 km northwest of the Rio Casma mouth. From a geological point of view, such a landscape dynamics is not surprising for valleys and coasts of the Peruvian desert. Late Pleistocene and Lower Holocene were, for example, earlier periods throughout which El Niños severely impacted coastal Peru, leaving signatures on ancient landscapes (see, e.g., References [19,85,93]).

Thirteen flood deposits younger than 3.2 ka are recorded along the stratigraphic section of Wells [17,82], six of which were deposited before the conquistadors first arrived in 1532. The upper two and the lower two of these latter are ^{14}C dated by means of incorporated organic material. The author, however, does not recognize the ca. 1100 event (i.e., The Chimu Flood) defined by Nials et al. [14] in the Moche Valley (see Section 5.2). She focus mainly on the recurrence of the hydrogeomorphic events rather than their strength and suggests, for the considered time interval, the increase in frequency of flooding and the occurrence of an event "*much larger than that which occurred during 1982–1983*" at least once every 1000 years. This hypothesis can be explained as follows: (a) the Mega El Niño events [19,80] actually exist and cause flood disasters in coastal Peru and possibly extreme climatic anomalies worldwide; or (b) the rainfall associated with the El Niño event is distributed such that extraordinary floods occur near Casma once every 1000 years [17]. Clearly, this question is crucial for the topic treated in the review.

Complementary evidence on the paleoflood record of the Casma district can be provided by archaeological data from Cerro Sechin site (Figure 1). However, only one laminated mud deposit ascribed to runoff water has been archaeologically dated (see Samaniego et al. [78]) and may be correlated to a terraced alluvial deposit ^{14}C dated at 1200 BCE by Wells [82]. Likely, new geoarchaeological research in this area would help in providing significant data on landscape change due to El Niño events.

5.5. Old Temple of Pachacamac and Urpi Kocha Lagoon (Lower Lurin Valley)

According to Franco and Paredes, the Old Temple of Pachacamac (Figure 4) was abandoned around 600 CE due to heavy rains that washed away and damaged the adobe walls of the building while runoff water deposited around thick layers of mud [90]. These authors argue that an unusual climate event triggered temple modifications and led to the development of new social trends and the introduction of architectural elements from the Andes Highlands. Later, Franco stated *"there is no doubt that the rains that caused this event correspond to a Mega Niño"* [134], which is now referred to also as the 6th-century El Niño event. The sedimentological and palynological study by Winsborough et al. on the near Urpi Kocha Lagoon (located 0.4 km west of Pachacamac) confirms such a suggestion [23]. These last authors found, in cores extracted from the bottom of the pool, evidence of three major floods associated with El Niño events in late pre-Hispanic time, the middle event dated between 436 and 651 CE (see Table 5 of Winsborough et al. [23], p. 611).

Figure 4. Pachacamac archaeological site; 2016 satellite image (12°15′08″–12°16′02″ S, 76°55′08″–76°53′31″ W). Old Temple and Urpi Kocha Lagoon are marked by a red circle and a blue circle, respectively. Archaeological and geological inferences on El Niño events were argued for by the authors [23,90,134].

The El Niño signature left at Pachacamac as recognized and interpreted by Franco and Paredes [90] is similar to the ones described by Samaniego et al. [78] and Wells [82] at Cerro Sechin (see Section 5.4) and by Nesbitt [120] at Caballo Muerto (see Section 5.3), respectively. It is apparent that, in late pre-Hispanic settlements, El Niño events may have left traces in the geoarchaeological record likely corresponding to landscape changes. Nevertheless, as carried out from further archaeological excavation, the adobe constructions of Pachacamac were seriously affected by repeated torrential rains up to the twentieth century, and especially by the 1925/1926 El Niño event [135]. The erosive processes caused by these rains have partially erased the traces of the most ancient events, making the reconstruction of the paleoflood record at the lower Lurin Valley more difficult.

The validity of the hypothesis that one, or even two, El Niño events affected the central coastal Peru around 600 CE is confirmed by the findings of Mauricio [136] at Huaca 20 site (Maranga archaeological complex, lower Rimac Valley, 30 km north-west of Pachacamac). Within such a site, now incorporated into the southern suburbs of Lima, the author describes two destructive floods separated by a phase of reconstruction during the Middle Horizon (^{14}C dated between 550 and 690 CE). Despite the limited size of the site, several geoarchaeological sections showing the above sequence are reported as basic data.

Other areas in the lower Lurin Valley offer opportunities for future research on the relationships between landscape modifications and El Niño events in the Initial Period and Early Horizon. Archaeological studies at the sites of Mina Perdida (5 km north-east of Pachacamac) and Manchay Bajo (10 km north-east of Pachacamac) have revealed the existence of debris flow, flood, and slack-water deposits in the respective geoarchaeological record [137,138]. A number of ^{14}C dating on organic remains extracted from the stratigraphic successions provide chronological references for the El Niño signatures.

5.6. Pampa de Palpa (Valleys of Rio Ica and Rio Grande)

The *pampa* southwest of Palpa (Figure 1) constitutes the northern margin of the Atacama desert, and was inhabited by several cultures during the pre-Hispanic times. It has been the object of different studies on the development of loess deposits, paleosoils, and alluvial terraces that provide insights about the long-term relationship between climate processes and desert landscape changes [4,101,106,118]. Moreover, this hyper-arid region includes relevant sites of interest for the present review and is next to the ceremonial center of Cahuachi (which will be discussed separately in Section 5.7).

Located at the lower valley of Rio Ica, the Samaca fluvial plain (Figure 1) is currently a riparian oasis that hosts numerous archaeological and paleobotanic remains. The latter are mainly constituted by partially fossilized trunks of the phreatophyte *Prosopis*, as established by Beresford-Jones et al. [103,104]. The geomorphological, archaeological, and paleoenvironmental evidence gathered by these authors allows them to state the occurrence of a major El Niño event "*at some time toward the end of the Early Intermediate Period, which spread a deep fluvial layer across the upper Samaca Basin, caused the river to cut some 5 m down into its floodplain and had catastrophic effects upon [a] canal system*" [103]. Such a result is consistent with the conclusions of other studies made in northern and central Peru [23,84,86]. However, it explicates part of the above landscape modification, as well as of the decline in the settlement. In fact, according to [103,104,109], human-induced gradual destruction of the *Prosopis* forest in pre-Hispanic time would have considerably increased the exposure to the flood hazard of landscape and settlement, making them more vulnerable to severe events. The deforestation process would culminate during the Middle Horizon, causing the final abandonment of the settlement [104].

Later, large urban settlements developed in the Late Intermediate Period (1000–1400 CE) within the *pampa* of Palpa, the most prominent example of what is the so-called *Ciudad Perdida de Huayuri* that, according to Hanzalova and Pavelka [117], was destroyed by an El Niño event. However, no indications or evidence of such an episode were found in other articles identified by citation database queries (Section 4, Tables 3 and 4). It must be observed that Eitel and their colleagues [4,106] posit an abandonment of such a "lost city" as a consequence of the depletion of water reserve, likely related to a climate change. The above questionable information in Ref. [117] may be due to a gap in the knowledge in the literature.

5.7. Cahuachi Ceremonial Center (Middle Nazca Valley)

According to Grodzicki [21,139], the archaeological site of Cahuachi (Figure 5) was affected by three catastrophic floods caused by El Niño between 2100 and 1000 BP. These studies support the major landscape changes ascribed to El Niño events in late pre-Hispanic time for the whole coastal desert of Peru. The second event would have even caused the collapse of the Nasca Culture (around 600 CE), while the third completely buried the ceremonial center [20,21]. In support of their hypothesis, the author describes conglomerate deposits that would result from deposition of exceptionally large, fluid debris flows. Finally, the landscape of Cahuachi would have been re-shaped (and the monumental building unearthed) throughout the last millennium. A necessary assumption for this argument is that abandonment and burial of the ceremonial center preceded the construction of the Nazca Lines [20]. However, such geoglyphs, created on the desert pavement by removing colored pebbles and leaving the underlying reg soil exposed, are mainly dated from 400 BCE to 600 CE [140–143].

Dealing with the second catastrophic flood inferred by Grodzicki [139], Silverman [6] observes that around 600 CE "*Cahuachi had ceased to function as the great early Nasca ceremonial center*". In their review on the occurrence of El Niño events, Machare and Ortlieb [56] consider the hydrogeomorphic events posited in References [20,21] no more than possible local climate change indicators. Nevertheless, the main question on the reliability of the Grodzicki's hypothesis comes from the research group headed by Eitel and Machtle [4,101,106]. Starting from the observation of "*the good state of preservation of geoglyphs, even where they cross valleys or erosion rills*", such authors "*disagree with ideas that catastrophic El Niño events destroyed the Nasca Civilization*" [4]. Again, as a result of extensive field surveys, they remark how the integrity "*of the line exemplarily illustrates that the valley floor has never been flooded since the Nasca period. Only a small channel in the foreground provides evidence for weak episodic runoff events during the past two millennia*" [106]. Finally, these authors also extend their conclusion to the contiguous *pampa* of Palpa (see Section 5.6).

Figure 5. Cahuachi ceremonial center; 2022 satellite image (14°48′41″–14°49′30″ S, 75°07′45″–75°06′11″ W). The monumental building area investigated by Grodzicki [20,21], Orefici [144], and Delle Rose et al. [26] is marked by a red line.

To further test the consistency of the hypothesis of Grodzicki, stratigraphic and petrographic analyses of upper bedrock and surficial cover of the ceremonial center were carried out by Delle Rose et al. [26]. The results of this study show that the conglomerate deposits interpreted by Grodzicki as signatures of El Niño events, belonging instead to the Tertiary–Quaternary clastic succession that forms the regional substratum. Moreover, such coarse-size sediments are a source of pebbles and cobbles, which form the desert pavement. The inconsistency of Grodzicki's hypothesis was likely due to knowledge gap in geological stratigraphy [26].

Recent archaeological excavations on *Templo Sur* have shed new light on hydrogeomorphic events that damaged the adobe brick buildings at Cahuachi during the development of the Nasca Culture. In fact, the inferences of Orefici [144] about two torrential rains that partially destroyed roof and walls of the temple next to the end of the fourth century CE, a time with no El Niño events reported in the reviewed literature, lead us to reconsider both the climatic pattern responsible for the events and the type of Earth surface response expected for the coastal desert. It must be noted that samples gathered at Cantalloc (20 km east of Cahuachi) by Pavelka et al. [122], considered by these authors remains of an ancient flood, are ^{14}C dated between 47 and 480 CE.

5.8. Moquegua Valley and Quebrada Miraflores

Geoarchaeological studies throughout Moquegua Valley and surrounding areas [22,24,51,84] were mainly aimed to argue the relationships between El Niño events and human responses rather than possible landscape processes [91]. Nevertheless, data exposed in the reviewed works allow one to argue for abrupt geomorphological changes to the desert environment. As a matter of fact, already thirty years ago, Moseley et al. [84] stated that *"serious landscape modification should indicate extreme ENSO conditions or ancient 'Mega-Niño' phenomena"*.

Three main depositional units confidently ascribed to El Niño events have been identified and dated by the authors. Two of them have a late pre-Hispanic age, while the third (named Chuza Unit) is dated at 1607–1608 CE [96]. The oldest ^{14}C dated unit (730–690 CE) is formed by debris flow deposits (identified within the dry Rio Muerto sub-basin [24] (see Figure 6)), and flood deposits recognized elsewhere [51]. As it signaled fast and extensive regolith mobilization, this unit constitutes evidence of a singular hydrogeomorphic event. For what concerns the regolith production, in a highly seismic region such as coastal Peru, it is increased by the shattering of the landscape during the frequent earthquakes that produce pervasive ground cracking and microfracturing of hill-slope materials [84,96].

Figure 6. Rio Muerto sub-basin; 2020 satellite image (17°17′58″–17°18′45″ S, 70°59′11″–70°57′34″ W). The sample point of the Miraflores Unit [51] is indicated by the red arrow (see text). Upstream, the cracked hill-slope area of regolith production may be observed.

The second depositional unit of interest for this review is a horizon marker referred to as the Miraflores Flood event [84]. It is formed by debris flow deposits and flood deposits indirectly dated between 1350 and 1370 CE by ice core data of the Quelccaya Ice Cap [22,145,146]. Such an age is confirmed by radiocarbon dates of samples taken at Rio Muerto [51] (Figure 6). At Quebrada Miraflores (Figure 7), the thickness of the marker reaches 1.2 m, which is a higher order of magnitude compared with any other flood deposits preserved in the geological record, including the Chuza Unit.

In this last area, mobilized sediments spread laterally out of the braided streams and up the ravine walls before descending to the sea across the coastal plain. Large clasts and boulders with a diameter of up to 3 m were also moved across the coastal plain by the Miraflores Flood [22]. Such a hydrogeomorphic event would also be implicated in the collapse or decline of the Chiribaya culture [111,147]. The imprint on the Peruvian desert surface of this 14th-century El Niño event is confirmed by zooarchaeological research. Rubinatto et al. posit *"that the abundance of anuran remains"* they found in Rio Muerto area may be related to the Miraflores Flood, since *"this event generated increased rainfall in the desert, creating conditions favorable for frogs and toads"* [130]. However, for a critical examination of such a hypothesis, these authors exhort further paleoenvironmental and zooarchaeological studies. In any case, the potential of Miraflores Unit as a horizon marker should be fully exploited in landscape reconstruction.

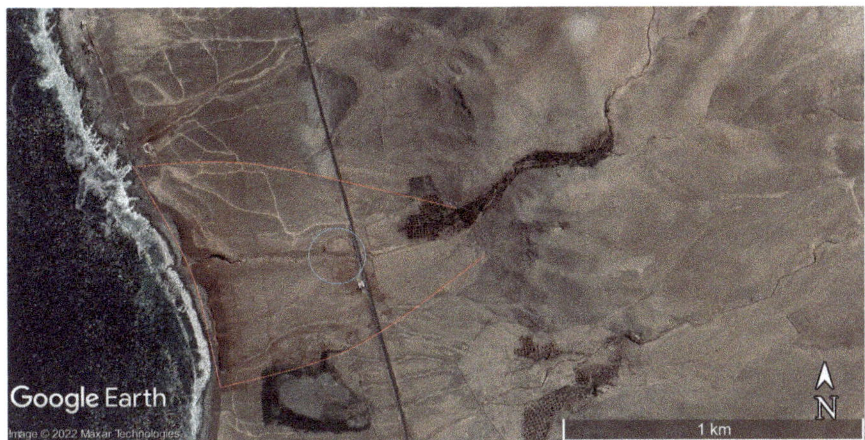

Figure 7. Quebrada Miraflores; 2021 satellite image (17°25′49″–17°26′36″ S, 71°22′22″–71°21′42″ W). The extent of the Miraflores Unit according to Satterlee et al. [22] is marked by a red line and the location of Chiribaya settlement by a blue circle (see text).

5.9. Discussion Summary

Landscape processes confidently ascribed to El Niño events are chronologically summarized in Table 5. They are considered responsible for changes in late pre-Hispanic coastal Peru to varying degrees.

Table 5. Chronology of late pre-Hispanic El Niño events argued as possibly responsible for landscape modifications.

Date	Landscape Process	Basin or Site	Reference
ca. 1450 CE [1]	formation of alluvial terraces	Rio Casma	[17]
1400–1450 CE [1]	mud deposition from runoff	Rio Moche	[133]
1360–1350 CE [1]	debris flow activation, flood sedimentation	Q. Miraflores, Rio Moquegua, Rio Muerto	[22,24,51,84]
1325 CE [1]	formation of alluvial terraces	Rio Casma	[17]
ca. 1100 CE [2]	flood sedimentation	Huacas de Moche	[14,15]
1008–995 CE [1]	mud deposition from runoff	Pachacamac	[23]
1000–650 CE [2]	flood sedimentation	Batan Grande	[16]
730–690 CE [1]	debris flow activation, flood sedimentation	Rio Muerto, Rio Moquegua	[24,51]
651–436 CE [1]	mud deposition from runoff	Urpi Kocka	[23]
ca. 600 CE [2]	mud deposition from runoff	Pachacamac	[18,90]
ca. 600 CE [2]	mud deposition from runoff	Huacas de Moche	[86]
600–550 CE [2]	flood sedimentation	Samaca	[103,104,109]
ca. 0 CE [1]	formation of alluvial terraces	Rio Casma	[17]
100 BCE [1]	debris flow activation	Rio Muerto	[51]
900–1100 BCE [2]	mud deposition from runoff	Caballo Muerto	[120]
ca. 1200 BCE [1]	formation of alluvial terraces	Rio Casma	[17,82]
ca. 1450–1600 BCE [1]	mud deposition from runoff	Caballo Muerto	[120]

[1] calibrated ^{14}C age; [2] date inferred from archaeological remains.

The above chronology may be tentatively compared with paleo El Niño proxies. By analyzing a high resolution marine sediment record about 60 km west of Lima, Rein et al. [148] argued maximum of El Niño activity during the third and second millennium BP, while El Niño events would have been persistently weak during the Medieval Climate Anomaly (MCA, ca. 900–1300 CE). Yan et al. [149], computing SOI (Section 2) from different precipitation proxy records in Pacific Ocean, found negative index during MCA, which indicates general El Niño-dominated conditions. However, such differences in climatic reconstructions may reflect the geographic complexities of the Pacific coastal regions [45]. As a matter of fact, SOI calculated for the Galapagos Islands (850 km off the coast of Ecuador) shows positive values during MCA in contrast with the ones of the western Pacific [149]. The oxygen isotopes ratio ($\delta^{18}O$) of ice core record from Quelccaya ice cap is depleted in ^{18}O between 1100 and 1300 CE, thus highlighting a warming trend over the Andes [150]. Such a finding is consistent with low El Niño activity during MCA [148,151]. Consequently, the hydrogeomorphic events that can be attributed to El Niño with the lowest uncertainties should be those that precede or follow the MCA.

As a result of the WoS and Scopus data mining and the subsequent full-text analysis, the first evaluable hydrogeomorphic signature is located at Caballo Muerto and ^{14}C-dated at about 3.5 ka BP [120]. However, it is in advance of the beginning of ENSO's modern periodicity (Section 2). Three more recent deposits dated around 3.0 ka and ascribed to El Niño events were also reported by this author (Table 5). To date, the magnitude of landscape response to these events along the middle Moche Valley is not yet investigated, while the only identified surface process is mud deposition from runoff over archaeological structures (Section 5.3). Moreover, the lower Lurin Valley also preserves geoarchaeological proxies [137,138] that can help shed light on the landscape modifications that occurred around the transition from the second to the first millennia BCE.

Around 1200 BCE, Casma Valley was affected by torrential rains, and thus, related proxies are preserved in both alluvial terrace sequences [82] and archaeological records of Cerro Sechin [78]. The hypothesis that both signatures were produced by the same El Niño event should be verified by further research (Section 5.4). According to Wells [17], the alluvial terrace system of the Casma Valley contains 13 flood deposits ascribable to late Holocene El Niños. Four ^{14}C-dated events occurred in pre-Hispanic time, but unfortunately, no date fits well with events identified in other basins or sites (Table 5). In contrast, one or more events transpiring about 600 CE have hit the valleys of Rio Moche (Huacas del Sol), Rio Lurin (Pachacamac), and Rio Ica (Samaca) (Sections 5.2, 5.5, and 5.6). Such evidence has led some authors to speak of a 6th-century Mega El Niño (see, e.g., [21,23,134]).

Nevertheless, the recognition of one or more events to the transition from the first to the second millennium CE at some sites of north-central coastal Peru (Table 5) has strengthened the paradigm of the 11th-century Mega El Niño (see, e.g., [21,56,152]). The seminal reconstruction of the Chimu Flood by Nials et al. [14,15] has greatly influenced the literature, but it should be tested by further research, and the related deposits should be absolute-dated. Furthermore, the age of the ethno-historical Naymlap Flood remains unclear [84,132]. It ranges from around 1000 CE (see, e.g., [16,23]) to around 1350 CE (see, e.g., [17,153,154]). As a matter of fact, Naymlap Flood *"may refer to different floods in different valleys at different times during the Late Intermediate Period"* [7].

If on the one hand, El Niño events had catastrophic consequences associated with floodplain stripping, irrigation system damage, and settlement destruction, on the other hand, they also provided cultural opportunities, triggered technological innovation, strengthened community resilience against hydrogeomorphological events, and played important roles in replenishing and maintaining groundwater resources throughout the coastal desert [51,88,138]. Nevertheless, extensive human intervention (desert reclamation, *Prosopis* deforestation) has likely caused some variations in landscape response, also increasing exposure to extreme events of settlements and infrastructures (Sections 5.2 and 5.6).

Authors have deeply investigated the landscape response and destructive effects of the Miraflores Flood, the last hydrogeomorphic event ascribed to a "Mega" El Niño in pre-Hispanic times [22,24,51,84,111,147] (Section 5.8). With the arrival of Spanish conquistadors, the production of historical documents begins, so the sources of knowledge on El Niño events are enriched with a fundamental element [34,56,81,155]. However, it is only with the systematic observations of the warm ocean current running along the central Pacific coast of South America (El Niño in its initial meaning, see Section 2) that all the necessary criteria to assess El Niño events are available.

A final consideration must be made on the limitation of the above finding due to the use of WoS and Scopus. Papers published in journals not indexed in these citation databases are obviously excluded. Moreover, as remarked upon in Section 3, documents written in English are favored over those written in other languages in data mining from WoS and Scopus [68]. Such problems may have partially affected this review. However, in the discussion on the main study areas (Section 5),documents written in Spanish and published in (not WoS- and/or Scopus-indexed) Peruvian journals and books ([131,134–136]) were also used, to the best of the author's knowledge. Clearly, the result may be hereafter improved and updated. On the other hand, data mining from other databases, systematic reading of papers written in Spanish and published in Peruvian journals, and the use of web search engines such as Google Scholar could be goals of future investigations.

6. Conclusions

Landscape modifications ascribed to El Niño events in late pre-Hispanic coastal Peru were mined and summarized by processing comprehensive databases (Section 3). Then, a review of the selected documents was performed in accordance with dependable guidelines (Section 4). The examination of the reliability of data contained in the reviewed documents allowed us to identify some critical questions in the treated topic and suggest several research goals (Section 5). The El Niño proxies discussed provide incomplete and heterogeneous paleoflood records for coastal Peru. The current state of geoarchaeological knowledge is not enough to define the temporal and spatial trend of the landscape effects of El Niño events during the late pre-Hispanic time. However, it gives some essential milestones (Section 5.9, Table 5).

Several landscape processes ascribable to El Niño events have been recognized by authors (Table 5). The less impressive one, mud deposition from runoff, has, however, sometimes been identified in archaeological sites, thus allowing valuable chronological attributions (Sections 5.3 and 5.5). Slack-water deposits found along the valleys of Rio La Leche and Rio Jequetepeque (Section 5.1), as well as other flood sediments recognized along the valleys of Rio Casma and Rio Samaca (Sections 5.4 and 5.6), and at the sites Huacas de Moche and Batan Grande (Sections 5.1 and 5.2), could have great potential in landscape reconstruction. The activation of debris flows has been recognized throughout Jequetepeque Valley (Section 5.1), Moche Valley (Section 5.3), Moquegua Valley, and the surrounding areas (Section 5.8). Within the catchment of Rio Muerto, regolith denudation and mobilization have shown a great size and extent (Section 5.8). The Miraflores Unit resulted in one of the major landscape changes for the whole coastal Peru, which was also involved in the end of the Chiribaya culture. It is apparently the better investigated proxy record confidently ascribed to an El Niño event (Section 5.8, Table 5).

Throughout the review, some possible research goals were highlighted that might be helpful in overcoming critical questions. They do not claim to be all-inclusive and are reported (not in the order of importance) in what follows: (a) to increase the number of the absolute dating of selected geoarchaeological records (see, e.g., the case of the deposits related to the Chimu Flood in the lower Moche Valley, Section 5.2); (b) to analyze stratigraphic successions that should preserve different proxies (see, e.g., the case of the alluvial fan within the Caballo Muerto Archaeological Complex, Section 5.3); (c) to investigate the landscape response of El Niño signatures discovered in archaeological sites (see the cases of Huaca Cortada and Cerro Sechin discussed in Sections 5.3 and 5.4, respectively); (d) to test the consistency of literature hypothesis with methods of various disciplines (see, e.g., the paleoecological and zooarchaeological study performed in the Rio Muerto catchment, Section 5.8); (e) to extract data from other databases also including documents written in Spanish and published in Peruvian journals (Section 5.9).

Funding: This research received no external funding.

Acknowledgments: I wish to thank G. Orefici, (the Centro de Estudios Arqueológicos Precolombinos (Peru)), for hospitality and valuable advice during research campaigns of 2012 and 2019 in Cahuachi.

Conflicts of Interest: The author declares no conflict of interest.

Appendix A. Works Excluded from the Discussion by Full-Text Analysis

As explained in Section 4, 20 documents selected from database search were not discussed in Section 5 because they were not appropriate for the review's purpose (Table A1).

Table A1. Works whose full text was analyzed with no data useful for the objective of the review.

Reference	Main Issue
[1]	Geomorphological effects of 1982/83 and 1997/98 El Niños
[61]	Morpho-pedological characterization of a Western Andean offshoot
[87]	Early-middle Holocene human-nature relationships
[95]	Late Quaternary evolution of a stretch of the Ecuador coast
[97]	Ancient peoples ability to mitigate El Niño effects
[98]	Debris flow burial episode of a Late Pleistocene site
[99]	Late pre-Hispanic desert reclamation and El Niño effects mitigation
[100]	Development of desert soil under ENSO conditions
[50]	Reliability of agricultural practices to face El Niño negative effects
[105]	Early-middle Holocene occupation history of a costal site
[107]	Introduction on methods and technologies in geoarchaeology
[108]	Geomorphological effects of El Niño on Western Andean Slope
[110]	Late Quaternary evolution of an Amazon Basin sector
[116]	Characterization of late Holocene coastal upwelling in Peru
[119,125]	El Niño influences on settlement pattern of a Western Andean offshoot
[121]	Paleobotanical analysis aimed to explore ancient desert reclamation
[123]	Reconstruction of the paleo-evolution of a coastal lagoon
[124]	Biological crusts effects on reg soil evolution
[127]	Late Quaternary environmental history of a Western Andean offshoot

References

1. Mettier, R.; Schlunegger, F.; Schneider, H.; Rieke-Zapp, D.; Schwab, M. Relationships between landscape morphology, climate and surface erosion in northern Peru at 5° S latitude. *Int. J. Earth Sci.* **2009**, *98*, 2009–2022. [CrossRef]
2. Morera, S.B.; Condom, T.; Crave, A.; Steer, P.; Guyot, J.L. The impact of extreme El Niño events on modern sediment transport along the western Peruvian Andes (1968–2012). *Sci. Rep.* **2017**, *7*, 11947. [CrossRef]
3. Rodriguez-Morata, C.; Ballesteros-Canovas, J.A.; Rohrer, M.; Espinoza, J.C.; Beniston, M.; Stoffel, M. Linking atmospheric circulation patterns with hydro-geomorphic disasters in Peru. *Int. J. Climatol.* **2018**, *38*, 3388–3404. [CrossRef]
4. Eitel, B.; Hecht, S.; Mächtle, B.; Schukraft, G.; Kadereit, A.; Wagner, G.A.; Kromer, B.; Unkel, I.; Reindel, M. Geoarchaeological evidence from desert loess in the Nazca–Palpa region, Southern Peru: Palaeoenvironmental changes and their impact on Pre-Columbian cultures. *Archaeometry* **2005**, *47*, 137–158. [CrossRef]
5. Jordan, T.E.; Kirk-Lawlor, N.E.; Blanco, N.P.; Rech, J.A.; Cosentino, N.J. Landscape modification in response to repeated onset of hyperarid paleoclimate states since 14 Ma, Atacama Desert, Chile. *Geol. Soc. Amer. Bull.* **2014**, *126*, 1016–1046. [CrossRef]
6. Silverman, H.; Proulx, D.A. *The Nasca*; Blackwell Publishers: Malden, MA, USA, 2002.
7. Huckleberry, G.; Caramanica, A.; Quilter, J. Dating the Ascope Canal System: Competition for Water during the Late Intermediate Period in the Chicama Valley, North Coast of Peru. *J. Field Archaeol.* **2018**, *31*, 17–30. [CrossRef]
8. Ortlieb, L.; Vargas G. Debris-flow deposits and El Niño impacts along the hyperarid southern Peru coast. In *El Niño in Peru: Biology and Culture over 10,000 Years*; Haas, J., Dillon, M.O., Eds.; Field Museum of Natural History: Chicago, IL, USA, 2003; pp. 24–51.
9. Vargas, G.; Rutllant, J.; Ortlieb, L. ENSO tropical–extratropical climate teleconnections and mechanisms for Holocene debris flows along the hyperarid coast of western South America (17°–24° S). *Earth Planet. Sci. Lett.* **2006**, *249*, 467–483. [CrossRef]
10. Rein, B. How do the 1982/83 and 1997/98 El Niños rank in a geological record from Peru? *Quat. Int.* **2007**, *161*, 56–66. [CrossRef]
11. Aceituno, P.; Prieto, M.D.R.; Solari, M.E.; Martinez, A.; Poveda, G.; Falvey, M. The 1877–1878 El Niño episode: Associated impacts in South America. *Clim. Chang.* **2009**, *92*, 389–416. [CrossRef]
12. Murga, J.C. Permanecer tras el desastre: La ciudad de Saña después de los Niños de 1578 y 1720. *Archaeobios* **2010**, *4*, 85–95. (In Spanish)
13. Sandweiss, D.H.; Andrus, C.F.T.; Kelley, A.R.; Maasch, K.A.; Reitz, E.J.; Roscoe, P.B. Archaeological climate proxies and the complexities of reconstructing Holocene El Niño in coastal Peru. *Proc. Natl. Acad. Sci. USA* **2020**, *117*, 201912242. [CrossRef]
14. Nials F.L.; Deeds, E.E.; Moseley, M.E.; Pozorski, S.E.; Pozorski, T.G.; Feldman, R. El Niño: The catastrophic flooding of coastal Peru. *Field Museum Nat. Inst. Bull.* **1979**, *50*, 4–14.
15. Nials F.L., Deeds, E.E.; Moseley, M.E.; Pozorski, S.E.; Pozorski, T.G.; Feldman, R. El Niño: The catastrophic flooding of coastal Peru. Part II. *Field Museum Nat. Inst. Bull.* **1979**, *50*, 4–10.
16. Craig, A.K.; Shimada, I. El Niño flood deposits at Batan Grande, northern Peru. *Geoarchaeology* **1986**, *1*, 29–38. [CrossRef]

17. Wells, L.E. Holocene history of the El Niño phenomenon as recorded in flood sediments of northern coastal Peru. *Geology* **1990**, *18*, 1134–1137. [CrossRef]
18. Paredes, P.; Ramos, J. Evidencias arqueologicas del "Niño" en las excavaciones de Pachacamac. In *Paleo ENSO Records*; Ortlieb, L., Machare, J., Eds.; Orstom-Concytec: Lima, Peru, 1992; pp. 225–233. (In Spanish)
19. Mörner, N.A. Present El Niño-ENSO events and past super-ENSO events. *Bull. Inst. Fr. Études Andines* **1993**, *22*, 3–12. [CrossRef]
20. Grodzicki, J. Los geoglifos de Nazca segun algunos datos geologicos. In *Paleo ENSO Records*; Ortlieb, L., Machare, J., Eds.; Orstom-Concytec: Lima, Peru, 1992; pp. 119–130. (In Spanish)
21. Grodzicki, J. *Nasca: Los Sintomas Geológicos del Fenómeno El Niño y Sus Aspectos Arquelógicos*; Warsaw University: Warsaw, Poland, 1994. (In Spanish)
22. Satterlee, D.R.; Moseley, M.E.; Keefer, D.K.; Tapia, J.E.A. The Miraflores El Niño disaster: Convergent catastrophes and prehistoric agrarian change in southern Peru. *Andean Past* **2000**, *6*, 95–116.
23. Winsborough, B.M.; Shimada, I.; Newsom, L.A.; Jones, J.G.; Segura, R.A. Paleoenvironmental catastrophies on the Peruvian coast revealed in lagoon sediment cores from Pachacamac. *J. Archaeol. Sci.* **2012**, *39*, 602–614. [CrossRef]
24. Magilligan, F.J.; Goldstein, P.S. El Niño floods and culture change: A late Holocene flood history for the Rio Moquegua, southern Peru. *Geology* **2001**, *29*, 431–434. [CrossRef]
25. Silverman, H. *Ancient Nasca Settlement and Society*; University of Iowa Press: Iowa City, IA, USA, 2002.
26. Delle Rose, M.; Mattioli, M.; Capuano, N.; Renzulli, A. Stratigraphy, Petrography and Grain-Size Distribution of Sedimentary Lithologies at Cahuachi (South Peru): ENSO-Related Deposits or a Common Regional Succession? *Geosciences* **2019**, *9*, 80. [CrossRef]
27. Moy, C.M.; Seltzer, G.O.; Rodbell, D.T.; Anderson, D.M. Variability of El Niño/Southern Oscillation activity at millennial timescales during the Holocene epoch. *Nature* **2002**, *420*, 162–164. [CrossRef]
28. Licciardi, J.M.; Schaefer, J.M.; Taggart, J.R.; Lund, D.C. Holocene glacier fluctuations in the Peruvian Andes indicate northern climate linkages. *Science* **2009**, *325*, 1677–1679. [CrossRef]
29. Meyer, I.; Wagner, S. The Little Ice Age in Southern South America: Proxy and Model Based Evidence. In *Past Climate Variability in South America and Surrounding Regions. Developments in Paleoenvironmental Research*; Vimeux, F., Sylvestre, F., Khodri, M., Eds.; Springer: Dordrecht, The Netherlands, 2009; Volume 14, pp. 395–412.
30. Moher, D.; Liberati, A.; Tetzlaff, J.; Altman, D.G.; PRISMA Group. Preferred reporting items for systematic reviews and meta-analyses: The PRISMA statement. *Biomed. J.* **2009**, *339*, b2535.
31. Page, M.J.; McKenzie, J.E.; Bossuyt, P.M.; Boutron, I.; Hoffmann, T.C.; Mulrow, C.D.; Shamseer, L.; Tetzlaff, J.M.; Akl, E.A.; Brennan, S.E.; et al. The PRISMA 2020 statement: An updated guideline for reporting systematic reviews. *Biomed. J.* **2021**, *372*, 71.
32. Satyamurty, P.; Nobre, C.; Silva Dias, P.L. South America. In *Meteorology of the Southern Hemisphere*; Monograph Series No. 27; Karoly, D.J., Vincent, D., Eds.; American Meteorological Society: Boston, MA, USA, 1998; pp. 119–140.
33. Rundel, P.; Villagra, P.E.; Dillon, M.; Roig-Juñent, S.; Debandi, G. Arid and Semi-Arid Ecosystems. In *The Physical Geography of South America*; Veblen, T.T., Young, K., Orme, A., Eds.; Oxford University Press: Oxford, UK, 2007; pp. 158–183.
34. Quinn, W.H.; Neal, V.T.; Antunez De Mayolo, S.E. El Niño occurrences over the past four and a half centuries. *J. Geophys. Res.* **1987**, *92*, 14449–14461 [CrossRef]
35. Sanabria, J.; Bourrel, L.; Dewitte, B.; Frappart, F.; Rau, P.; Solis, O.; Labat, D. Rainfall along the coast of Peru during strong El Niño events. *Int. J. Climatol.* **2017**, *38*, 1737–1747. [CrossRef]
36. Rodriguez-Morata, C.; Diaz, H.F.; Ballesteros-Canovas, J.A.; Rohrer M.; Stoffel, M. The anomalous 2017 coastal El Niño event in Peru. *Clim. Dyn.* **2019**, *52*, 5605–5622. [CrossRef]
37. Son, R.; Wang, S.Y.S.; Tseng, W.L.; Barreto Schuler, C.W.; Becker, E.; Yoon, J.H. Climate diagnostics of the extreme floods in Peru during early 2017. *Clim. Dyn.* **2020**, *54*, 935–945. [CrossRef]
38. Trenberth, K.E. The definition of El Niño. *Bull. Am. Meteorol. Soc.* **1997**, *78*, 2771–2777. [CrossRef]
39. Trenberth, K.E.; Caron, J.M. The Southern Oscillation revisited: Sea level pressure, surface temperatures, and precipitations. *J. Clim.* **2000**, *13*, 4358–4365. [CrossRef]
40. Jadhav, J.; Panickal, S.; Marathe, S.; Ashok, K. On the possible cause of distinct El Niño types in the recent decades. *Sci. Rep.* **2015**, *5*, 17009. [CrossRef]
41. Hu, Z.Z.; Huang, B.; Zhu, J.; Kumar, A.; McPhaden, M.J. On the variety of coastal El Niño events. *Clim. Dyn.* **2019**, *52*, 7537–7552. [CrossRef]
42. Wang, C.; Deser, C.; Yu, J.Y.; DiNezio, P.; Clement, A. El Niño and Southern Oscillation (ENSO): A review. In *Coral Reefs of the Eastern Pacific*; Glymm, P., Manzello, D., Enochs, I., Eds.; Springer: Berlin/Heidelberg, Germany, 2017; pp. 85–106.
43. Lin, J.; Qian, T. A New Picture of the Global Impacts of El Nino-Southern Oscillation. *Sci. Rep.* **2019**, *9*, 17543. [CrossRef]
44. Sandweiss, D.H.; Richardson, J.B.; Reitz, E.J.; Rollins, H.B.; Maasch, K.A. Geoarchaeological evidence from Peru for a 5000 year B.P. onset of El Nino. *Science* **1996**, *273*, 1531–1533. [CrossRef]
45. Cobb, K.M.; Westphal, N.; Sayani, H.R.; Watson, J.T.; Di Lorenzo, E.; Cheng, H.; Edwards, R.L.; Charles, C.D. Highly variable El Niño-Southern Oscillation throughout the Holocene. *Science* **2013**, *339*, 67–70. [CrossRef]
46. Waylen, P.R.; Caviedes, C.N. El Niño and annual floods in coastal Peru. In *Catastrophic Flooding*; Mayer, L., Nash, D., Eds.; Allen and Unwin: Boston, MA, USA, 1987; pp. 57–78.

47. Tapley, T.D.; Waylen, P.R. Spatial variability of annual precipitation and ENSO events in western Peru. *Hydrol. Sci. J.* **1990**, *35*, 429–446. [CrossRef]
48. Bird, B.; Abbott, M.B.; Vuille, M.; Rodbell, D.T.; Stansell, N.D.; Rosenmeier, M.F. A 2300-year-long annually resolved record of the South American summer monsoon from the Peruvian Andes. *Proc. Natl. Acad. Sci. USA* **2011**, *108*, 8583–8588. [CrossRef]
49. Giraldez, L.; Silva, Y.; Zubieta, R.; Sulca, J. Change of the Rainfall Seasonality Over Central Peruvian Andes: Onset, End, Duration and Its Relationship With Large-Scale Atmospheric Circulation. *Climate* **2020**, *8*, 23. [CrossRef]
50. Manners, R.B.; Magilligan, F.J.; Goldstein, P.S. Floodplain Development, El Niño, and Cultural Consequences in a Hyperarid Andean Environment. *Ann. Am. Assoc. Geogr.* **2007**, *97*, 229–249. [CrossRef]
51. Magilligan, F.J.; Goldstein, P.S.; Fisher, G.B.; Bostick, B.C.; Manners, R.B. Late Quaternary hydroclimatology of a hyper-arid Andean watershed: Climate change, floods, and hydrologic responses to the El Niño-Southern Oscillation in the Atacama Desert. *Geomorphology* **2008**, *101*, 14–32. [CrossRef]
52. Hanley, D.E.; Bourassa, M.A.; O'Brien, J.J.; Smith, S.R.; Spade, E.R. A Quantitative Evaluation of ENSO Indices. *J. Clim.* **2003**, *16*, 1249–1258. [CrossRef]
53. Khalil, A.F.; Kwon, H.H.; Lall, U.; Miranda, M.J.; Skees, J. El Niño–Southern Oscillation–based index insurance for floods: Statistical risk analyses and application to Peru. *Water Resour. Res.* **2007**, *43*, W10416. [CrossRef]
54. Wang, Z.; Wu, D.; Chen, X.; Qiao, R. ENSO indices and analyses. *Adv. Atmos. Sci.* **2013**, *30*, 1491–1506. [CrossRef]
55. Sandweiss, D.H.; Maasch, K.A.; Burger, R.L.; Richardson, J.B.; Rollins, H.B.; Clement, A. Variation in Holocene El Niño frequencies: Climate records and cultural consequences in ancient Peru. *Geology* **2001**, *29*, 603–606. [CrossRef]
56. Machare, J.; Ortlieb, L. Registros del fenomeno El Niño en el Perù. *Bull. Inst. Fr. Études Andines* **1993**, *22*, 35–52. (In Spanish) [CrossRef]
57. Río Chicama Bloquea Panamericana Norte. (In Spanish). Available online: https://www.elperuano.pe/noticia/53379-rio-chicama-bloquea-panamericana-norte (accessed on 16 November 2022).
58. Osterkamp, W.R.; Friedman, J.M. The disparity between extreme rainfall events and rare floods—With emphasis on the semi-arid American West. *Hydrol. Process.* **2000**, *14*, 2817–2829. [CrossRef]
59. Tooth, S.; Nanson, G.C. Equilibrium and nonequilibrium conditions in dryland rivers. *Phys. Geogr.* **2000**, *21*, 183–211. [CrossRef]
60. Matmon, A.; Simhai, O.; Amit, R.; Haviv, I.; Porat, N.; Mcdonald, E.; Benedetti, L.; Finkel, R. Desert pavement-coated surfaces in extreme deserts present the longest-lived landforms on Earth. *Geol. Soc. Amer. Bull.* **2009**, *121*, 688–697. [CrossRef]
61. Calderoni, G.; Lazzarotto, L.; Rodolfi, G.; Terribile, F. Evolution of the coastal reliefs of southern Peru (Lomas) as suggested by the soil-landform relationships: A case study from Mejìa site (Department of Arequipa). *Geogr. Fis. Din. Quat.* **2002**, *25*, 23–43.
62. Sparavigna, A.C. Peruvian Transverse Dunes in the Google Earth Images. *Philica* **2014**, *474*, hal-01389724f. Available online: https://hal.archives-ouvertes.fr/hal-01389724/document (accessed on 14 October 2022).
63. Delle Rose, M. The geology of Cahuachi. In *Ancient Nasca World. New Insights from Science and Archaeology*; Lasaponara, R., Masini, N., Orefici, G., Eds.; Springer: Basel, Switzerland, 2016; pp. 47–64.
64. Magirl, C.S.; Griffiths, P.G.; Webb, R.H. Analyzing debris flows with the statistically calibrated empirical model LAHARZ in southeastern Arizona, USA. *Geomorphology* **2010**, *119*, 111–124. [CrossRef]
65. Aguilar, G.; Cabre, A.; Fredes, V.; Villela, B. Erosion after an extreme storm event in an arid fluvial system of the southern Atacama Desert: An assessment of the magnitude, return time, and conditioning factors of erosion and debris flow generation. *Nat. Hazards Earth Syst. Sci.* **2020**, *20*, 1247–1265. [CrossRef]
66. Kochel, R.C.; Baker, V.R. Paleoflood hydrology. *Science* **1982**, *215*, 353–361. [CrossRef]
67. Salisbury, L. Web of Science and Scopus: A Comparative Review of Content and Searching Capabilities. *Charlest. Advis.* **2009**, *11*, 5–18.
68. Mongeon, P., Paul-Hus, A. The journal coverage of Web of Science and Scopus: A comparative analysis. *Scientometrics* **2016**, *106*, 213–228. [CrossRef]
69. Pranckute, R. Web of Science (WoS) and Scopus: The Titans of Bibliographic Information in Today's Academic World. *Publications* **2021**, *9*, 12. [CrossRef]
70. Frodeman, R. Geological reasoning: Geology as an interpretive and historical science. *Geol. Soc. Am. Bull.* **1995**, *107*, 960–968. [CrossRef]
71. Dott, R.H. What is unique about geological reasoning? *Geol. Soc. Am. Today* **1998**, *10*, 15–18.
72. Raab T.; Frodeman, R. What is it like to be a geologist? A phenomenology of geology and its epistemological implications. *Philos. Geogr.* **2002**, *5*, 69–81. [CrossRef]
73. Hodder, I. Interpretative archaeology and its role. *Am. Antiq.* **1991**, *56*, 7–18. [CrossRef]
74. Fogelin, L. Inference to the Best Explanation: A Common and Effective Form of Archaeological Reasoning. *Am. Antiq.* **2007**, *72*, 603–625. [CrossRef]
75. Lucas, G. Interpretation in Archaeological Theory. In *Encyclopedia of Global Archaeology*; Smith, C., Ed.; Springer: New York, NY, USA, 2014; pp. 4013–4022.
76. Butzer, K.W. Challenges for a cross-disciplinary geoarchaeology: The intersection between environmental history and geomorphology. *Geomorphology* **2008**, *101*, 402–411. [CrossRef]
77. Hill, C.L.; Rapp, G. Geoarchaeology. In *Encyclopedia of Global Archaeology*; Smith, C., Ed.; Springer: New York, NY, USA, 2014; pp. 3008–3017.

78. Samaniego, L.; Vergara, E.; Bischof, H. New evidence on Cerro Sechin, Casma Valley, Peru. In *Early Ceremonial Architecture in the Andes*; Donnan, C.B., Ed.; Dumbarton Oaks Research Library and Collection: Washington, DC, USA, 1985; pp. 165–190.
79. Sandweiss, D.H. The beach ridges at Santa, Peru: El Niño, uplift, and prehistory. *Geoarchaeology* **1986**, *1*, 17–28. [CrossRef]
80. Rollins, H.B.; Richardson, J.B.; Sandweiss, D.H. The birth of El Niño: Geoarchaeological evidence and implications. *Geoarchaeology* **1986**, *1*, 3–15. [CrossRef]
81. DeVries, T.J. A review of geological evidence for ancient El Niño activity in Peru. *J. Geophys. Res.* **1987**, *92*, 471–479. [CrossRef]
82. Wells, L.E. An alluvial record of El Niño events from northern coastal Peru. *J. Geophys. Res.* **1987**, *92*, 14463–14470. [CrossRef]
83. Moseley, M.; Richardson, J.B. Doomed by natural disaster. *Archaeology* **1992**, *45*, 44–45.
84. Moseley, M.; Tapia, J.; Satterlee, D.S.; Richardson, J.B. Flood events, El Niño events, and tectonic events. In *Paleo ENSO Records*; Ortlieb, L., Machare, J., Eds.; Orstom-Concytec: Lima, Peru, 1992; pp. 207–212. (In Spanish)
85. Ortlieb, L.; Machare, J. Former El Niño events: Records from western South America. *Glob. Planet. Chang.* **1993**, *7*, 181–202. [CrossRef]
86. Uceda, C.S.; Canziani Amico, J. Evidencias de grandes precipitaciones en diversas etapas constructivas de la Huaca de la Luna, costa norte del Perú. *Bull. Inst. Fr. Études Andines* **1993**, *22*, 313–343.
87. Keefer, D.; Moseley, M.; Richardson, J.; Satterlee, D.; Day Lewis, A. Early Maritime Economy and El Niño Events at Quebrada Tacahuay, Peru. *Science* **1998**, *281*, 1833–1835. [CrossRef] [PubMed]
88. Wells, L.E.; Noller, J.S. Holocene coevolution of the physical landscape and human settlement in northern coastal Peru. *Geoarchaeology* **1999**, *14*, 755–789. [CrossRef]
89. Veit, H. Klima- und Landschaftswandel in der Atacama. *Geographische Rundschau* **2000**, *52*, 4–9. (In German)
90. Franco, R.; Paredes, P. El Templo Viejo de Pachacamac: Nuevos aportes al studio del Horizonte Medio. *Bol. Arqueol. PUCP* **2000**, *4*, 607–630. (In Spanish)
91. Van Buren, M. The archaeology of El Niño events and others "natural" disasters. *J. Archaeol. Method Theory* **2001**, *8*, 129–149. [CrossRef]
92. Huckleberry, G.; Billman, B.R. Geoarchaeological insights gained from surficial geologic mapping, middle Moche Valley, Peru. *Geoarchaeology* **2003**, *18*, 505–521. [CrossRef]
93. Keefer, D.K.; Moseley, M.E.; deFrance, S.D. A 38,000-year record of floods and debris flows in the Ilo region of southern Peru and its relation to El Niño events and great earthquakes. *Palaeogeogr. Palaeoclimatol. Palaeoecol.* **2003**, *194*, 41–77. [CrossRef]
94. Dillehay, T.; Kolata, A.L.; Pino, M.Q. Pre-industrial human and environment interactions in northern Peru during the late Holocene. *Holocene* **2004**, *14*, 272–281. [CrossRef]
95. Federici, P.R.; Rodolfi, G. Geomorphological Features and Evolution of the Ensenada de Atacames (Provincia de Esmeraldas, Ecuador). *J. Coast. Res.* **2004**, *20*, 700–708. [CrossRef]
96. Keefer, D.K.; Moseley, M.E. Southern Peru desert shattered by the great 2001 earthquake: Implications for paleoseismic and paleo-El Niño-Southern Oscillation records. *Proc. Natl. Acad. Sci. USA* **2004**, *101*, 10878–10883. [CrossRef]
97. Brooks, W.E.; Willett, J.C.; Kent, J.D.; Vasquez, V.; Rosales, T. The Muralla Pircada: An ancient Andean debris flow retention dam, Santa Rita B archaeological site, Chao Valley, northern Peru. *Landslides* **2005**, *2*, 117–123. [CrossRef]
98. deFrance, S.D.; Keefer, D.K. Quebrada Tacahuay, Southern Peru: A Late Pleistocene Site Preserved by a Debris Flow. *J. Field Archaeol.* **2005**, *30*, 385–399. [CrossRef]
99. Zaro, G.; Alvarez, A.U. Late Chiribaya agriculture and risk management along the arid Andean coast of southern Perú, A.D. 1200–1400. *Geoarchaeology* **2005**, *20*, 717–737. [CrossRef]
100. Fabre, A.; Gauquelin, T.; Vilasante, F.; Ortega, A.; Puig, H. Phosphorus content in five representative landscape units of the Lomas de Arequipa (Atacama Desert-Peru). *Catena* **2006**, *65*, 80–86. [CrossRef]
101. Machtle, B.; Eitel, B.; Kadereit, A.; Unkel, I. Holocene environmental changes in the northern Atacama desert, southern Peru (14°30′S) and their impact on the rise and fall of Pre-Columbian cultures. *Z. Geomorphol. Suppl.* **2006**, *142*, 47–62.
102. Andrus, C.F.T.; Sandweiss, D.H.; Reitz, E.J. Climate Change and Archaeology: The Holocene History Of El Niño On The Coast Of Peru. In *Case Studies in Environmental Archaeology. Interdisciplinary Contributions to Archaeology*; Reitz, E.J., Scudder, S.J., Scarry, C.M., Eds.; Springer: New York, NY, USA, 2008; pp. 143–157.
103. Beresford-Jones, D.G.; Arce Torres, S.; Whaley, O.Q.; Chepstow-Lusty, A.J. The role of Prosopis in ecological and landscape change in the Samaca Basin, lower Ica Valley, south coast Peru from the Early Horizon to the Late Intermediate Period. *Latin Am. Antiqu.* **2009**, *20*, 303–332. [CrossRef]
104. Beresford-Jones, D.G.; Lewis, H.; Boreham, S. Linking cultural and environmental change in Peruvian prehistory: Geomorphological survey of the Samaca Basin, Lower Ica Valley, Peru. *Catena* **2009**, *78*, 234–249. [CrossRef]
105. deFrance, S.D.; Grayson, N.; Wise, K. Documenting 12,000 Years of Coastal Occupation on the Osmore Littoral, Peru. *J. Field Archaeol.* **2009**, *34*, 227–246. [CrossRef]
106. Eitel, B.; Machtle, B. Man and Environment in the Eastern Atacama Desert (Southern Peru): Holocene Climate Changes and Their Impact on Pre-Columbian Cultures. In *New Technologies for Archaeology*; Reindel, M., Wagner, G.A., Eds.; Springer: Berlin/Heidelberg, Germany, 2009; pp. 17–37.
107. Reindel, M.; Wagner, G.A. Introduction—New Methods and Technologies of Natural Sciences for Archaeological Investigations in Nasca and Palpa, Peru. In *New Technologies for Archaeology*; Reindel, M., Wagner, G.A., Eds.; Springer: Berlin/Heidelberg, Germany, 2009; pp. 1–13.

108. Abbuhl, L.M.; Norton, K.P.; Schlunegger, F.; Kracht, O.; Aldahan, A.; Possnert, G. El Niño forcing on 10Be-based surface denudation rates in the northwestern Peruvian Andes? *Geomorphology* **2010**, *123*, 257–268. [CrossRef]
109. Beresford-Jones, D.G. *The Lost Woodlands of Ancient Nasca: A Case-Study in Ecological and Cultural Collapse*; Oxford University Press: Oxford, UK, 2011; pp. 1–208.
110. Bernal, C.; Christophoul, F.; Darrozes, J.: Soula, J.C.; Baby, P.; Burgos, J. Late Glacial and Holocene avulsions of the Rio Pastaza Megafan (Ecuador–Peru): Frequency and controlling factors. *Int. J. Earth. Sci.* **2011**, *100*, 1759–1782. [CrossRef]
111. Goldstein, P.S.; Magilligan, F.J. Hazard, risk and agrarian adaptations in a hyperarid watershed: El Niño floods, streambank erosion, and the cultural bounds of vulnerability in the Andean Middle Horizon. *Catena* **2011**, *85*, 155–167. [CrossRef]
112. Gayo, E.M.; Latorre, C.; Jordan, T.E.; Nester, P.L.; Estay, S.; Ojeda, K.F.; Santoro, C.M. Late Quaternary hydrological and ecological changes in the hyperarid core of the northern Atacama Desert (21° S). *Earth Sci. Rev.* **2012**, *113*, 120–140. [CrossRef]
113. Huckleberry, G.; Hayashida, F.; Johnson, J. New Insights into the Evolution of an Intervalley Prehistoric Irrigation Canal System, North Coastal Peru. *Geoarchaeology* **2012**, *27*, 492–520. [CrossRef]
114. Sandweiss, D.H.; Kelley, A.R. Archaeological Contributions to Climate Change Research: The Archaeological Record as a Paleoclimatic and Paleoenvironmental Archive. *Annu. Rev. Anthropol.* **2012**, *41*, 371–391. [CrossRef]
115. Sandweiss, D.H.; Quilter, J. Collation, Correlation, and Causation in the Prehistory of Coastal Peru. In *Sudden Environmental Change: Answers From Archaeology*; Cooper, J., Sheets, P., Eds.; University Press of Colorado: Boulder, CO, USA, 2012; pp. 117–141.
116. Etayo-Cadavid, M.F.; Andrus, C.F.T.; Jones, K.B.; Hodgins, G.W.L.; Sandweiss, D.H.; Uceda-Castillo, S.; Quilter, Q. Marine radiocarbon reservoir age variation in Donax obesulus shells from northern Peru: Late Holocene evidence for extended El Niño. *Geology* **2013**, *41*, 599–602. [CrossRef]
117. Hanzalova, K.; Pavelka, K. Documentation and virtual reconstruction of historical objects in Peru damaged by an earthquake and climatic events. *Adv. Geosci.* **2013**, *35*, 67–71. [CrossRef]
118. Machtle, B.; Eitel, B. Fragile landscapes, fragile civilizations—How climate determined societies in the pre-Columbian south Peruvian Andes. *Catena* **2013**, *103*, 62–73. [CrossRef]
119. Kalicki, P.; Kalicki, T.; Kittel, P. The Influence of El Niño on Settlement Patterns in Lomas de Lachay, Central Coast, Peru. *Interdiscip. Archaeol.* **2014**, *5*, 147–160. [CrossRef]
120. Nesbitt, J. El Niño and second-millennium BC monument building at Huaca Cortada (Moche Valley, Peru). *Antiquity* **2016**, *351*, 638–653. [CrossRef]
121. Caramanica, A.; Koons, M.L. Living on the edge. Pre-Columbian habitation of the desert periphery of the Chicama valley, Perú. In *The Archaeology of Human-Environment Interactions: Strategies for Investigating Anthropogenic Landscapes, Dynamic Environments, and Climate Change in the Human Past*; Contreras, D., Ed.; Routledge: London, UK, 2016; pp. 141–164.
122. Pavelka, K.; Matouskova, E.; Faltynova, M. Dating of Artefacts from Nasca Region and Falsification of "Mandala" Geoglyph. In Proceedings of the 16th International Multidisciplinary Scientific GeoConference SGEM, Albena, Bulgaria, 30 June–6 July 2016; Volume 1, pp. 165–172.
123. Christol, A.; Wuscher, P.; Goepfert, N.; Mogollon, V.; Bearez, P.; Gutierrez, B.; Carre, M. The Las Salinas palaeo-lagoon in the Sechura Desert (Peru): Evolution during the last two millennia. *Holocene* **2017**, *27*, 26–38. [CrossRef]
124. Wang, F.; Michalski, G.; Luo, H.; Caffee, M. Role of biological soil crusts in affecting soil evolution and salt geochemistry in hyper-arid Atacama Desert, Chile. *Geoderma* **2017**, *307*, 54–64. [CrossRef]
125. Kalicki, P.; Kalicki, T.; Kittel, P. Fog, mountain and desert: Human-environment interactions in Lomas de Lachay, Peru. In *People in the Mountains*; Pelisiak, A., Nowak, M., Astalos, C., Eds.; Archaeopress Publishing: Oxford, UK, 2018; pp. 207–223.
126. Caramanica, A.; Huaman, L.H.; Morales, C.R.; Huckleberry, G.; Castillo, L.J.; Quilter, J. El Niño resilience farming on the north coast of Peru. *Proc. Natl. Acad. Sci. USA* **2020**, *117*, 24127–24137. [CrossRef] [PubMed]
127. Kalicki, T.; Kalicki, P. Fluvial activity in the Lomas de Lachay during the upper Pleistocene and Holocene. *Geomorphology* **2020**, *357*, 107087. [CrossRef]
128. Uceda, S.; Gayoso, H.; Castillo, F.; Rengifo, C. Climate and Social Changes: Reviewing the Equation with Data from the Huacas de Moche Archaeological Complex, Peru. *Lat. Am. Antiq.* **2021**, *32*, 705–722. [CrossRef]
129. Sandweiss, D.H.; Maasch, K.A. Climatic and Cultural Transitions in Lambayeque, Peru, 600 to 1540 AD: Medieval Warm Period to the Spanish Conquest. *Geosciences* **2022**, *12*, 238. [CrossRef]
130. Rubinatto Serrano, J.R.; Vallejo-Pareja, M.C.; deFrance, S.D.; Baitzel, S.I.; Goldstein, P.S. Contextual, Taphonomic, and Paleoecological Insights from Anurans on Tiwanaku Sites in Southern Peru. *Quaternary* **2022**, *5*, 16. [CrossRef]
131. Castillo, L.J.; Christopher B.D. Los mochicas del norte y los mochicas del sur: Una Perspectiva desde el Valle del Jequetepeque. In *Vicus*; Makowsky, K., Ed.; Banco de Crédito del Perú: Lima, Peru, 1994; pp. 142–181. (In Spanish)
132. Rowe, J.H. Absolute chronology in the Andean area. *Am. Antiq.* **1945**, *10*, 265–284. [CrossRef]
133. Prieto, G.; Verano, J.W.; Goepfert, N.; Kennett. D.; Quilter, J.; LeBlanc, S.; Fehren-Schmitz, L.; Forst, J.; Lund, M.; Dement, B.; et al. A mass sacrifice of children and camelids at the Huanchaquito-Las Llamas site, Moche Valley, Peru. *PLoS ONE* **2019**, *14*, e0211691. [CrossRef]
134. Franco, R.G. Poder religioso, crisis y prosperidad en Pachacamac: Del Horizonte medio al intermedio tardío. *Bull. Inst. Fr. Etudes Andines* **2004**, *33*, 465–506. (In Spanish)

135. Makowski, K. Pachacamac y la politica imperial Inca. In *El Inca y la Huaca. La Religion del Poder y el Poder de la Religion en el Mundo Andino Antiguo*; Curatola Petrocchi, M., Szeminski, J., Eds.; Pontificia Universidad Catolica del Peru: Lima, Peru, 2016; pp. 153–208. (In Spanish)
136. Mauricio, A.C. Ecodinamicas Humanas en Huaca 20: Reevaluando el impacto de El Niño a finales del periodo Intermedio Temprano. *Bol. Arqueol. PUCP* **2014**, *18*, 159–190. (In Spanish)
137. Burger, R.L.; Salazar-Burger, L. A Sacred Effigy from Mina Perdida and the Unseen Ceremonies of the Peruvian Formative. *Res. Anthropol. Aesthet.* **1998**, *33*, 28–53. [CrossRef]
138. Burger, R.L. El Niño, early Peruvian civilization, and human agency: Some thoughts from the Lurin Valley. *Fieldiana Bot.* **2003**, *43*, 90–107.
139. Grodzicki, J. Las catástrofes ecológicas en la Pampa de Nazca a fines del Holoceno y el fenómeno El Niño. In Proceedings of th El Fenómeno El Niño: A Través de las Fuentes Arqueológicas y Geológicas, Warsaw, Poland, 18–19 May 1990; pp. 64–102. (In Spanish)
140. Silverman, H.; Browne, D. New Evidence for the Date of the Nazca Lines. *Antiquity* **1991**, *65*, 208–220. [CrossRef]
141. Dorn, R.I.; Clarkson, P.B.; Nobbs, M.F.; Loendorf, L.L.; Whitley, D.S. New Approach to the Radiocarbon Dating of Rock Varnish, with Examples from Drylands. *Ann. Am. Assoc. Geogr.* **1992**, *82*, 136–151. [CrossRef]
142. Rink, W.J.; Bartoll, J. Dating the geometric Nasca lines in the Peruvian desert. *Antiquity* **2004**, *79*, 390–401. [CrossRef]
143. Stanish, C.; Tantalean, H.; Nigra, B.T.; Griffin, L. A 2,300-year-old architectural and astronomical complex in the Chincha Valley, Peru. *Proc. Natl. Acad. Sci. USA* **2014**, *111*, 7218–7223. [CrossRef]
144. Orefici, G. Recent discoveries in Cahuachi: The Templo Sur. In *Ancient Nasca World. New Insights from Science and Archaeology*; Lasaponara, R., Masini, N., Orefici, G., Eds.; Springer: Basel, Switzerland, 2016; pp. 363–374.
145. Thompson, L.G.; Mosley-Thompson, E.; Morales Arnao, B. El Niño-Southern Oscillation Events Recorded in the Stratigraphy of the Tropical Quelccaya Ice Cap, Peru. *Science* **1984**, *226*, 50–53. [CrossRef]
146. Thompson, L.G.; Mosley-Thompson, E.; Bolzan, J.F.; Koci, B.R. A 1500-year record of tropical precipitation in ice cores from the Quelccaya Ice Cap, Peru. *Science* **1985**, *229*, 971–973. [CrossRef]
147. Reycraft, R.M. Long-term Human Response to El Niño in South Coastal Peru. In *Environmental Disaster and the Archaeology of Human Response*; Bawden, G., Reycraft, R.M., Eds.; Maxwell Museum of Anthropology: Albuquerque, NM, USA, 2000; pp. 99–119.
148. Rein, B.; Luckge, A.; Reinhardt, L.; Sirocko, F.; Wolf, A.; Dullo, W.C. El Niño variability off Peru during the last 20,000 years. *Paleoceanography* **2005**, *20*, PA4003. [CrossRef]
149. Yan, H.; Sun, L.; Wang, Y.; Huang, W.; Qiu, S.; Yang, C. A record of the Southern Oscillation Index for the past 2000 years from precipitation proxies. *Nat. Geosci.* **2011**, *4*, 611–614. [CrossRef]
150. Thompson, L.; Mosley-Thompson, E.; Davis, M.; Zagorodnov, V.; Howat, I.; Mikhalenko, V.; Lin, P.N. Annually Resolved Ice Core Records of Tropical Climate Variability Over the Past 1800 Years. *Science* **2013**, *340*, 945–950. [CrossRef]
151. Rein, B.; Luckge, A.; Sirocko, F. A major Holocene ENSO anomaly during the Medieval period. *Geophys. Res. Lett.* **2004**, *31*, L17211. [CrossRef]
152. Shimada, I.; Wagner, U. Craft Production on the Pre-Hispanic North Coast of Peru: A Holistic Approach and its Results. In *Archaeology as Anthropology: Theoretical and Methodological Approaches*; Skibo, J., Grave, M., Stark, M., Eds.; University of Arizona Press: Tucson, AZ, USA, 2007; pp. 163–197.
153. Kosok, P. *Life, Land and Water in Ancient Peru*; Long Island University Press: New York, NY, USA, 1965; 264p.
154. Pozorski, T.; Pozorski, S. The Impact of the El Niño Phenomenon on Prehistoric Chimu Irrigation Systems of the Peruvian Coast. In *El Niño in Peru: Biology and Culture Over 10,000 Years*; Dillon, M.O., Haas, J., Eds.; Fieldiana, Botany New Series: Chicago, IL, USA, 2003; pp. 71–90.
155. Ortlieb, L. The Documented Historical Record of El Niño Events in Peru: An Update of the Quinn Record (Sixteenth through Nineteenth Centuries). In *El Niño and the Southern Oscillation: Multiscale Variability and Global and Regional Impacts*; Diaz, H., Markgraf, V., Eds.; Cambridge University Press: Cambridge, UK, 2000; pp. 207–296.

MDPI
St. Alban-Anlage 66
4052 Basel
Switzerland
www.mdpi.com

Land Editorial Office
E-mail: land@mdpi.com
www.mdpi.com/journal/land

www.ingramcontent.com/pod-product-compliance
Lightning Source LLC
LaVergne TN
LVHW071254170726
843515LV00006BA/1405

* 9 7 8 3 0 3 6 5 9 9 0 9 0 *